Safety Evaluation of Dam and Geotechnical Engineering

Safety Evaluation of Dam and Geotechnical Engineering

Editors

Rui Pang
Binghan Xue
Yantao Zhu
Xiang Yu
Xiaoying Lin

MDPI

Basel • Beijing • Wuhan • Barcelona • Belgrade • Novi Sad • Cluj • Manchester

Editors

Rui Pang
Dalian University of Technology
Dalian
China

Binghan Xue
Zhengzhou University
Zhengzhou
China

Yantao Zhu
Hohai University
Nanjing
China

Xiang Yu
Zhengzhou University
Zhengzhou
China

Xiaoying Lin
Zhengzhou University
Zhengzhou
China

Editorial Office
MDPI AG
Grosspeteranlage 5
4052 Basel, Switzerland

This is a reprint of articles from the Special Issue published online in the open access journal *Water* (ISSN 2073-4441) (available at: https://www.mdpi.com/journal/water/special_issues/DamSafety_Evaluation).

For citation purposes, cite each article independently as indicated on the article page online and as indicated below:

Lastname, A.A.; Lastname, B.B. Article Title. *Journal Name* **Year**, *Volume Number*, Page Range.

ISBN 978-3-7258-1905-8 (Hbk)
ISBN 978-3-7258-1906-5 (PDF)
doi.org/10.3390/books978-3-7258-1906-5

Contents

water

MDPI

Article

Behavior of Porewater Pressures in an Earth Dam by Principal Component Analysis

Seong-Kyu Yun [1], Jiseong Kim [2], Eun-Sang Im [3] and Gichun Kang [4,*]

1 Engineering Research Institute, Gyeongsang National University, Jinju 52828, Korea; tjdrb330@gnu.ac.kr
2 Department of Cadastre & Civil Engineering, Vision College of Jeonju, Jeongju 55069, Korea; kimjs@jvision.ac.kr
3 Water Energy & Infrastructure Research Center, K-Water, Daejeon 34045, Korea; esim89@kwater.or.kr
4 Department of Civil Engineering, College of Engineering, Gyeongsang National University, Jinju 52828, Korea
* Correspondence: gkang@gnu.ac.kr; Tel.: +82-55-772-1792

Abstract: This study deals with the utilization of the pore pressure meter for evaluating the stability of a dam through the correlation between the porewater pressure installed in the fill dam and the water level of the dam. To this end, principal components analysis was performed on a total of 18 porewater pressure meters, and the main components were classified into three groups: internal (Group A), external (Group B), and upper (Group C), on the basis of the seepage line formed within the dam body. The coefficient of correlation between the porewater pressure and water level was found to be 0.86 to 1.00, indicating a strong positive linear relationship. This means that the maintenance of the dam is possible through the pore pressure meter present in Group A. Furthermore, the regression analysis for porewater pressures and water levels resulted in a linear regression model with the coefficient of determination (R^2) of Group A being between 0.74 and 0.99. In particular, R^2 between the power water pressure installed at the base of the dam and the water level was more than 0.99. Therefore, it was shown that the prediction of the porewater pressure is possible by using the relationships with the water level, making it possible to determine the safety of the dam by comparing it with the currently measured values.

Keywords: fill dam; principal components analysis; porewater pressure; seepage line; regression analysis

Citation: Yun, S.-K.; Kim, J.; Im, E.-S.; Kang, G. Behavior of Porewater Pressures in an Earth Dam by Principal Component Analysis. *Water* **2022**, *14*, 672. https://doi.org/10.3390/w14040672

Academic Editors: Rui Pang, Binghan Xue, Xiang Yu and Yantao Zhu

Received: 21 January 2022
Accepted: 17 February 2022
Published: 21 February 2022

1. Introduction

Among the national infrastructure facilities, dams serve important functions such as that of electricity production and flood control, as well as water supply for living, industry, and agriculture. Dams in the Republic of Korea are more than 18,000 in number, and medium-sized and small dams have been built and managed since the 1960s. Due to the construction of dams, which started from Japanese colonial era, multi-purpose dams and water supply dams that have existed for more than 30 years account for more than 60%. The aging of these dams is affecting their structural stability. In this regard, dam accidents can cause huge human/economic losses. Therefore, recently, studies on the utilization of dam measurement data have been conducted frequently for the safety (Pang et al., 2020 [1]) and maintenance of dams (K-water, 2019) [2].

According to the International Commission on Large Dams, about 150,000 cases of dam collapse and accidents have been reported worldwide, and more than 2000 since the 12th century and more than 200 after the 20th century have been reported to have caused casualties involving more than 238,000 people (ICOLD, 1995) [3]. Outside the country, more than 200 dams have collapsed in Italy, the United States, and France in the 1900s, causing more than 11,000 casualties (Jansen, 1983) [4]. In Korea, more than 100 people died in the collapse of the Hyogiri Dam in Namwon, North Jeolla Province, in 1961 (Chang et al., 1998) [5]. Yeoncheon Dam, located in Yeoncheon-gun, Gyeonggi Province, caused huge economic damage to Paju and Pocheon, as well as surrounding areas, due to dam collapse in two instances due to overflowing caused by heavy rains in the summers of 1996 and 1998. Therefore, in order to cope with the threat of safety of the dam, the main agent managing the dam must assess the stability of the dam by installing various measuring instruments and conducting real-time or regular stability evaluations (Kang et al., 2018) [6].

In particular, the porewater pressure meter, which is installed and operated during construction of the fill dam, is an important measurement item for monitoring the barrier role of the dam along with the water seepage. In a study by the U.S. Commission on Large Dams (USCOLD, 1975) [7], 77 cases of collapse of rockfill dams in the USA were analyzed, with 44% of the damage being found to be caused by leaks and piping through the dam's body or foundation, and finding that the porewater pressure meter is an important measurement item for monitoring this type of destruction. For this reason, research on the use of porewater pressure is being actively conducted (Wang et al., 2018 [8]). The porewater pressure meter is only used in Korea to check the presence or absence of a stable barrier role after construction. The reason is that the porewater pressure meter buried inside the dam has a relatively short lifespan compared to the external instruments, since it is difficult to maintain due to aging caused by wet conditions in the meter and breaking caused by deformation of the dam due to time passing after completion. However, thanks to the recent development of measuring technology nationally and abroad, the installation and operation of instruments have been stable, and the importance of the porewater pressure meter, which is a major measurement item of the fill dam, has emerged (Kang et al., 2020) [9].

In this study, we intended to analyze the correlation between the porewater pressure installed in the fill dam and the water level of the dam, and conducted a stability evaluation to present a plan for the utilization of the porewater pressure meter for future dam safety.

2. Region Subject to the Study

2.1. Dam Subject to the Study

Figure 1 shows the overall view of Gampo Dam located in the Republic of Korea. It is a central core-type rockfill dam with a total capacity of 2.39 million m^3, a height of 35 m, a dam extension of 108 m, and a volume of 190,000 m^3. The dam was completed in 2007, and 13 years have passed since impoundment.

Figure 1. Study fill dam.

2.2. Installation Status of Measuring Instruments

The measuring instruments of the dam to be studied were installed for stability and behavior analysis during construction, after completion, and impoundment, and they detect stability and behavior of the dam through the analysis of measured data. Their purpose is to be utilized for operation and maintenance of the dam for long-term use. In general, in the case of the fill dam, various instruments such as porewater pressure meter, earth pressure gauge, inclinometer, and differential settlement gauge are installed. In the case of the porewater pressure meter subject to this study, a total of 18 locations are installed in the base, core, and filter sections of the dam, as shown in Figure 2, in order to determine the appropriateness of the penetration outflow through the variation of the water pressure after impoundment. Measured data were automatically measured from 1 June 2009 after impoundment, and this study analyzed measurement data for around 10 years until 10 June 2019.

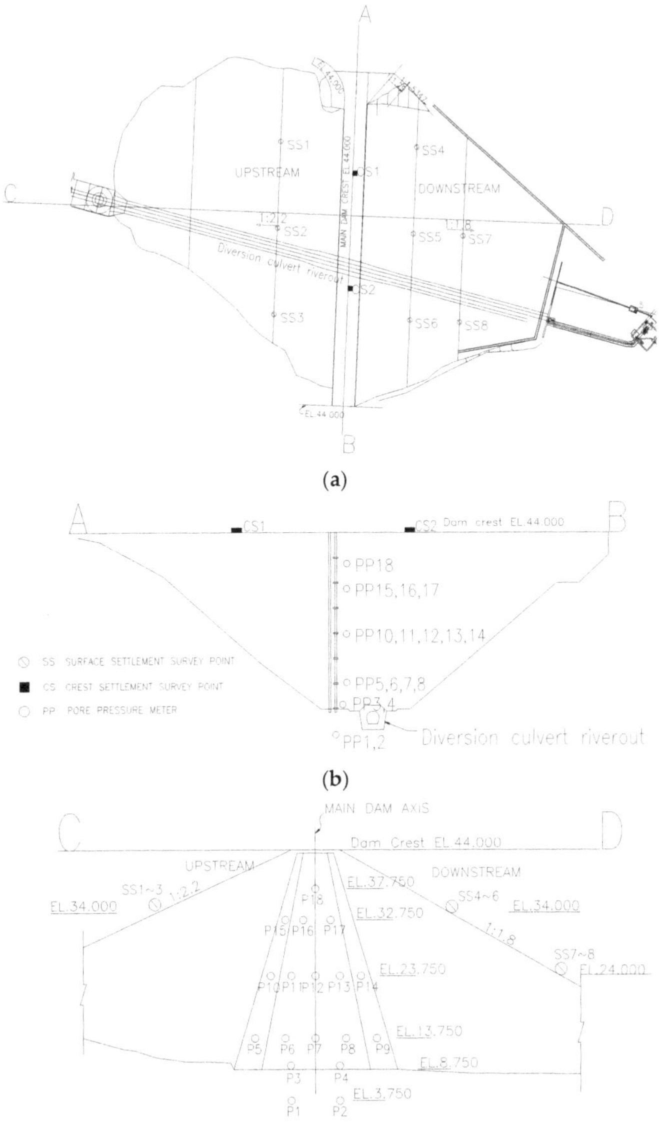

(a)

(b)

(c)

Figure 2. Location of porewater pressure transducers: (**a**) on the floor plan; (**b**) on the cross-section (A–B); (**c**) on the cross-section (C–D).

3. Measurement Status

3.1. *Water Level*

Figure 3 shows changes in water level and rainfall over time in the dam subject to the study. The missing rate of the water level during the data collection period was 0%, indicating that the data are well managed. As shown in Figure 3, it can be seen that the water level rises when rainfall increases rapidly, and that it is operated and managed within the normal high-water level (EL.40 m). The average of the water level gauge was 35.68 m, and its standard deviation was 2.97 m, while the average rainfall was 3.27 mm (std 12.13 mm), and the maximum rainfall was 234 mm.

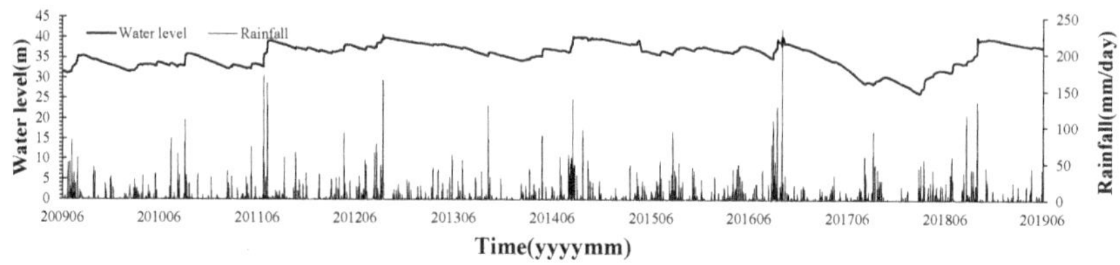

Figure 3. Water level and rainfall.

3.2. *Porewater Pressure*

The measurement results from 1 June 2009 to 10 June 2019 were analyzed. The average missing rate of the porewater pressure meter during the data collection period was 7.5%, and the missing frequency for each point was similar. The pore pressure meter point PP09 was excluded from the analysis as it became inoperable after completion. Figure 4 are representations of the water level and the porewater pressure at point PP01–PP18. Each analysis result can be represented by Table 1. It shows that the water pressure is high at PP01 and PP02, which are located at the bottom, affected by the seepage line formed inside the dam's body. Located in the core part of the dam, PP12 showed relatively low water pressure, which indicates that PP12 may be at the boundary of the seepage line. In the case of PP04 and PP08 installed on the downstream side of the dam's core part, it can be seen that relatively large values work in the early stages of impoundment and become smaller over time. It may be determined that the water pressure worked as the seepage line was formed during the early impoundment and the water pressure became lower as the dam became stabilized. PP13 and PP14 are determined to be located at the upper part of the seepage line, measured to have negative porewater pressure with averages of −26 kPa and −45 kPa, respectively. In the case of PP16 to PP18 installed at the top of the upper side of the dam, it was shown that the negative (−) porewater pressure mainly worked. It is installed near the water level and is considered to be the negative (−) porewater pressure caused by the unsaturated ground when the low water level becomes lower.

Table 1. Pore pressure gauge technical statistics analysis result.

PP	Mean (kPa)	Standard Deviation (kPa)	Range		
			Minimum (kPa)	Medium (kPa)	Maximum (kPa)
PP0001	3.33	0.25	2.53	3.39	3.73
PP0002	2.53	0.22	1.82	2.58	2.88
PP0003	1.38	0.22	0.78	1.38	1.93
PP0004	0.74	0.25	0.32	0.67	1.27
PP0005	2.11	0.30	1.18	2.19	2.55
PP0006	1.84	0.25	0.97	1.90	2.21
PP0007	1.39	0.17	0.72	1.42	1.64
PP0008	0.70	0.23	0.33	0.64	1.37
PP0010	0.89	0.30	−0.09	0.96	1.36
PP0011	0.65	0.27	−0.25	0.71	1.09
PP0012	0.32	0.19	−0.34	0.36	0.58
PP0013	−0.26	0.07	−0.45	−0.27	−0.07
PP0014	−0.46	0.04	−0.53	−0.46	−0.36
PP0015	−0.19	0.23	−0.62	−0.15	0.25
PP0016	−0.40	0.75	−11.22	−0.30	0.27
PP0017	−0.25	0.28	−4.71	−0.22	0.05
PP0018	−0.30	0.10	−2.45	−0.31	−0.13

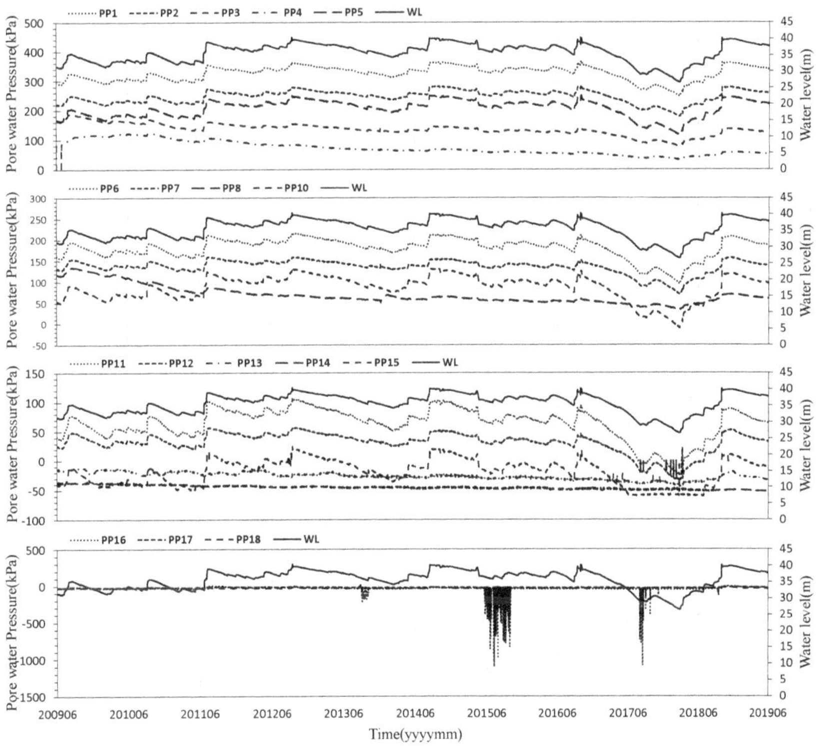

Figure 4. Relationship between porewater pressure and water level.

4. Principal Component Analysis (PCA)

Principal component analysis is one of the methods utilized in multivariate analysis. It is a method of reducing high-dimensional data to low-dimensional data. The concept was first proposed in the 1900s, similarly to the principal axis theorem theory (Person, 1901) [10], and was later established by Hotelling (1933) [11]. Unlike multidimensional scaling, which can be applied only with distance matrices between objects (Kwon, 2016) [12], principal component analysis is widely used in various disciplines such as the humanities, economics, business, and engineering (Kwon et al., 2020) [13] due to its advantages of understanding potential characteristics of variables and its enabling of spatial representation of consumer perceptions and preferences in addition to data reduction (Kim, 2016) [14].

Principal component analysis is a method of summarizing and analyzing a small number of comprehensive characteristics using linear combinations of independent variables while minimizing the loss of information. It is a technique that is often used to identify factors behind a particular idea or when it is more reasonable to deal with specific ideas and their background factors comprehensively rather than independently (Gwak and Kim, 2016 [15]; Park and Rhee, 2012 [16]). For example, it would be more reasonable and rational to identify measured items used for safety monitoring of dams by integrating them into several common factors rather than to identify them in detail by instrument.

Principal component analysis, which is a linear combination similar to regression analysis, can be expressed in the following formula (Lee, 2012 [17]; Lee and Nho, 2015 [18], Nho, 2007 [19]):

$$\begin{aligned} z_1 &= a_{11}x_1 + a_{12}x_2 + \cdots + a_{1p}x_p \\ z_2 &= a_{21}x_1 + a_{22}x_2 + \cdots + a_{2p}x_p \\ z_n &= a_{n1}x_1 + a_{n2}x_2 + \cdots + a_{np}x_p \end{aligned} \tag{1}$$

where $z_1 \sim z_n$, $a_{11} \sim a_{np,}$ *and* $x_1 \sim x_p$ are the principal component, the coefficient of principal component analysis, and the independent variable, respectively.

4.1. Calculation of the Principal Components

The linear combination of regression analysis and principal component analysis is similar, but for regression analysis linear combinations, it is a linear combination that minimizes the independent variable linear combination deviation for the dependent variable. On the other hand, the linear combination of principal component analysis is different, as it is the linear combination minimizing the deviation by the linear combination of independent variables. Therefore, the principal component deviation is calculated as the vertical line, which is the shortest connection line to the linear combination line with the variable. The linear combination minimizing this deviation (R) is shown in Equation (2), which is interpreted as a linear combination minimizing the loss of information of the variables (Lee, 2012 [17]; Lee and Nho, 2015 [18], Nho, 2007 [19]).

$$R = \frac{|a_{11}x_1 + a_{12}x_2 + \cdots + a_{1p}x_p|}{\sqrt{a_{11}^2 + a_{12}^2 + a_{1p}^2}} \tag{2}$$

In order for the loss of information in principal component analysis to be minimized, the value of simultaneous equation composed by the assumption of "$a_{11} + a_{12} + \cdots + a_{1p} = 1$" is changed into the Lagrange function (λ), and the partial differential value of the squared deviation is set as zero. The formula for this is expressed in Equations (3) and (4) (Lee, 2012 [17]).

$$\mathrm{F} = R\,(a_{11}, a_{12}, \ldots, a_{1p}) - \lambda\,(a_{11}, a_{12}, \ldots, a_{1p}) \tag{3}$$

$$\frac{\partial F}{\partial a_{11}} = 0,\ \frac{\partial F}{\partial a_{12}} = 0,\ \ldots, \frac{\partial F}{\partial a_{1p}} = 0 \tag{4}$$

In the regular equation of the principal component analysis, the information loss minimization is calculated with the Lagrange function λ value. That is, if there are n

independent variables, then there are n principal components that aggregate them. When the Lagrange function λ value reaches its maximum, the information loss becomes the lowest, which is shown as the maximum of the principal component variance according to the relationship found in Equation (5) (Lee, 2012 [17]).

$$[Sum\ of\ independent\ variable\ variance] = \left[\frac{Sum\ of\ squared\ information\ loss}{measured\ number} - 1\right] + [Principal\ component\ variance] \quad (5)$$

In the principal component analysis, the primary principal component of eigenvector $(a_{11}, a_{12}, \ldots a_{1p})$ according to the maximum eigenvalue λ_1 can be expressed as $z_1 = a_{11}x_1 + a_{12}x_2 + \cdots + a_{1p}x_p$, and the proportion of this primary principal component can be expressed as Equation (6) (Lee, 2012 [17]).

$$\mathrm{P} = \frac{\lambda_1}{\lambda_1 + \lambda_2 + \cdots + \lambda_p} \quad (6)$$

4.2. Analysis of Principal Components and Score of Principal Components

It is not easy to accurately interpret what the principal component means when the measured variables are abbreviated to a small number of principal components while minimizing the loss of information of the measured variables. Principal component analysis is performed according to existing studies or empirical results, and they are generally interpreted in an abstract and conceptual sense, not as measurable concepts, as the meanings are given subjectively by the research analyst on the basis of the principal component calculation coefficient of the measured variables. The analysis of the principal components is conducted on the basis of the interpretation of the coefficients of the formula comprising the principal components, and the researcher properly interprets the names and contents of the principal components in consideration of the characteristics and coefficients of the measurement variables.

If there are n independent variables, there are also n principal components that aggregate them, and there is no prescribed method for determining the number of principal components. Generally, the criteria for determining the number of principal components are divided into the case on the basis of the eigenvalue 1.0 of the Lagrange function, or the case where the number of components are determined at a level of 80% of cumulative proportion of the principal components cumulated from the maximum proportion of the principal components (Lee, 2012 [17]). In order highly correlated measured variables to be summarized into a small number of principal components, the principal component score for each measurement case can be calculated after calculating the coefficient of the formula and organizing the principal component calculation formula. The analysis of the principal components for the measurement case is performed according to the principal component score, and the formula for calculating the principal component score of the principal component i is a functional formula according to the measurement of measured variables of the principal component coefficient and a functional formula according to the average value of measured variables of the principal component coefficient, which is expressed as Equation (7) (Lee, 2012 [17]).

$$Z\ (x_1, x_2, \ldots, x_p) = a_{i1}x_1 + a_{i2}x_2 + \cdots a_{ip}x_p - \left(a_{i1}x_1^* + a_{i2}x_2^*, \ldots, a_{ip}x_p^*\right) \quad (7)$$

where $a_{i1}, a_{i2} \ldots a_{ip}$ represents the coefficient of principal component analysis, and $x_1^*, x_2^*, \ldots, x_p^*$ represents the average value of variables.

5. Result of Porewater Pressure Analysis

5.1. Linear Interpolation of Missing Data

Figure 5 shows the raw data of PP01 and the time series data after linear interpolation. The average missing rate was 7.5%. The missing point showed a downward curve with a value of 0, as shown in the figure, with the measurement recorded as "0". As shown in the figure, according to the time series distribution of the porewater pressure meter, it was determined that unusual behaviors such as severe vertical vibration or large amount of outliers occurring locally were not observed in the time series distribution, and therefore linear interpolation using Spline was conducted. In other words, linear interpolation was conducted for missing sections by generating Spline, connecting local peaks extracted from inflection point analysis and synthesizing the extracted raw data and the generated Spline.

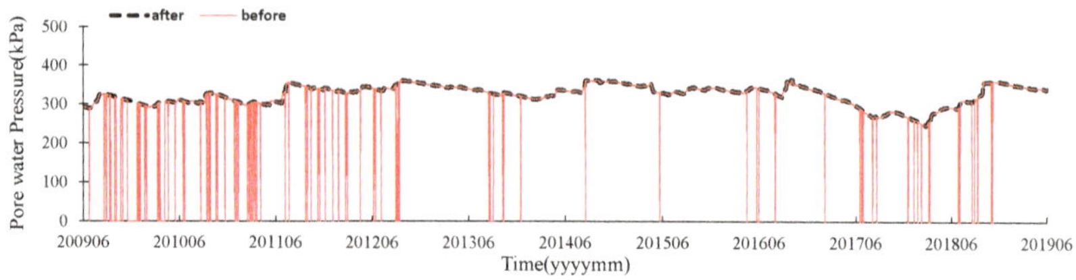

Figure 5. Linear interpolation for porewater pressure.

5.2. Principal Component Analysis

In this study, principal component analysis was performed to determine the appropriate group on the basis of the statistical similarity of the porewater pressure meters installed in the dam body under study. Although the criteria for how many principal components to be adopted have not been theoretically determined, the porewater pressure meter was summarized in three principal components by applying the generally applied grouping criteria (Lee, 2012 [17]): (1) 1 or higher correlation matrix eigenvalue (latent), (2) cumulative proportion of more than 70~80%. Figure 6 represents the component chart according to the three selected principal components (Table 2). The results of the group classification of the porewater pressure meter performed from the component score coefficient matrix results are shown in Table 3. According to the component scores in Table 3, PP01, PP02, PP03, PP05, PP06, PP07, PP10, PP11, PP12, and PP15 were classified as Group A; PP04, PP08, PP13, and PP14 as Group B; and PP16, PP17, and PP18 as Group C.

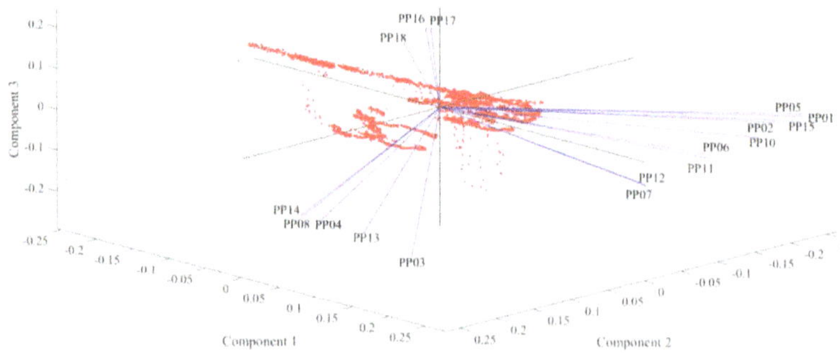

Figure 6. Results of the PCA.

Table 2. Latent and cumulative scores for PCA.

Component	Latent	Cumulative (%)
Comp.1	9.79	60
Comp.2	4.23	80
Comp.3	2.10	90

Table 3. Coefficient matrix for component scores and group distributions for porewater pressures.

	Comp.1	Comp.2	Comp.3	Group
PP0001	0.28	−0.23	0.01	A
PP0002	0.29	−0.22	0.00	A
PP0003	0.25	0.29	−0.14	A
PP0004	0.16	0.39	−0.14	B
PP0005	0.28	−0.22	0.01	A
PP0006	0.31	−0.10	−0.03	A
PP0007	0.32	0.00	−0.06	A
PP0008	0.13	0.41	−0.15	B
PP0010	0.30	−0.17	−0.01	A
PP0011	0.31	−0.10	−0.04	A
PP0012	0.32	0.00	−0.07	A
PP0013	0.20	0.33	−0.13	B
PP0014	0.12	0.41	−0.15	B
PP0015	0.28	−0.19	0.01	A
PP0016	0.09	0.14	0.59	C
PP0017	0.11	0.14	0.59	C
PP0018	0.12	0.21	0.45	C

Figure 7 shows the location of three group-specific porewater pressure meters classified on the basis of principal component analysis. As shown in the figure, the porewater pressure meters classified as Group A were located primarily in the upper/lower part of the upstream dam, whereas those classified as Group B were mainly distributed in the downstream part/intermediate part of the dam, and those classified as group C were mainly distributed in the upper part of the upstream dam. They were classified as internal, external, and upper on the basis of the seepage line formed within the dam body.

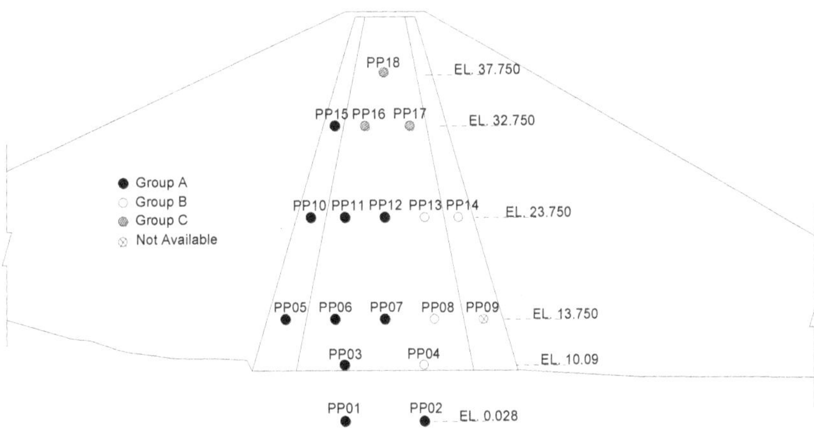

Figure 7. Distributions of porewater pressure transducers by PCA.

5.3. Correlation Analysis by Group

As for the correlation analysis for the porewater pressure meter, the instrument-specific correlation analysis within the group and the correlation analysis between the classified group and water level were conducted. Correlation analysis in Group A (PP01, PP02, PP03, PP05, PP06, PP07, PP10, PP11, PP12, PP15), Group B (PP04, PP08, PP13, PP14), and Group C (PP16, PP17, PP18) is shown in Table 4. The correlation coefficient in Group A showed a strong positive correlation, except for PP03. From these results, we believe that individual instruments in Group A may have been complementary to each other when they missed the data or exhibited mechanical abnormal behavior. For Group B and Group C, the correlation coefficients between the instruments in the group were 0.83 to 0.92 and

0.67 to 0.90, respectively, indicating relatively high positive correlations but no correlations with other groups. Furthermore, the relationship between the porewater pressure meter and the water level can be seen in Group A, with a correlation coefficient of 0.86 to 1.00, except for PP03, indicating a strong positive linear relationship with the water level. In addition, Group A can be classified into three groups according to the size of the correlation coefficient on the basis of the results of correlation analysis with water level. Group A-a (PP01, PP02, PP05) showed the highest correlation with the correlation coefficient, with a water level of 0.99, while Group A-b (PP06, PP10, PP11, PP15) had the correlation of 0.94–0.96, and Group A-c (PP07, PP12) had the correlation of 0.86–0.87. In other words, it was shown to have a high correlation with water level and to perform similar behaviors. On the other hand, Group B and Group C showed low correlations with water levels of −0.09 to 0.22 and 0.14 to 0.18, respectively, indicating that they did not correlate with water levels. Therefore, it is believed that the risk factors of dam safety such as increased penetration pressure due to the creation of flow paths inside the dam's body could be detected in advance from the porewater pressure meter in Group A, which showed no change in water level and significant change in porewater pressure.

Table 4. Relationships between water level and porewater pressures in the groups.

Group		Porewater Pressure Gauge, PP	Correlation Coefficient, r	
			Correlation Coefficient within Group	Correlation Coefficient between Group and Water Level
A	a	PP01, PP02, PP05	>0.99	>0.99
	b	PP06, PP10, PP11, PP15	0.94~0.99	0.94~0.96
	c	PP07, PP12	0.99	0.86~0.87
	*	PP03,	0.40~0.78	0.38
B		PP04, PP08, PP13, PP14	0.83~0.92	−0.09~0.22
C		PP16, PP17, PP18	0.67~0.90	0.14~0.18

* It couldn't be classified as any group because reliability of the instrument was degraded.

Figure 8 shows a comparison of the measured porewater pressure and water level for each group. As shown in Figure 8a,c,e, the size of the porewater pressure in Group A varied depending on the location of the installation, but it tended to be similar to the water level. In other words, the trend of water level was consistent with the porewater pressure, represented in the z-score distribution chart of Group A shown in Figure 8b,d,f. This indicates that Group A had a very high correlation with water level as well as with the correlation between instruments in the group, as mentioned above. On the other hand, as shown in Figures 9 and 10, Group B and Group C did not show much of a correlation with water levels. Figure 11 shows changes in the seepage line of the dam prepared from the measurement results of the porewater pressure installed in the dam following the changes in water level. As shown in Figure 11, Group A-a existed inside the line, and Group A-b was included inside the line when the water level was 37.75 m. If the water level was relatively low at EL. 26.55 m, PP11 and PP12 would be outside the boundary of the seepage line, and therefore the porewater pressure would not normally be measured, whereas PP11 and PP12 would be included inside the seepage line in a normal high water level of EL.40 m. In other words, it is believed that the porewater pressure is related to the path of the seepage line caused by changes in water level.

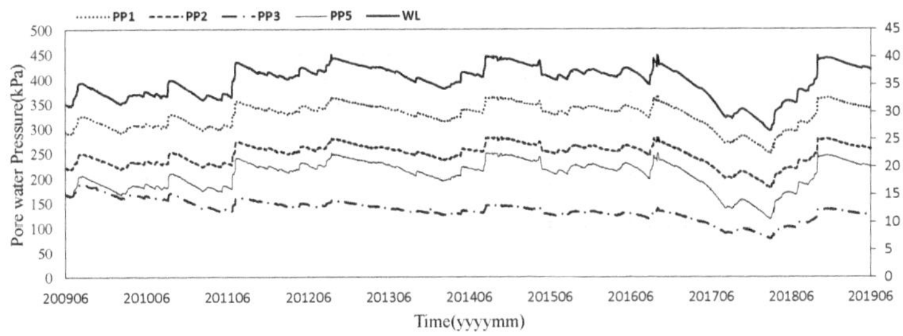

(**a**)

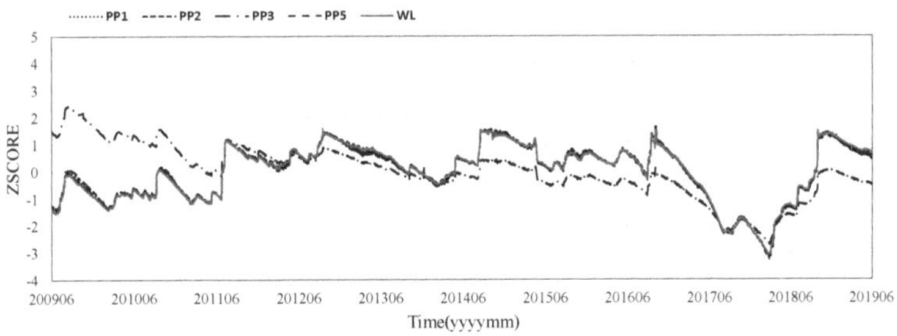

(**b**)

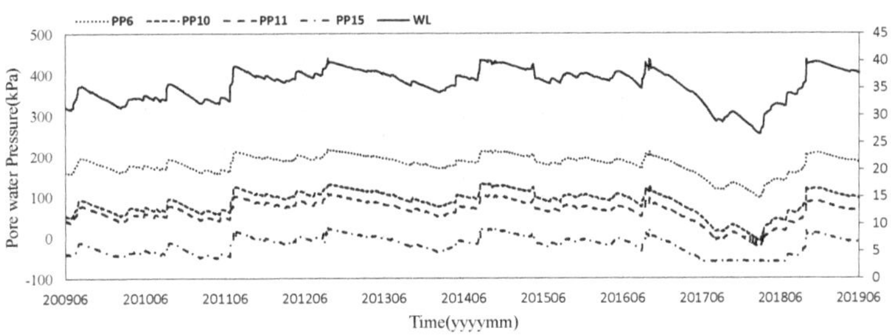

(**c**)

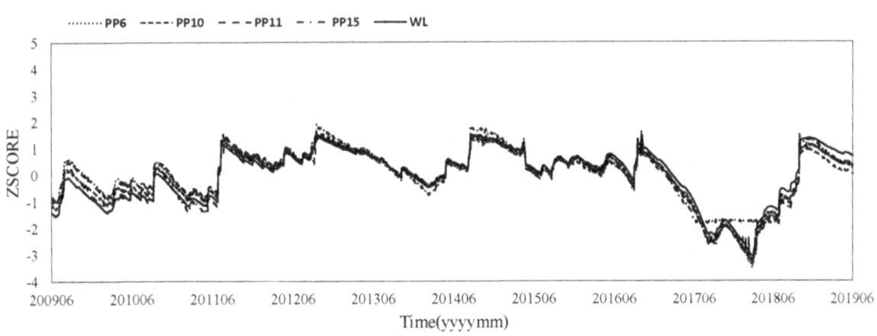

(**d**)

Figure 8. *Cont.*

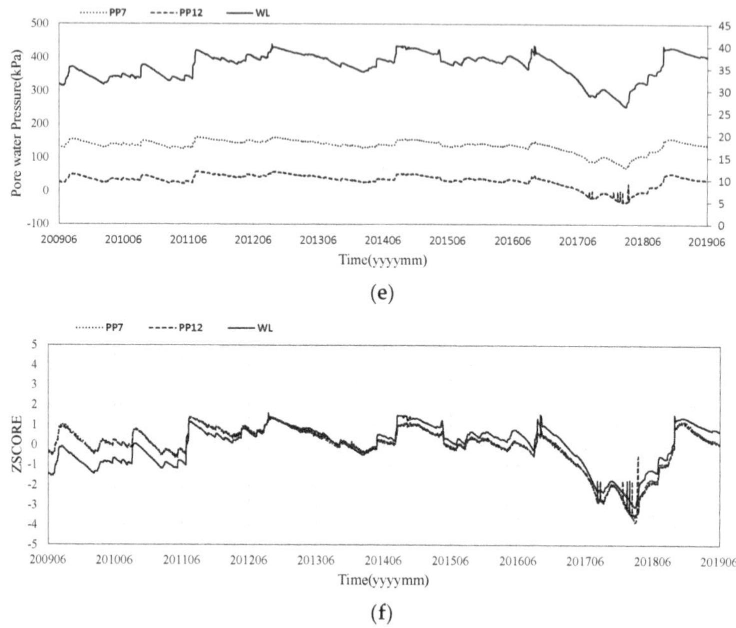

Figure 8. Relationships between water level and porewater pressure for group A: (**a**) comparisons with water level and Group A-a; (**b**) z-score for group A-a with water level; (**c**) comparisons with water level and Group A-b; (**d**) z-score for group A-b with water level; (**e**) comparisons with water level and Group A-c; (**f**) z-score for group A-c with water level.

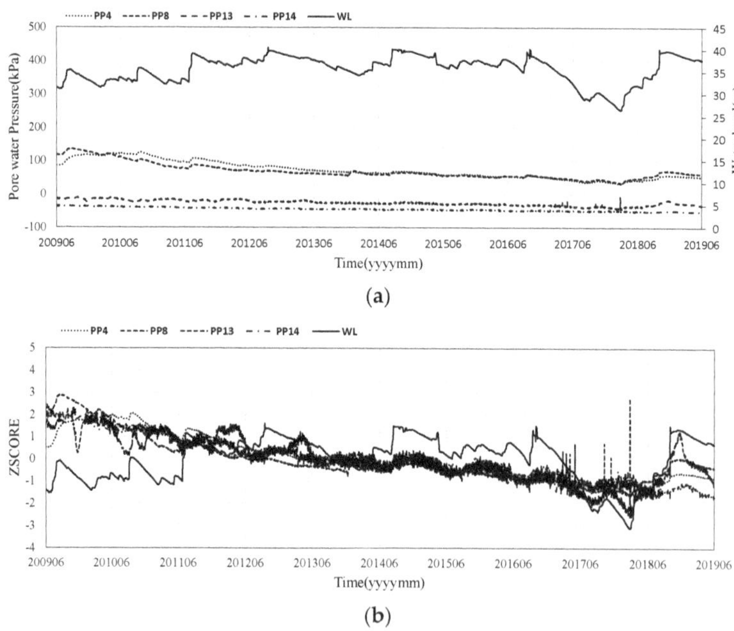

Figure 9. Relationships between water level and porewater pressure for group B: (**a**) comparisons with water level and Group B; (**b**) z-score for Group B with water level.

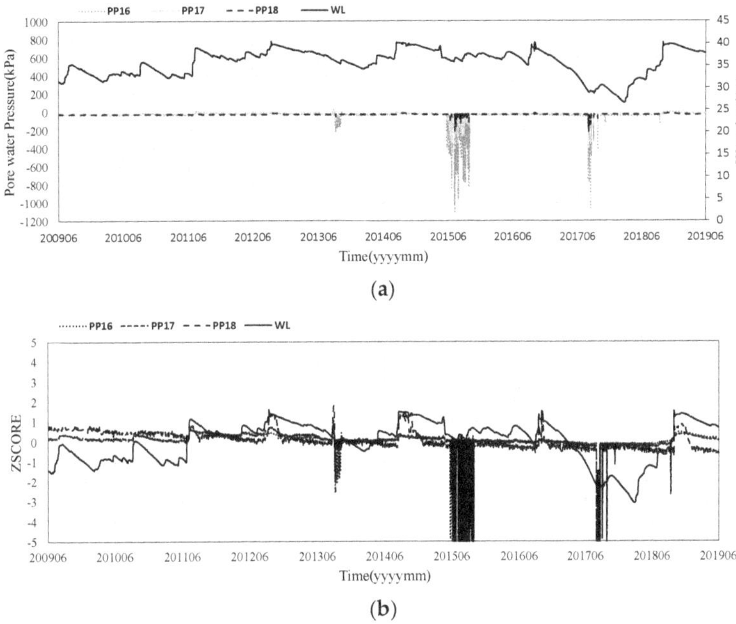

Figure 10. Relationships between water level and porewater pressure for Group C: (**a**) comparisons with water level and Group C; (**b**) z-score for Group C with water level.

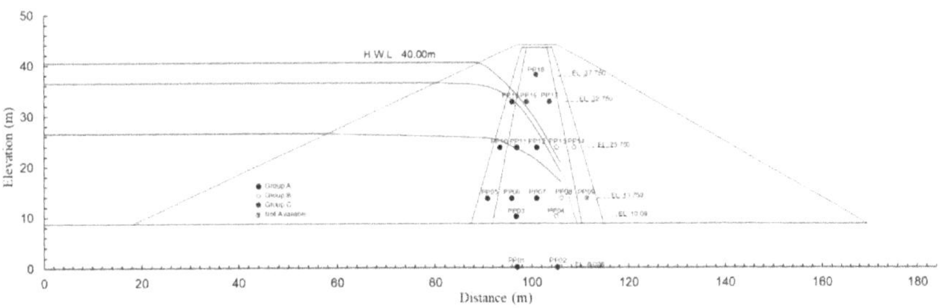

Figure 11. Expected phreatic line drawn from measured porewater pressures.

For PP03, the porewater pressure over time tended to be similar to water level, as shown in Figure 8a. However, as shown in Figure 8b, the z-score of PP03 was relatively large at the beginning after its completion and relatively small after about 2014, and therefore it was determined that the reliability of the instrument was degraded. That is, it was classified as Group A by the principal component analysis, but as shown in Table 4, the correlation coefficients between the instruments and the water levels in the group were 0.40~0.78 and 0.38, respectively, and was excluded from the regression analysis of following section.

5.4. Regression Analysis

Since the porewater pressure in Group A showed high coefficient of correlation with water level, we propose a model that can predict the porewater pressure according to the changes in water level by conducting a regression analysis on water level and porewater pressure meter. The regression analysis was performed for the case where the water level was set as an explanatory variable, and each porewater pressure was set as a response variable. Table 5 summarizes the results of the development of the primary linear regression model with the porewater pressure meter and water level. Figure 12 illustrates the results of the development of the primary linear regression model, showing only the results for the highly correlated Group A. As shown in Figure 12a,b,d, PP01, PP02, and PP05 were classified as Group A-a with an R^2 of 0.99 or more, drawing on highly related relation formula. The R^2 of Group A-b was 0.8927 to 0.9676, indicating relatively high correlation. On the other hand, the R^2 of Group A-c was 0.7495 to 0.7429, indicating a relatively low correlation with the water level among Group A, but the group showed a stable distribution of each measured value around the developed regression line. For Groups B and C, the R^2 was distributed from 0.001 to 0.0497, and 0.0187 to 0.0331, respectively, indicating no correlation with water level. As summarized in Table 5, the determination coefficients of the primary linear regression model with the porewater pressure meter and water level were analyzed to satisfy the appropriate explanatory power for Group A only, and the porewater pressure was determined to be predictable with changes in the water level. In other words, it is possible to predict porewater pressure when the water level is fixed as an explanatory variable and to determine the safety of the dam by comparing it with the currently measured value. In addition, an appropriateness test was conducted through an F-test that was performed for each regression model developed, which was also analyzed to satisfy statistical goodness of fit in Group A, aside from Group B and Group C. In addition, the variability of the measured value tended to increase as the installation position of the porewater pressure meter became higher due to the effect of the water level. It is determined that changes in the seepage line of the dam body due to the fluctuation of the water level have a significant impact on the measured value.

Table 5. Regression analysis between water level and porewater pressures.

y	x	y = α + βx		R^2	F-Test		
		α	β		F	*p*-Value	Test
PP0001	WL	30.37	8.297	0.996	9.8×10^5	0.000	ok
PP0002		−6.3588	7.1313	0.995	3.75×10^5	0.000	ok
PP0003		35.734	2.8063	0.146	-	-	NG
PP0005		−140.71	9.7398	0.997	1.06×10^6	0.000	ok
PP0006		−97.607	7.7965	0.911	3.76×10^4	0.000	ok
PP0007		−41.599	4.9886	0.749	1.09×10^4	0.000	ok
PP0010		−261.76	9.7793	0.968	1.09×10^5	0.000	ok
PP0011		−240.91	8.544	0.893	3.04×10^4	0.000	ok
PP0012		−157.17	5.2975	0.743	1.06×10^4	0.000	ok
PP0015		−276.24	7.2211	0.918	4.08×10^4	0.000	ok

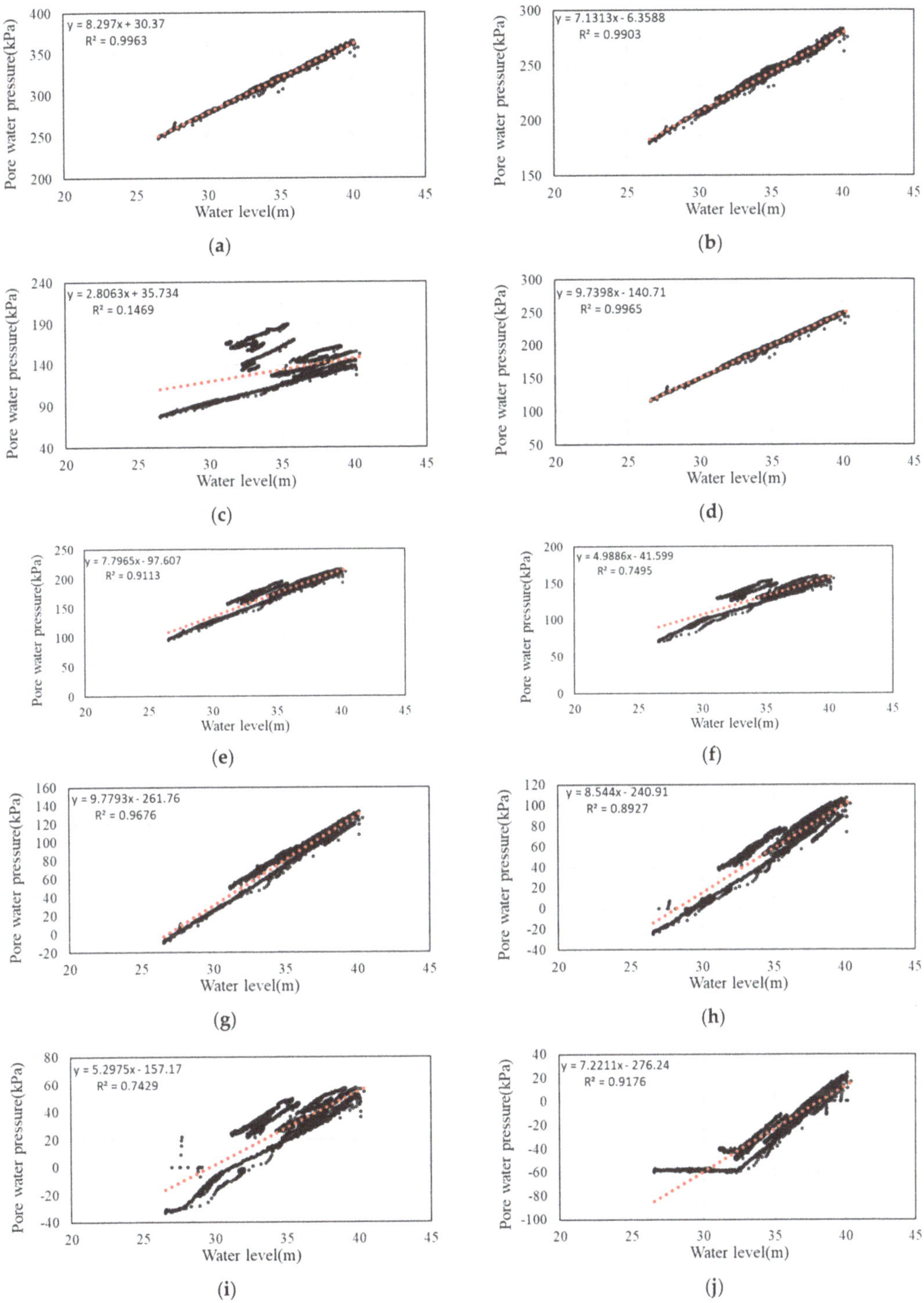

Figure 12. Regression analysis between water level and porewater pressures of Group A. (**a**) PP01; (**b**) PP02; (**c**) PP03; (**d**) PP05; (**e**) PP06; (**f**) PP07; (**g**) PP10; (**h**) PP11; (**i**) PP12 and (**j**) PP15.

6. Conclusions

In this study, the following results were obtained through analysis of the correlation with the dam water level for the porewater pressure meter utilized to predict leakage and piping of the fill dam.

1. As the result of linear interpolation for missing porewater pressure and principal component analysis, we determined the three groups: middle/lower part of upstream dam (Group A), middle part of downstream dam (Group B), and upper part of the upstream dam (Group C). Similar behaviors were shown between porewater pressure meters within a group.
2. The correlation analysis within Group A present inside the seepage line showed a correlation of 0.94 or higher, which is considered to be complementary.
3. The primary linear regression analysis of Group A, satisfying a significant correlation between water level and porewater pressure, showed the determination coefficient (R^2) in the range of 0.75 to 0.99, satisfying high explanatory power, and statistical goodness of fit was also found to be significant at the significance level of 5%.
4. Through the regression analysis, we found the R^2 of Group A-a and Group A-b to be 0.99 or more at 0.8927 and 0.9676, respectively, showing a relatively high correlation. The proposed regression analysis can predict the porewater pressure and the seepage line at constant water level.
5. Therefore, it is possible to predict the porewater pressure when there is a change in the porewater pressure at a constant dam water level, as well as when the low water level is fixed as an explanatory variable and pre-detection of the threat to dam safety due to leakage or piping inside the body is possible. Thus, the comparison with currently measured values can determine the safety of the dam.

From the results of this study, we expect that maintenance by the porewater pressure meter in Group A is required and that an early warning system can be operated using the model formula derived above for real-time safety monitoring of dam safety.

Author Contributions: Conceptualization, methodology, G.K. and S.-K.Y.; modeling tests and data curation, J.K. and E.-S.I.; writing—original draft preparation, S.-K.Y. and G.K.; writing—review and editing, G.K. and S.-K.Y.; supervision, G.K.; funding acquisition, G.K. All authors have read and agreed to the published version of the manuscript.

Funding: This research was supported by the Basic Science Research Program through the National Research Foundation of Korea (NRF) funded by the Ministry of Education (no. 2020R1I1A3067248).

Institutional Review Board Statement: Not applicable.

Informed Consent Statement: Not applicable.

Data Availability Statement: The data presented in this study are available on request from the corresponding author. The data are not publicly available due to privacy reasons.

Conflicts of Interest: The authors declare no conflict of interest.

References

1. Rui, P.; Xu, B.; Zhou, Y.; Zhang, X.; Wang, X. Fragility analysis of high CFRDs subjected to mainshock-aftershock sequences based on plastic failure. *Eng. Struct.* **2020**, *206*, 110152.
2. K-water. *A Study on the Combined Analysis Method of Measurement Data for Dam Safety and Maintenance (II)*; K-water: Daejeon, Korea, 2019.
3. ICOLD. *Bulletin 99, Dam Failures, Statistical Analysis*; International Commission on Large Dams: Paris, France, 1995; p. 77.
4. Jansen, R.B. *Dams and Public Safety*; United States Department of the Interior: Washington, DC, USA, 1983; p. 94.
5. Chang, P.W.; Park, Y.K.; Woo, C.W. Evaluation of Degradation and Safety of Small Agricultural Reservoir-Case Study for Kosam Reservoir in Kyungki Provice. *J. Korean Soc. Agric. Eng.* **1998**, *40*, 49–56.
6. Kang, G.-C.; Kim, S.-J.; Ryou, M.; Ahn, K.; Kim, J. Case Study on Real-time Monitoring System of Dam Displacement Based on SAMP (Single Antenna-Multi Point) GPS. *Korean Cadastre Inf. Assoc.* **2018**, *20*, 3–13. [CrossRef]
7. USCOLD. *Lessons of Dam Incidents*; Committee on Dam Safety of the United States Committee on Large Dams: New York, NY, USA, 1975.

8. Wang, Y.; Le, Y.; Yin, T.; Yu, L.; Huo, Z. Stability Analysis of Partially Submerged Landslide with the Consideration of the Relationship between Porewater Pressure and Seepage Force. *Geofluids* **2018**, 1–9. [CrossRef]
9. Kang, G.-C.; Kim, D.-H.; Yoon, S.-M.; Jang, B.-S.; Kim, J.-S. Assessing the Stability of Fill Dams by Relationship between Water Level and Porewater Pressure. *J. Korean Geotech. Soc.* **2020**, *36*, 5–15.
10. Pearson, K.L., III. On lines and planes of closest fit to systems of points in space. *Lond. Edinb. Dublin Philos. Mag. J. Sci.* **1901**, *2*, 559–572. [CrossRef]
11. Hotelling, H. Analysis of a complex of statistical variables into principal components. *J. Educ. Psychol.* **1933**, *24*, 417. [CrossRef]
12. Kwon, Y.; Jang, K.; Jang, I.G. Multidimensional Scaling Analysis of Inter-regional Public Transit Services: Focusing on Inter-regional Railways. *J. Korean Soc. Railw.* **2016**, *19*, 243–250. [CrossRef]
13. Kwon, Y.; Kim, S.; Byun, J. Cognitive Perception of an Eco-friendly Public Transportation: Using Principal Component Analysis. *Korea Inst. Intell. Transp. Syst.* **2020**, *19*, 71–82. [CrossRef]
14. Kim, S.H. *Multivariate Data Anlysis*; Bob-Munsa: Seoul, Korea, 2016.
15. Gwak, B.-R.; Kim, I.-K. Characterization of Water Quality in Changnyeong-Haman Weir Section Using Statistical Analyses. *J. Korean Soc. Environ. Eng.* **2016**, *38*, 71–78. [CrossRef]
16. Park, J.S.; Rhee, K.H. Study on characteristics of water quality variation in the Yeongsan River using Multivariate Analysis. *J. Korean Soc. Water Sci. Technol.* **2012**, *20*, 61–72.
17. Lee, J. *Matlab Recipes for Statistical Analysis*; A-Jin: Seoul, Korea, 2012.
18. Lee, H.; Nho, S. *Advanced Statistical Analysis*; Moonwoo: Seoul, Korea, 2015.
19. Nho, H. *Multivariate Analysis for Excel and SPSS*; Hyungseul: Paju, Korea, 2007.

water

MDPI

Article

Comparison of the Seismic Responses of an Arch Dam under Excitation from the Design Response Spectrum in the New and Old Chinese National Standards

Binghan Xue [1,2], Jing Wang [1,2,*], Na Li [1], Chao Zhang [2] and Jianguo Chen [2,3]

1 Engineering Research Center on Dike Safety and Disease Prevention of the Ministry of Water Resources, Yellow River Institute of Hydraulic Research, Zhengzhou 450003, China; xuebinghan@zzu.edu.cn (B.X.); zhuofan8227@163.com (N.L.)
2 School of Water Conservancy Engineering, Zhengzhou University, Zhengzhou 450001, China; czhang@zzu.edu.cn (C.Z.); chenjianguo198414@163.com (J.C.)
3 Guangxi Key Laboratory of Water Engineering Materials and Structures, Guangxi Institute of Water Resources Research, Nanning 530000, China
* Correspondence: wangjing@hky.yrcc.gov.cn

Abstract: The new Chinese national standard, *Standard for the seismic design of hydraulic structures* (GB51247-2018), has been published. Compared with the *Specifications for the seismic design of hydraulic structures* (SL203-1997), the standard design response spectrum curve was revised in the new national standard. In order to compare the seismic responses of an arch dam under excitation from the design response spectrum in the new and old standards, the dynamic calculation of a 240 m high arch dam is carried out by a three-dimensional finite element method. In the dynamic calculation, the B-differentiable equation is used to simulate the tension motion of arch dam contraction joints, and the multi-transmitting boundary method and the Westergaard added mass method are used to simulate the dam–infinite foundation and dam–reservoir interactions, respectively. The results show that the dynamic stress responses of the arch dam under excitation from the design response spectrum in the new standard are increased compared with those of the old standard. The seismic safety of an arch dam may decrease under excitation from the design response spectrum in the new standard. Thus, the seismic validation on built arch dams should be carried out by using the new standard when it is possible.

Keywords: arch dam; standard for seismic design; design response spectrum; seismic response

Citation: Xue, B.; Wang, J.; Li, N.; Zhang, C.; Chen, J. Comparison of the Seismic Responses of an Arch Dam under Excitation from the Design Response Spectrum in the New and Old Chinese National Standards. *Water* **2022**, *14*, 832. https://doi.org/10.3390/w14050832

Academic Editor: Helena M. Ramos

Received: 10 December 2021
Accepted: 3 March 2022
Published: 7 March 2022

1. Introduction

The seismic design standard is a special technical standard, which is constantly revised and improved based on existing scientific knowledge, economic conditions, and accumulated aseismic experience and data. In order to guide the design and construction of hydraulic structures in earthquake zones, the *Standard for the seismic design of hydraulic structures* (GB51247-2018) [1] has been published as a new national standard based on extensive investigation of the status of the seismic design of hydraulic structures in China and referring to the seismic design methods and standards of hydraulic structures in other country. Compared with the *Specifications for the seismic design of hydraulic structures* (SL203-1997) [2], the standard design response spectrum curve was revised in the new national standard.

The design response spectrum in various seismic design standards reflects the statistical law of the seismic acceleration response spectrum with different magnitudes and epicentral distances [3–5], which is an important basic parameter in seismic calculation using the dynamic time–history method [6]. The seismic safety of arch dams is very important, because any potential failure of arch dams may induce a major disaster [7]. In the seismic validation of arch dams, the acceleration time–history generated from the design response spectrum is usually used to calculate the dynamic responses of arch dams. Thus, it is necessary for dam designers to understand the difference between the seismic responses of arch dams under excitation from the design response spectrums in the new and old standards.

In order to compare the seismic responses of arch dams under excitation from the design response spectrum in the new and old standards, the dynamic calculation of a 240 m high arch dam was carried out by the three-dimensional finite element method. In the dynamic calculation, B-differentiable equations [8–10] are used to simulate the tension motion of arch dam contraction joints; the multi-transmitting boundary method [11] and Westergaard added mass method [12] are used to simulate the dam–infinite foundation and dam–reservoir interactions, respectively.

The rest of this paper is organized as follows. The design response spectrums in the old and new standards are compared in Section 2. The computational method, mode, and results of the arch dam–reservoir–foundation system are presented in Sections 3–5 respectively. Some conclusions are drawn in Section 6.

2. Comparison of Design Response Spectrum in Old and New Standards

In the standard for the seismic design of hydraulic structures, the product of the maximum amplification coefficient spectrum and the peak ground acceleration is used to represent the maximum absolute acceleration response spectrum. The maximum value of the standard design response spectrum β_{max}, the characteristic period T_g, and the attenuation index γ are the three main parameters that determine the design response spectrum. Taking an arch dam built on a Class I type site [13] as an example, the design response spectrums in the new and old standards are compared.

(1) The maximum value β_{max} of the design response spectrum suitable for seismic design of arch dams are set as 2.5 in the new and old standards.
(2) In the old standard, the value of the characteristic period T_g mainly considered the influence of site type, the characteristic period of the design response spectrum was suggested to be 0.2 s for arch dams built on a class I site; The new standard takes into account the effects of site type, epicentral distance, and magnitude; hence, the characteristic period of the design response spectrum is suggested to be 0.3 s for arch dams built on a class I site.
(3) The design response spectrum reflects the statistical law of the response spectrum of ground motion acceleration with different magnitudes and epicentral distances and, in fact, reflects the attenuation relation of the response spectrum. The attenuation index γ of the design response spectrum in the old standard was derived from the seismic intensity transformation in the seismic hazard analysis of the site and was suggested to be 0.9. The seismic response spectrum proposed in the new standard is based on the latest next generation attenuation relation (NGA) in the USA [14], which is a normalized mean response spectrum. The attenuation index γ is suggested to be 0.6.

Thus, as shown in Figure 1, the differences between the design response spectrum in the new standard and the old standard are mainly reflected in the characteristic period T_g and the attenuation index γ for arch dams built on a Class I site.

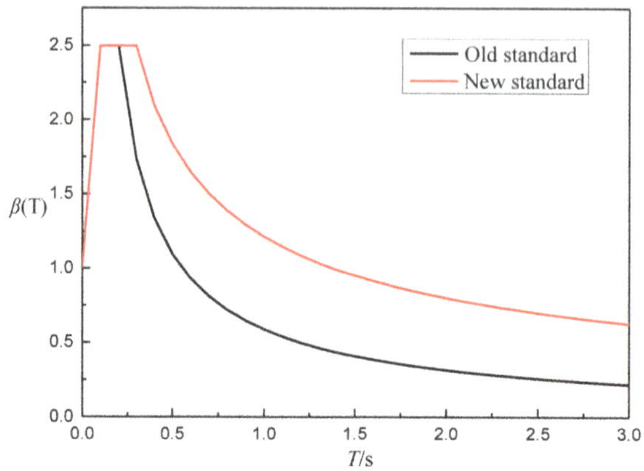

Figure 1. Comparison of the design response spectrum in the new and old standards (Class I site).

3. Computational Method

3.1. B-Differentiable Equations

Under the action of strong earthquakes, arch dam contraction joints may open, close, and have frictional dislocation, which can be solved as a contact problem [15,16]. A method of B-differentiable equations [8] is used to solve the three-dimensional dynamic friction contact between dam blocks induced by an earthquake. The following is a brief introduction of the method of B-differentiable equations for the three-dimensional elastic frictional contact problem, taking the contact of two bodies (represented by body 1 and body 2, respectively) as an example.

Based on the assumption of small deformation and small strain, the point–point contact model is adopted for the contact surface after FE discretization, that is, the nodes of the two contact surfaces are one-to-one corresponding along the normal direction of the contact surface, forming a plurality of contact pairs. The contact conditions at ith contact pair can be expressed in the form of B-differentiable equations as follows:

$$H_2^i\left(d\mathbf{u}_c^i, d\mathbf{P}^i\right) = \min\left\{r\Delta u_n^i, P_n^i\right\} = 0 \tag{1}$$

$$H_3^i\left(d\mathbf{u}_c^i, d\mathbf{P}^i\right) = P_a^i - \lambda P_a^i(r) = 0 \tag{2}$$

$$H_4^i\left(d\mathbf{u}_c^i, d\mathbf{P}^i\right) = P_b^i - \lambda P_b^i(r) = 0 \tag{3}$$

where

$$P_a^i(r) = P_a^i - r\Delta du_a^i \tag{4}$$

$$P_b^i(r) = P_b^i - r\Delta du_b^i \tag{5}$$

$$\lambda = \min\left\{\frac{\mu\min\{P_n^i, 0\}}{\sqrt{\left(P_a^i(r)\right)^2 + \left(P_b^i(r)\right)^2}}, 1\right\} \tag{6}$$

where μ is the friction coefficient. $\Delta u_n^i, \Delta du_a^i, \Delta du_b^i$ denote the increment of the normal relative displacement and the tangential relative displacement of the ith contact pair, respectively. P_n^i, P_a^i, P_b^i represents the normal and tangential contact forces of the ith contact pair, respectively. Equations (1)–(3) is non-differentiable causing by the operator min. The contact equations in the form of B-differentiable equations can be solved by the B-differentiable damped Newton method [8].

3.2. Multi-Transmitting Boundary Method

The radiation damping effect of an infinite foundation and the input mechanism of ground motion should be considered in the dam–infinite foundation interaction model [17]. In the direct method [18,19], an artificial boundary condition is applied to the outer boundary of the finite domain model, which can simulate the propagation of the scattered waves caused by the vibration of the arch dam. The multi-transmitting formula [11] as an artificial boundary condition is based on the plane wave assumption. The multi-transmitting formula is suitable to simulate the far field condition of a homogeneous infinite foundation, and has the characteristics of space–time decoupling which greatly reduce the amount of calculation. Because the local artificial boundary multi-transmitting formula does not consider the coupling effect of boundary nodes, the artificial boundary should be selected far enough away from the structure. Thus, the multi-transmitting boundary method is used to simulate the dam–infinite foundation interaction.

3.3. Westergaard Added Mass Method

The dam–reservoir interaction is simulated by the Westergaard added mass method [12,20]. The added mass of hydrodynamic pressure is calculated according to Westergaard's formula. The Westergaard added mass model has been recognized as too conservative according to experimental and numerical analysis [21]. Thus, the Westergaard formula (7) is reduced by 25%.

$$m_w(h) = \frac{7}{8}\rho\sqrt{H_0 h} \tag{7}$$

where $m_w(h)$ is the added mass at the node, with depth h, ρ is the mass density of water, H_0 is the depth of reservoir, and h is the depth of the node.

3.4. Prediction–Correction Explicit Integration Method

A prediction–correction explicit integration method is used to solve the dynamic equilibrium equation of an arch dam–reservoir–infinite foundation system. In the prediction–correction explicit integration method, the dynamic equilibrium equation can be written as:

$$\mathbf{M}\ddot{\mathbf{u}}^{t+dt} + \mathbf{C}\tilde{\dot{\mathbf{u}}}^{t+dt} + \mathbf{K}\tilde{\mathbf{u}}^{t+dt} = \mathbf{F}^{t+dt} + \mathbf{P}_c^{t+dt} \tag{8}$$

where $\mathbf{M}$, $\mathbf{C}$, and $\mathbf{K}$ denote the mass, damping, and stiffness matrices respectively. $\mathbf{F}^{t+dt}$ and $\mathbf{P}_c^{t+dt}$ represent the external load and the contact force on the contact surface at time $t + dt$, respectively. $\ddot{\mathbf{u}}^{t+dt}$, $\tilde{\dot{\mathbf{u}}}^{t+dt}$, and $\tilde{\mathbf{u}}^{t+dt}$ represent the acceleration, predicted velocity and predicted displacement at time $t + dt$, respectively. $\tilde{\dot{\mathbf{u}}}^{t+dt}$ and $\tilde{\mathbf{u}}^{t+dt}$ can be obtained from the known displacement, velocity, and acceleration at time t. In Equation (8), $\ddot{\mathbf{u}}^{t+dt}$ and $\mathbf{P}_c^{t+dt}$ are unknown variables, which can be obtained by simultaneously solving Equation (8) and Equations (1)–(3).

4. Computational Model

The QBT hydropower plant is being constructed upstream of the Burqin River in northwest China. The QBT dam is a concrete hyperbolic arch dam with a volume of 3.78 million m^3. The maximum height is 240 m. The chord length of the crest is 600 m. The thickness is 14 m at the crest, and 65 m at the base. The normal storage and the lowest generating level of the reservoir are 235 m and 150 m, respectively. The level of the Earthquake Intensity at the QBT dam site is VII. The Earthquake Intensity indicates the degree of the earthquake's impact on the ground and buildings. The geological conditions at the QBT dam site are complex. There are several faults intersecting interlayer shear weakness zones at the dam abutment. The level of the seismic fortification is Class A The level of seismic fortification is determined on the basis of the comprehensive consideration of the seismic environment, the degree of importance of the construction project, the allowable risk level, the safety target to be achieved, and the national economic bearing capacity. In this paper, the effect of the faults on the seismic stability of the dam–foundation system is not taken into consideration.

4.1. Finite Element Model

In order to satisfy the requirements for calculation accuracy and efficiency, as shown in Figure 2, the dimensions of the FE model of the QBT arch dam–foundation system were extended by one time of the dam height in the transverse direction, vertical direction, and river direction. A 3D solid element with 8 nodes was used to create the mesh of the model. In order to accurately simulate the propagation of a seismic wave, the size of the FE mesh of the foundation was not more than 20 m in the vertical direction, which satisfied the fact that the maximum element size should be no fewer than 10 elements per wavelength. There were 162,428 nodes and 148,695 elements in the FE model of the arch dam–foundation system, and the number of dam elements and nodes were 4518 and 8596, respectively. Three layers of grids were divided along the thickness direction of the arch dam, and 34 contraction joints with 2288 contact pairs were simulated, as shown in Figure 3. The interface between the dam and the foundation was neglected. The contraction joints were simulated with B-differentiable equations method. The friction coefficient was 1.0. The response of the dam was calculated using our in-house code.

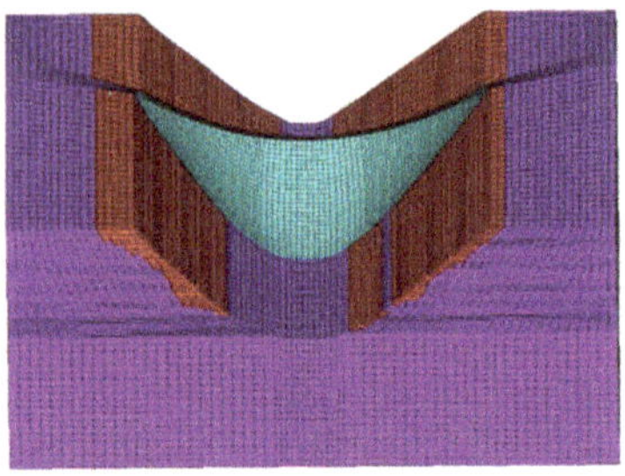

Figure 2. The FE model of the arch dam–foundation system.

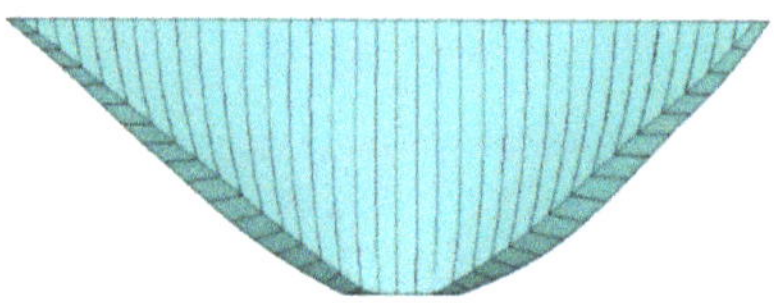

Figure 3. Distribution of the arch dam contraction joints.

4.2. Material Parameters

Concrete and bedrock were assumed to be isotropic linear elastic materials. The physical and mechanical parameters of concrete and bedrock are listed in Table 1. Bedrock II (the red zone in Figure 2) and Bedrock III (the violet zone in Figure 2) represented the rock classification.

Table 1. Material parameters.

Parameter Material	Mass Density (kg/m^3)	Elasticity Modulus (GPa)	Poisson's Ratio	Linear Expansivity (/°C)	Thermal Conductivity (W/m·°C)	Specific Heat (kJ/(kg·°C))
Concrete	2400	21	0.167	1.0×10^{-5}	3.0	970
Bedrock II	2755	13.5	0.24	1.0×10^{-5}	2.67	840
Bedrock III	2700	10.0	0.26	1.0×10^{-5}	2.67	840

4.3. Loading Conditions

The static load includes the self-weight of the dam, the hydrostatic pressure, the sediment pressures and the temperature load. The normal storage and the lowest generating level of the reservoir are 235 m and 150 m, respectively. The sediment depth is 57.5 m. The buoyant unit weight and frictional angle of sediment are 9.0 kN/m^3 and 12°, respectively. The sediment pressure was treated as hydrostatic pressure [18] and was applied on the upstream surface of the dam. The static load was firstly applied on the arch dam–foundation system. Then, the dynamic load was applied to shake the dam–foundation system.

In the phase of dynamic calculation, the effect of sediment was neglected and the dam–foundation system damping was incorporated using Rayleigh damping. The damping ratio of the dam–foundation system was assumed to be 5%. The peak ground acceleration of the design earthquake was 0.357 g, when the exceedance probability of the ground motion was 2% in a 100-year period. Using the design response spectrum in the new and old standards as the target spectrum, artificial seismic waves with a duration of 20 s were fitted as the input of ground motion. In the dynamic analysis of the dam–reservoir–foundation system, it was assumed that the seismic wave was incident vertically, and the combined effects of the horizontal ground motion (along and across rivers) and vertical ground motion were simultaneously considered. The vertical ground motion was taken as 2/3 of that in the horizontal direction.

5. Numerical Results

5.1. Dynamic Characteristics

Under the condition of the normal storage level of the reservoir, as shown in Table 2, the base frequency of the arch dam was 1.2 Hz, and the natural frequency of the arch dam was relatively dense in the range of 1–5 Hz (natural vibration period 0.2 s–1 s). In the range of this period, the difference between the new and old standards was significant. Thus, the artificial waves generated by different design response spectrum had a certain influence on the tensile and compressive principal stress of the dam body and the distribution range of high stresses.

Table 2. The first 20 order frequencies and the corresponding vibration modes.

Order	Frequency (Hz)	Vibration Mode in the Arch Direction
1	1.1960	antisymmetry
2	1.3093	symmetry
3	1.7188	symmetry
4	2.2163	antisymmetry
5	2.5623	symmetry
6	2.7981	symmetry
7	2.9636	antisymmetry
8	3.2813	antisymmetry
9	3.4350	antisymmetry
10	3.5391	symmetry
11	3.8739	symmetry
12	4.1229	symmetry
13	4.3710	symmetry
14	4.6033	antisymmetry
15	4.7468	antisymmetry
16	4.8678	antisymmetry
17	5.0934	symmetry
18	5.3399	antisymmetry
19	5.5150	symmetry
20	5.6603	symmetry

5.2. Dynamic Stress Distribution

Under the excitation from the design response spectrum in the new standard, as shown in Figure 4, the maximum tensile principal stress of the arch dam was 7.35 Mpa, which appeared at the upstream surface of the dam heel. The high stress zones appeared at the upstream surface of the dam heel and the middle and upper elevation of the downstream surface. The local ultra-high tensile stress zones (greater than the dynamic tensile strength of C40 concrete 4.89 Mpa) appeared at the upstream surface of the dam heel, which extended about 10 m from the bottom of the dam upward along the elevation direction, and about 1/4 of the section thickness of the dam body along the dam thickness direction. The tensile stress in other parts did not exceed the dynamic tensile strength of the concrete (4.89 MPa).

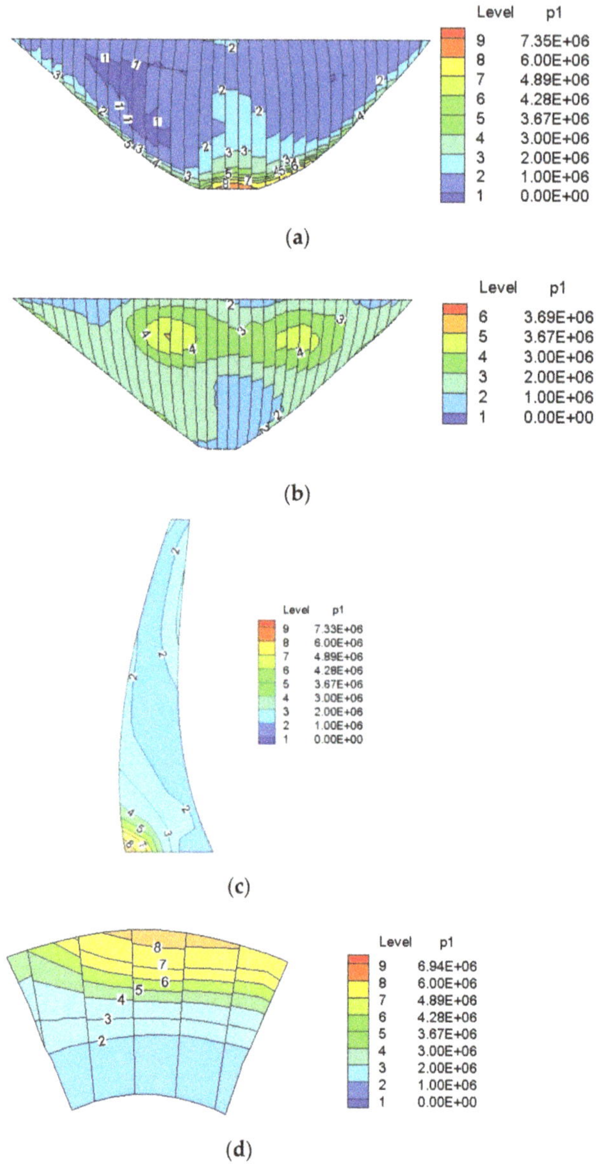

Figure 4. Envelopes of maximum tensile principal stresses using the new standard (Pa). (**a**) Upstream surface, (**b**) downstream surface, (**c**) section of arch crown beam, and (**d**) arch section at dam bottom elevation.

Under excitation from the design response spectrum in the new standard, as shown in Figure 5, the maximum compressive principal stress of the arch dam was 15.3 Mpa, which appeared at the top of arch crown beam. The high stress zones were mainly concentrated in the area near the top of the arch crown beam and the local area near the boundary of the dam–foundation at the middle and lower elevation.

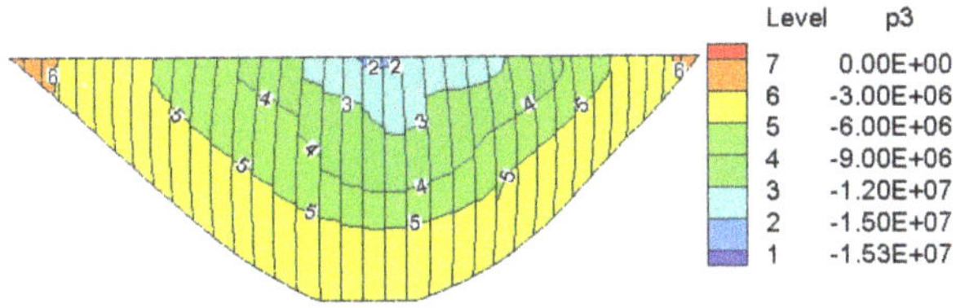

(a)

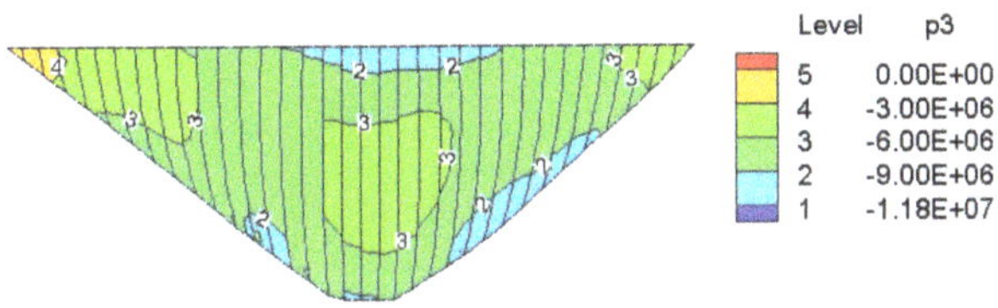

(b)

Figure 5. Envelopes of maximum compressive principal stresses using the new standard (Pa). (**a**) Upstream surface and (**b**) downstream surface.

Under excitation from the design response spectrum in the old standard, as shown in Figure 6, the maximum tensile principal stress of the arch dam was 5.19 Mpa, which appeared at the upstream surface of the dam heel. The distribution of the maximum tensile principal stresses were similar to those in Figure 4. However, the range of the high stress zones and the local ultra-high tensile stress zones were less than those in Figure 4.

Under excitation from the design response spectrum in the old standard, as shown in Figure 7, the maximum compressive principal stress of the arch dam was 12.1 Mpa, which appeared near the top of the arch crown beam. The distributions of the maximum compressive principal stresses were similar to those in Figure 5. However, the ranges of the high stress zone and the local ultra-high tensile stress zone were less than those in Figure 5.

As shown in Table 3, the maximum tensile principal stress of the arch dam using the design response spectrum in the new and old standards was 7.35 MPa and 5.19 MPa, respectively. The maximum compressive principal stress of the arch dam using the design response spectrum in the new and old standards was 15.3 MPa and 12.1 MPa, respectively. The maximum tensile and compressive principal stresses of the arch dam using the design response spectrum in the new standard were significantly increased compared with those using the design response spectrum in the old standard.

Table 3. The maximum tensile and compressive principal stresses.

MPS / Standard	TS-US (MPa)	TS-DS (MPa)	TS-ACB (MPa)	TS-DBE (MPa)	CS-US (MPa)	CS-DS (MPa)
New standard	7.35	3.69	7.33	6.94	15.3	11.8
Old standard	5.19	2.08	5.19	4.86	12.1	10.8

Notes: MPS means maximum principle stress; TS means tensile stress; CS means compressive stress; US means upstream surface; DS means downstream surface; ACB means arch crown beam; and DBE means dam bottom elevation.

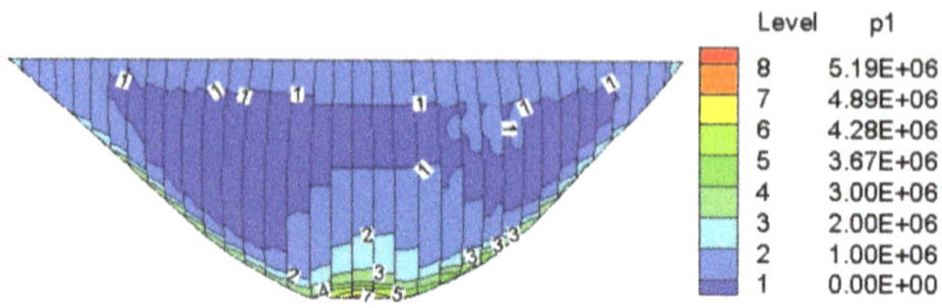

(**a**)

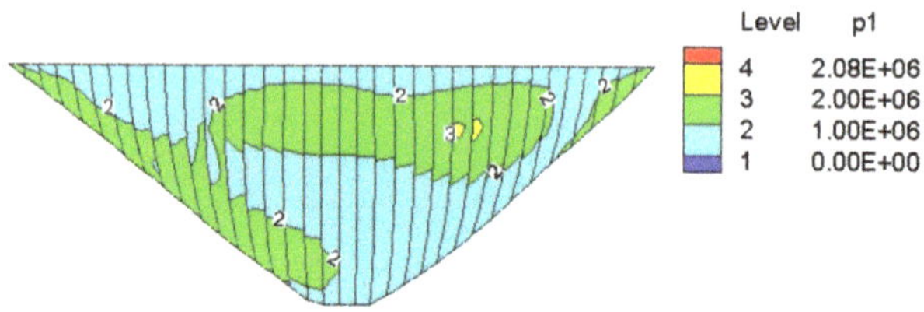

(**b**)

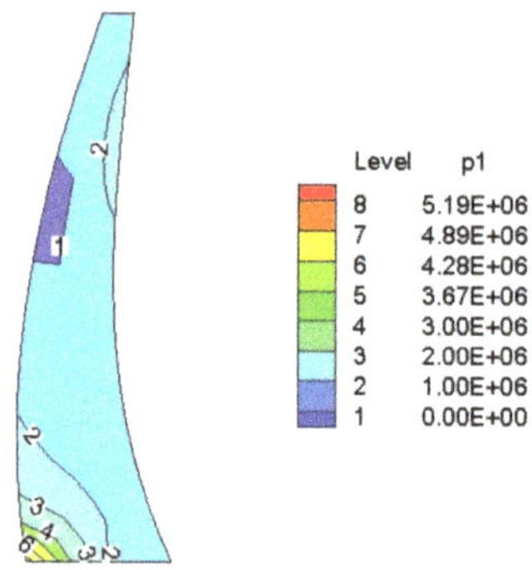

(**c**)

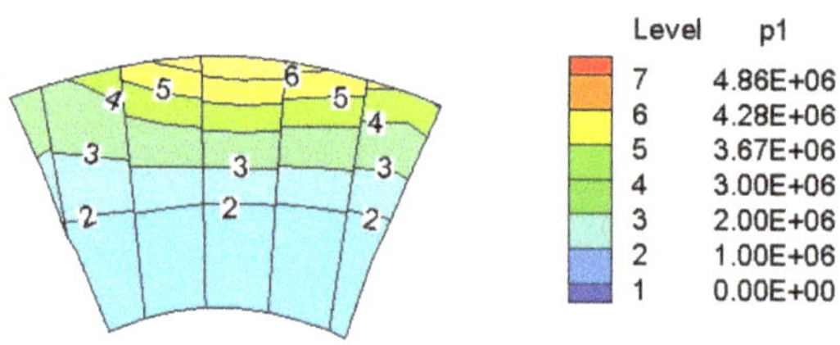

(**d**)

Figure 6. Envelopes of maximum tensile principal stresses using the old standard (Pa). (**a**) Upstream surface, (**b**) downstream surface, (**c**) section of arch crown beam, and (**d**) arch section at dam bottom elevation.

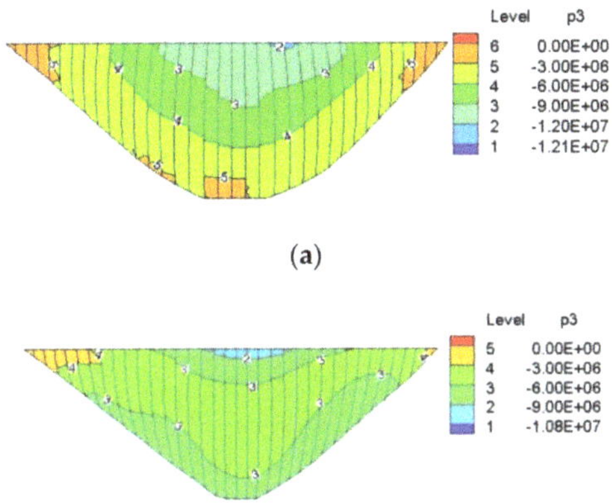

(a)

(b)

Figure 7. Envelopes of maximum compressive principal stresses using the old standard (Pa). (**a**) Upstream surface and (**b**) downstream surface.

5.3. Contraction Joint Opening

As shown in Figures 8 and 9, the ranges of the contraction joints opening zones under excitation from the design response spectrum in the new standard were greater than those under excitation from the design response spectrum in the old standard. The maximum values of the contraction joints opening under excitation from the design response spectrum in the new and old standards were 5.18 cm and 3.26 cm, respectively.

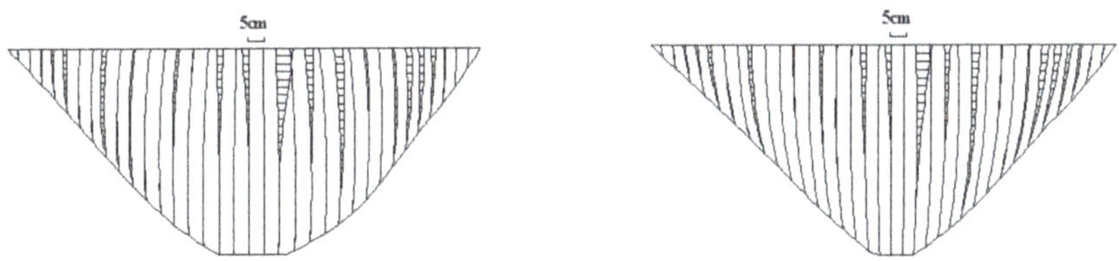

(a) (b)

Figure 8. Envelopes of the contraction joint opening using the new standard (cm). (**a**) Upstream surface and (**b**) downstream surface.

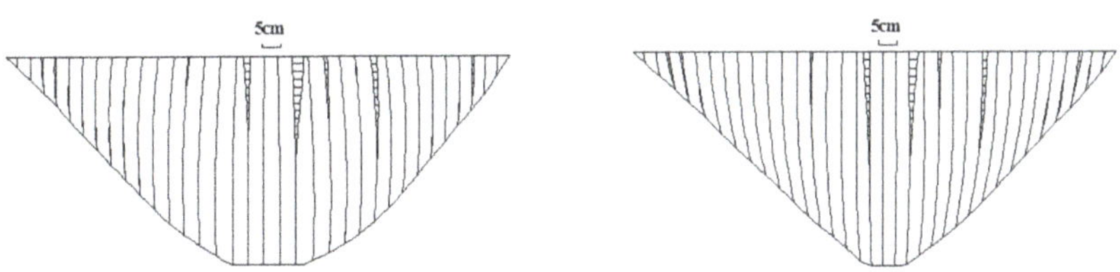

(a) (b)

Figure 9. Envelopes of the contraction joint opening using the old standard (cm). (**a**) Upstream surface and (**b**) downstream surface.

6. Conclusions

In order to compare the seismic responses of an arch dam under excitation from the design response spectrums in new and old standards, the dynamic calculation of a 240 m high arch dam was carried out by a 3D finite element method.

(1) The differences between the design response spectrums in the new standard and the old standard were mainly reflected in the characteristic period T_g and the attenuation index γ for arch dams at a Class I site.
(2) The high stress and ultra-high stress of the arch dam basically occurred at the same zone using the design response spectrum in the old and new standards. The dynamic responses including the maximum principal stress, the distribution range of high stress, the maximum value of the contraction joints opening, and the range of the contraction joints opening zone using the design response spectrum in the new standard were greater than those using the design response spectrum in the old standard.
(3) The seismic safety of the arch dam may decrease under excitation from the design response spectrum in the new standard. Thus, seismic validation on the built arch dams should be carried out when possible.

Author Contributions: Conceptualization, B.X. and J.W.; methodology, B.X.; validation, N.L. and J.W.; formal analysis, B.X.; investigation, C.Z.; data curation, B.X.; writing—original draft preparation, B.X.; writing—review and editing, J.C.; visualization, B.X.; supervision, N.L.; project administration, J.W.; funding acquisition, B.X. and J.W. All authors have read and agreed to the published version of the manuscript.

Funding: This research was supported by the National Natural Science Foundation of China with grant No. 52109169, the Natural Science Foundation of Henan with grant No. 212300410279, the Open Fund of Engineering Research Center on Dike Safety and Disease Prevention of the Ministry of Water Resources with grant No. LSDP202103, and the Open Research Fund of Guangxi Key Laboratory of Water Engineering Materials and Structures, Guangxi Institute of Water Resources research with grant No. GXHRI-WEMS-2020-12.

Institutional Review Board Statement: Not applicable.

Informed Consent Statement: Not applicable.

Data Availability Statement: The data are available from the corresponding author upon request.

Conflicts of Interest: The authors declare no conflict of interest.

References

1. *GB 51247-2018*; Standard for Seismic Design of Hydraulic Structures. China Planning Press: Beijing, China, 2018.
2. *SL 203-1997*; Specifications for Seismic Design of Hydraulic Structures. China Water & Power Press: Beijing, China, 1997.
3. Hariri-Ardebili, M.A.; Saouma, V.E. Probabilistic seismic demand model and optimal intensity measure for concrete dams. *Struct. Saf.* **2016**, *59*, 67–85. [CrossRef]
4. Pang, R.; Xu, B.; Zhou, Y. Seismic time-history response and system reliability analysis of slopes considering uncertainty of multi-parameters and earthquake excitations. *Comput. Geotech.* **2021**, *136*, 104245. [CrossRef]
5. Wang, X.; Xue, B.; Xu, B.; Pang, R. Role of strong motion duration on seismic responses of high concrete faced rockfill dams. *Structures* **2021**, *32*, 1092–1102. [CrossRef]
6. Chen, H. Seismic safety of high concrete dams. *Earthq. Eng. Eng. Vib.* **2014**, *13*, 1–16. [CrossRef]
7. Wang, R.; Chen, L.; Zhang, C. Seismic design of Xiluodu ultra-high arch dam. *Water Sci. Eng.* **2018**, *11*, 288–301. [CrossRef]
8. Christensen, P.W.; Klarbring, A.; Pang, J.S.; Strömberg, N. Formulation and comparison of algorithms for frictional contact problems. *Int. J. Numer. Methods Eng.* **1998**, *42*, 145–173. [CrossRef]
9. Lin, G.; Xue, B.; Hu, Z. A mortar contact formulation using scaled boundary isogeometric analysis. *Sci. China Phys. Mech. Astron.* **2018**, *61*. [CrossRef]
10. Xue, B.; Du, X.; Wang, J.; Yu, X. A scaled boundary finite-element method with B-Differentiable equations for 3D frictional contact problems. *Fractal Fract.* **2022**, *6*, 133. [CrossRef]
11. Liao, Z. Extrapolation non-reflecting boundary conditions. *Wave Motion* **1996**, *24*, 117–138. [CrossRef]
12. Westergaard, H.M. Water pressures on dams during earthquakes. *Trans. ASCE* **1933**, *98*, 418–432. [CrossRef]
13. *GB 18306-2015*; Seismic Ground Motion Parameters Zonation Map of China. China Standards Press: Beijing, China, 2015.
14. Abrahamson, N.; Silva, W. Summary of the Abrahamson & Silva NGA Ground-Motion Relations. *Earthq. Spectra* **2012**, *24*, 67–97.
15. Omidi, O.; Lotfi, V. Seismic plastic–damage analysis of mass concrete blocks in arch dams including contraction and peripheral joints. *Soil Dyn. Earthq. Eng.* **2017**, *95*, 118–137. [CrossRef]
16. Khassaf, S.I. Effect of contraction joints on the structural performance of arch dam. *Int. J. GEOMATE* **2020**, *19*, 219–227. [CrossRef]
17. Xu, Q.; Chen, J.; Li, J.; Zhao, C.F. Influence of seismic input on response of Baihetan arch dam. *J. Cent. South Univ. Engl. Ed.* **2014**, *21*, 2437–2443. [CrossRef]

18. Pan, J.; Xu, Y.; Jin, F.; Wang, J. Seismic stability assessment of an arch dam-foundation system. *Earthq. Eng. Eng. Vib.* **2015**, *14*, 517–526. [CrossRef]
19. Saouma, V.; Miura, F.; Lebon, G.; Yagome, Y. A simplified 3D model for soil-structure interaction with radiation damping and free field input. *Bull. Earthq. Eng.* **2011**, *9*, 1387–1402. [CrossRef]
20. Chopra, A.K. Hydrodynamic pressures on dams during earthquake. *J. Eng. Mech. Div.* **1967**, *93*, 205–224. [CrossRef]
21. Chen, H.; Hou, S.; Yang, D. Study on arch dam-reservoir water interaction under earthquake condition. *J. Hydraul. Eng.* **1989**, *7*, 29–39. (In Chinese)

water

MDPI

Article

A Novel Calculation Method of Hydrodynamic Pressure Based on Polyhedron SBFEM and Its Application in Nonlinear Cross-Scale CFRD-Reservoir Systems

Jianjun Xu [1], He Xu [1,*], Dongming Yan [2], Kai Chen [3] and Degao Zou [3]

[1] Power China Huadong Engineering Corporation Limited, Hangzhou 311122, China; xu_jj@hdec.com

[2] College of Civil Engineering and Architecture, Zhejiang University, Hangzhou 310058, China; dmyan@zju.edu.cn

[3] School of Hydraulic Engineering, Dalian University of Technology, Dalian 116024, China; chenkai@dlut.edu.cn (K.C.); zoudegao@dlut.edu.cn (D.Z.)

* Correspondence: bieshuxuhe@163.com; Tel.: +86-150-4243-1787

Abstract: Hydrodynamic pressure is an important factor that cannot be ignored in the seismic safety evaluation of dams. However, when the polyhedron-scaled boundary finite element method is used to simulate dams in a cross-scale dynamic analysis, polygonal surfaces often appear on the upstream face of dams, which is difficult to deal with for conventional methods of hydrodynamic pressure. In this paper, a three-dimensional calculation method of hydrodynamic pressure based on the polyhedron-scaled boundary finite element method is proposed, in which polygon (triangle, quadrilateral, pentagon, hexagon, heptagon, octagon, etc.) semi-infinite prismatic fluid elements are constructed using the mean-value shape function. The proposed method, with a high efficiency, overcomes the limitation of conventional methods in which only quadrangle or triangle boundary faces of elements are permitted. The accuracy of the proposed method is proved to be high when considering various factors. Furthermore, combined with the polyhedron-scaled boundary finite element method for a solid dam, the proposed method for reservoir water is used to develop a nonlinear dynamic coupling method for cross-scale concrete-faced rockfill dam-reservoir systems based on the polyhedron SBFEM. The results of the numerical analysis show that when the hydrodynamic pressure is not considered, the error of rockfill dynamic acceleration and displacement could reach 15.4% and 12.7%, respectively, and the error of dynamic face slabs' stresses could be 24.9%, which is not conducive to a reasonable seismic safety evaluation of dams.

Keywords: scaled boundary finite element method (SBFEM); hydrodynamic pressure; polyhedral element; dam reservoir interaction; concrete faced rockfill dam (CFRD)

Citation: Xu, J.; Xu, H.; Yan, D.; Chen, K.; Zou, D. A Novel Calculation Method of Hydrodynamic Pressure Based on Polyhedron SBFEM and Its Application in Nonlinear Cross-Scale CFRD-Reservoir Systems. *Water* **2022**, *14*, 867. https://doi.org/10.3390/w14060867

Academic Editor: Georg Umgiesser

Received: 20 January 2022
Accepted: 9 March 2022
Published: 10 March 2022

1. Introduction

The prevalence of the scaled boundary finite element method (SBFEM) [1] can be attributed to its unique advantages, and as a result, SBFEM is being applied to an ever-expanding range of numerical computation analyses. Initially, SBFEM was conceived for the field of computational elasticity [2–4]. As SBFEM became more mature, Zhang and Wegner [5,6] spearheaded SBFEM-based dynamic coupling analysis between the three-dimensional (3D) infinite foundation and structure, and in their studies, the substructure method was used to improve computational efficiency and analyze the wave motion of seismic waves. Subsequently, further details of the dynamic interaction between structure and foundation were studied [7–10] by SBFEM, which automatically satisfied the infinite radiation condition. SBFEM also demonstrated its superiority over traditional numerical methods for modeling fracture mechanics, as similarity centers of SBFEM can be directly placed at a crack tip to allow a straightforward accurate simulation of the singular stress distribution at the crack tip with no need for very fine meshes, which is essential for the conventional finite element method (FEM). Song et al. [11], acting as a pioneer, inspired researchers to launch research efforts towards fracture mechanics problems using SBFEM [12–14]. At the same time, SBFEM has also obtained some achievements in the Cosserat continuum analysis [15], crack propagation [16,17], the analysis of sandwich plates [18,19], image-based analyses [20–22], acoustics [23], the contact problem [24], and electromagnetism [25].

Recently, Chen et al. [26–29] constructed 3D polyhedral elements based on SBFEM, which facilitated breakthroughs in the cross-scale non-linear static and dynamic analysis of engineering structures. Compared to conventional methods, the polyhedral element supports not only triple or quadrilateral surfaces, but also polygonal surfaces, and thus, it can readily process complex geometries. Having the ability for cross-scale computation and economizing efforts during pre-processing work, the octree polyhedral element [28,29] allows sparse and dense grids to be connected to each other quickly and smoothly, and above all, considerable degrees of freedoms (DOFs) can be eliminated. The polyhedron SBFEM has been applied to the elasto-plastic analysis of large structures, such as concrete gravity dams [26] and earth-rock dams [27,28]. In addition, this polyhedron SBFEM could effectively evaluate the safety of dams in strong earthquake hazard areas, where there may be earthquakes with a high Richter scale and seismic intensity. However, polygonal surfaces often appear on the (upstream) face of dams when the octree technique is employed to divide the grids as shown in Figure 1. It is difficult to use conventional methods, which support only triangle and quadrilateral grids, to solve and compute the hydrodynamic pressure of a reservoir with polygonal grids under an earthquake. In the analysis of the seismic safety evaluation of dams, the hydrodynamic pressure is an important influencing factor that cannot be ignored [30–33]. Considering the important influence of hydrodynamic pressure on dam response, it is an urgent problem, which until now remains to be solved, for a cross-scale dynamic dam analysis system based on a polyhedron SBFEM.

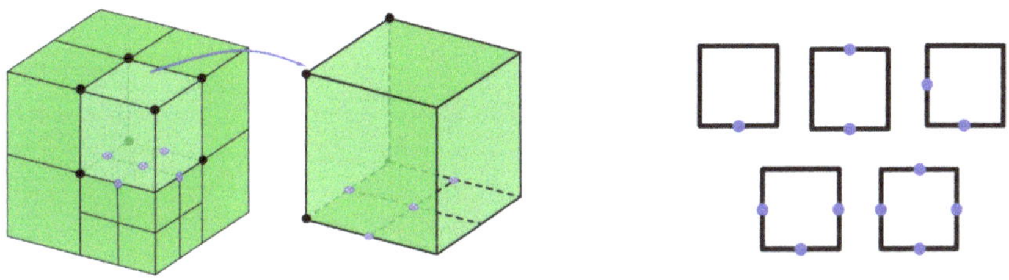

(a) (b)

Figure 1. Octree polyhedral elements. (**a**) Octree mesh; (**b**) Polygonal surfaces.

In the analysis of the interaction between fluid and structure [34–37], the calculation method of hydrodynamic pressure on dams has always been one of the hot research topics. At present, much research on the numerical analysis of the dam–reservoir dynamic interaction under earthquake conditions has emerged. The scope of research covers arch dams [38–41], gravity dams [41–44], and concrete face rockfill dams (CFRDs) [45–49]. The hydrodynamic pressure computational methods used in research include FEM, the boundary element method (BEM) and SBFEM. FEM is widely used in the computation of hydrodynamic pressure in complex reservoirs. However, many nodal DOFs need to be introduced, especially for large scale 3D models of actual projects, which can dramatically increase the calculation amount when simulating dynamic coupling of dam–reservoir systems. Lin et al. [50] realized an efficient SBFEM-based solution of hydrodynamic pressure in a 3D reservoir by only discretizing the two-dimensional (2D) interfaces between the reservoir water and the dam's upstream face, thus saving many DOFs, improving computational efficiency, and facilitating large-scale numerical analysis. Using this method [50] to simulate a reservoir, Xu et al. [45,48] further developed a nonlinear dynamic coupling method for CFRD-reservoir systems based on the FEM-SBFEM approach. Fortunately, this method [45,50] is suitable for constructing polyhedral fluid elements and can be seamlessly integrated with a (octree) cross-scale dam analysis system based on polyhedron SBFEM.

Based on previous research, this study proposes a novel 3D calculation method of hydrodynamic pressure based on polyhedron SBFEM (PSBFEM). Using the mean-value shape function [51], polygonal semi-infinite prismatic elements, i.e., polyhedral fluid elements, are constructed. The proposed method could directly make use of 2D grids on the upstream face of a dam model to automatically generate the 3D mesh of the semi-infinite reservoir, which simplifies the pre-treatment process. More importantly, the computational cost of hydrodynamic pressure is less compared with FEM for reducing the dimension of discretizing by one. The proposed method provides an accurate and efficient analysis tool for calculating hydrodynamic pressure in a cross-scale dynamic dam–reservoir analysis system based on polyhedron SBFEM. It should be noted that the 3D PSBFEM-based hydrodynamic pressure calculation method is not only efficient for a cross-scale or multi-scale dam analysis system with polyhedral or octree meshing, but it is also suitable for a traditional FEM-based dam analysis system. This method is also convenient for formulation and implementation in the analysis program.

Furthermore, a nonlinear dynamic coupling method for cross-scale CFRD-reservoir systems based on the polyhedron SBFEM has been developed in this study, in which the proposed method of hydrodynamic pressure is adopted to simulate a reservoir, and the polyhedron SBFEM is used to model the elastic–plastic CFRD and foundation system. In the end, the seismic safety of the CFRD is evaluated using the developed dynamic coupling method. The coupling method may have a significant prospect of practical application in hydraulic structure engineering.

2. A Calculation Method of Hydrodynamic Pressure and Polyhedral Fluid Element

The following section firstly provides a brief theoretical derivation of the hydrodynamic pressure computation method based on polyhedron SBFEM, and secondly, the basic formulas for the mean-value shape function on polygons is introduced. Finally, polygonal prismatic fluid elements are constructed.

2.1. Computation Method of Hydrodynamic Pressure Based on Polyhedron SBFEM

It is assumed that the reservoir water is an ideal fluid, which is incompressible, undisturbed, and not viscous. Under seismic excitation, the hydrodynamic pressure in front of the dam satisfies the Laplace equation:

$$\nabla^2 p = 0 \tag{1}$$

Ignoring the micro-amplitude gravity wave, the free surface S_0 boundary condition of reservoir water is:

$$p = 0 \tag{2}$$

The boundary condition on the water face S_1 of dam satisfies:

$$\partial p / \partial n = -\rho \ddot{u}_n \tag{3}$$

The boundary condition at the interface S_2 between the reservoir and the river–valley satisfies:

$$\partial p / \partial n = -\rho \ddot{v}_n \tag{4}$$

In the above equations, ∇^2 is the Laplacian operator, p is the hydrodynamic pressure, n is the normal direction of the interface, ρ represents the fluid density, and $\ddot{u}_n$ and $\ddot{v}_n$ are the normal accelerations of the dam–reservoir interface and the river–valley interface, respectively. Since the whole semi-infinite reservoir water in front of the dam is discretized by SBFEM, the radiation boundary condition at infinity S_3 of the reservoir is satisfied automatically, the theory of which is expounded as below.

Taking a pentagonal grid on the dam upstream face as an example as shown in Figure 2, the similarity center O is selected at the downstream infinity of the dam, and on the grounds of SBFEM thought, the semi-infinite prismatic fluid element is generated using the 2D surface grid of the dam upstream face. By this means, the 3D model of a reservoir, as indicated in Figure 3, consisting of a series of semi-infinite polyhedral elements, is discretized automatically only in two dimensions by utilizing the element grid of the dam on the upstream surface, which means that there are limited DOFs, and there is no need to divide the reservoir grid additionally.

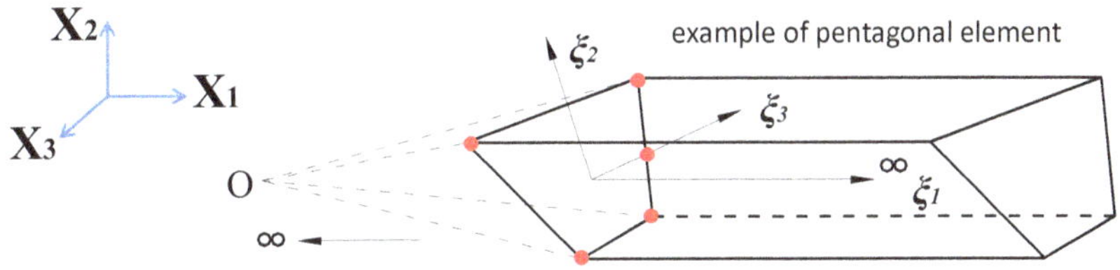

Figure 2. Typical polygonal scale boundary finite element of fluid.

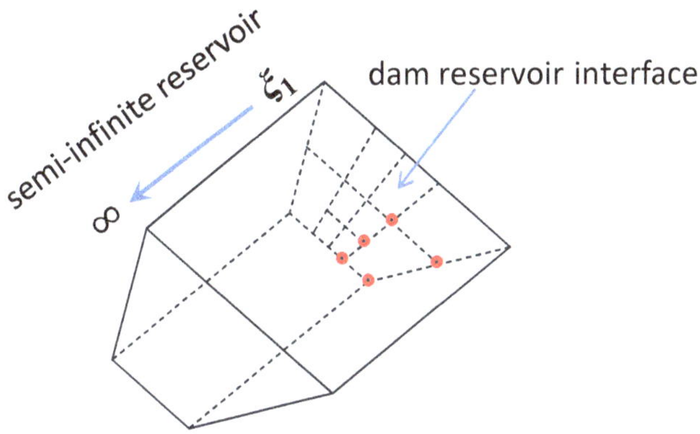

Figure 3. Reservoir model discretized by SBFEM.

The range of the local radial coordinate ξ_1 is [0, +∞] from upstream face of the dam to infinity of the reservoir. Furthermore, the range of local circumferential coordinates ξ_2 and ξ_3 is [−1, +1]. By using the scaled boundary coordinate transformation, the coordinates of the global Cartesian coordinate system at any point (X_1, X_2, X_3) in the reservoir can be expressed by the local scaled boundary coordinates (ξ_1, ξ_2, ξ_3). Here, ξ_1 serves as a factor of proportionality, as follows:

$$\begin{aligned} X_1(\xi_1,\xi_2,\xi_3) &= x_1(\xi_2,\xi_3) + \xi_1 = [N(\xi_2,\xi_3)]\{x_1\} + \xi_1 \\ X_2(\xi_1,\xi_2,\xi_3) &= x_2(\xi_2,\xi_3) = [N(\xi_2,\xi_3)]\{x_2\} \\ X_3(\xi_1,\xi_2,\xi_3) &= x_3(\xi_2,\xi_3) = [N(\xi_2,\xi_3)]\{x_3\} \end{aligned} \tag{5}$$

where (x_1, x_2, x_3) represent global coordinates of a node on a reservoir grid at the dam upstream face (ξ_1 = 0), and $[N(\xi_2,\xi_3)]$ denotes the polygon mean-value shape function, which is compatible with the polygon mesh and particularly presented in Section 2.2 below. With the help of the interpolation function $[N(\xi_2,\xi_3)]$, the hydrodynamic pressure $p(\xi_1, \xi_2, \xi_3)$ at any point in a polygon element can be expressed by the hydrodynamic pressure $\{p(\xi_1)\}$ at nodes of the fluid element as:

$$p(\xi_1,\xi_2,\xi_3) = [N(\xi_2,\xi_3)]\{p(\xi_1)\} \tag{6}$$

The differential operations $\bigtriangledown$ in the global coordinate system can be transformed into the local scaled boundary coordinate system by the Jacobian matrix [*J*] as follows:

$$[J] = \begin{bmatrix} X_{1,\xi_1} & X_{2,\xi_1} & X_{3,\xi_1} \\ X_{1,\xi_2} & X_{2,\xi_2} & X_{3,\xi_2} \\ X_{1,\xi_3} & X_{2,\xi_3} & X_{3,\xi_3} \end{bmatrix} = \begin{bmatrix} 1 & 0 & 0 \\ [N(\xi_2,\xi_3),_{\xi_2}]\{x_1\} & [N(\xi_2,\xi_3),_{\xi_2}]\{x_2\} & [N(\xi_2,\xi_3),_{\xi_2}]\{x_3\} \\ [N(\xi_2,\xi_3),_{\xi_3}]\{x_1\} & [N(\xi_2,\xi_3),_{\xi_3}]\{x_2\} & [N(\xi_2,\xi_3),_{\xi_3}]\{x_3\} \end{bmatrix} \tag{7}$$

$$\{\ \tfrac{\partial}{\partial X_1}\ \ \tfrac{\partial}{\partial X_2}\ \ \tfrac{\partial}{\partial X_3}\ \}^T = [J]^{-1}\{\ \tfrac{\partial}{\partial \xi_1}\ \ \tfrac{\partial}{\partial \xi_2}\ \ \tfrac{\partial}{\partial \xi_3}\ \}^T = \left\{b^1\right\}\frac{\partial}{\partial \xi_1} + \left\{b^2\right\}\frac{\partial}{\partial \xi_2} + \left\{b^3\right\}\frac{\partial}{\partial \xi_3} \tag{8}$$

where $[J]^{-1} = [\{b^1\}\ \{b^2\}\ \{b^3\}]$.

Through the weight function w, the weak form of the integral equation is formulated as Equation (9) by applying the method of the weighted residual to Equations (1)–(4).

$$\int_V \nabla w \nabla p dV + \rho \int_{S_1} w\ddot{u}_n dS + \rho \int_{S_2} w\ddot{v}_n dS = 0 \tag{9}$$

Substituting Equations (5)–(8) into Equation (9) and after a series of derivative processes, the SBFEM governing equation (Equation (10)) and boundary condition equation (Equation (11)) of the hydrodynamic pressure can be obtained:

$$\left[E^0\right]\{p(\xi_1)\},_{\xi_1\xi_1} + (\left[E^1\right]^T - \left[E^1\right])\{p(\xi_1)\},_{\xi_1} - \left[E^2\right]\{p(\xi_1)\} - \rho\left[C^0\right]\{\ddot{v}_n\} = 0 \tag{10}$$

$$(\left[E^0\right]\{p(\xi_1)\},_{\xi_1} + \left[E^1\right]^T p(\xi_1) + [M^1]\{\ddot{u}_n\}|_{\xi_1=0} = 0 \tag{11}$$

where the coefficient matrices $[E^0]$, $[E^1]$, $[E^2]$, $[C^0]$, and $[M^1]$ only depend on geometry information of the mesh on the upstream face of the dam and are expressed as follows:

$$\left[B^1\right] = \left\{b^1\right\}[N], \qquad \left[B^2\right] = \left\{b^2\right\}[N],_{\xi_2} + \left\{b^3\right\}[N],_{\xi_3} \tag{12}$$

$$\left[M^1\right] = \rho \int_{-1}^{1}\int_{-1}^{1} [N]^T[N] A d\xi_2 d\xi_3 \tag{13}$$

$$\left[E^0\right] = \int_{-1}^{1}\int_{-1}^{1} [B^1]^T[B^1]|J| d\xi_2 d\xi_3 \tag{14}$$

$$\left[E^1\right] = \int_{-1}^{1}\int_{-1}^{1} [B^2]^T[B^1]|J| d\xi_2 d\xi_3 \tag{15}$$

$$\left[E^2\right] = \int_{-1}^{1}\int_{-1}^{1} [B^2]^T[B^2]|J| d\xi_2 d\xi_3 \tag{16}$$

$$A = \sqrt{(x_{2,\xi_2}x_{3,\xi_3} - x_{3,\xi_2}x_{2,\xi_3})^2 + (x_{3,\xi_2}x_{1,\xi_3} - x_{1,\xi_2}x_{3,\xi_3})^2 + (x_{1,\xi_2}x_{2,\xi_3} - x_{2,\xi_2}x_{1,\xi_3})^2} \tag{17}$$

$$\left[C^0\right] = \int_\Gamma [N]^T[N] d\Gamma \tag{18}$$

where $[N]$ represents the polygon mean-value shape function $[N(\xi_2,\xi_3)]$. and Γ in Equation (18) represents the projection of the intersecting line between the dam upstream face and S_2 on the (X_2, X_3) plane.

This section may be divided by subheadings. It should provide a concise and precise description of the experimental results, their interpretation, as well as the experimental conclusions that can be drawn.

$$d\Gamma = \sqrt{x_{2,\xi_2}{}^2 + x_{3,\xi_2}{}^2} d\xi_2 \Big|_{\xi_3=-1} \tag{19}$$

An auxiliary variable $\{q(\xi_1)\}$ is introduced to solve the governing equation of the hydrodynamic pressure (Equation (10)) analytically.

$$\{q(\xi_1)\} = \left[E^0\right]\{p(\xi_1)\}_{,\xi_1} + [E^1]^T\{p(\xi_1)\} \tag{20}$$

where $\{q(\xi_1)\}$ is the nodal force resulting from the hydrodynamic pressure. Then, the governing equation (Equation (10)) can be transformed with a first-order ordinary differential equation (Equation (22)) by making use of the new variables and expressions (Equation (21)).

$$\{X(\xi_1)\} = \left\{ \begin{array}{c} \{p(\xi_1)\} \\ \{q(\xi_1)\} \end{array} \right\}, \quad \{F_0\} = \left\{ \begin{array}{c} 0 \\ -\rho\left[C^0\right]\{\ddot{v}_n\} \end{array} \right\} \tag{21}$$

$$\{X(\xi_1)\}_{,\xi_1} = [Z]\{X(\xi_1)\} + \{F_0\} \tag{22}$$

in which the Hamilton coefficient matrix [Z] is expressed as follows:

$$[Z] = \left[\begin{array}{cc} -[E^0]^{-1}[E^1]^T & [E^0]^{-1} \\ [E^2] - [E^1][E^0]^{-1}[E^1]^T & [E^1][E^0]^{-1} \end{array} \right] \tag{23}$$

The eigenvalue problem of the Hamilton matrix [Z] is to be solved first.

$$[Z][\Phi] = [\Phi][\Lambda] \tag{24}$$

The eigenvalue matrix $[\Lambda]$ and eigenvector matrix $[\Phi]$ of the matrix [Z] are written in partitioned form:

$$[\Lambda] = \left[\begin{array}{cc} [\lambda_i] & 0 \\ 0 & [-\lambda_i] \end{array} \right], [\Phi] = \left[\begin{array}{cc} [\Phi_{11}] & [\Phi_{12}] \\ [\Phi_{21}] & [\Phi_{22}] \end{array} \right] \tag{25}$$

in which $[\lambda_i]$ is the diagonal matrix, and the real part of $\lambda_i \geq 0$.

The matrix $[A]$, which is the inverse of the eigenvector matrix $[\Phi]$, is solved and partitioned secondly.

$$[A] = [\Phi]^{-1}, \quad [A] = \left[\begin{array}{cc} [A_{11}] & [A_{12}] \\ [A_{21}] & [A_{22}] \end{array} \right] \tag{26}$$

Finally, by taking the boundary condition (Equation (11)) into account and executing a series of solution procedures, the hydrodynamic pressure of the reservoir acting on the dam upstream face due to an earthquake can be expressed as:

$$\{p(\xi_1 = 0)\} = -[\Phi_{12}]\left[\Phi_{22}\right]^{-1}[M^1]\{\ddot{u}_n\} - \left([\Phi_{12}]\left[\Phi_{22}\right]^{-1}[B_1] - [B_2]\right)\rho\left[C^0\right]\{\ddot{v}_n\} \tag{27}$$

where

$$[B_1] = [\Phi_{21}]\left[\lambda_i^{-1}\right][A_{12}] + [\Phi_{22}]\left[-\lambda_i^{-1}\right][A_{22}] \tag{28}$$

$$[B_2] = [\Phi_{11}]\left[\lambda_i^{-1}\right][A_{12}] + [\Phi_{12}]\left[-\lambda_i^{-1}\right][A_{22}] \tag{29}$$

It can be seen from Equation (27) that the hydrodynamic pressure consists of the following two components: the hydrodynamic pressure caused by the dam upstream face vibration $\{\ddot{u}_n\}$ and that caused by the vibration of the river valley $\{\ddot{v}_n\}$ in the reservoir.

2.2. Polyhedral Scaled Boundary Finite Element of Fluid

2.2.1. Polygon Mean-Value Shape Function

The mean-value shape function [29,51–53] can be used for the interpolating of polygonal elements. The application of the function is straightforward, convenient. and efficient.

Initially, this method was employed to analyze and solve FE polygonal elements with an acceptable computational efficiency and accuracy [52]. In this paper, the shape function formulated for general polygonal elements is briefly introduced. A more detailed description can be found in [29,52]. The mean-value coordinate system is shown in Figure 4, and the interpolation function with linear accuracy is expressed as:

$$N_i(x) = \frac{w_i(x)}{\sum_{j=1}^{n} w_j(x)} \tag{30}$$

$$w_i(x) = \frac{tan(\alpha_{i-1}/2) + tan(\alpha_i/2)}{\|x - x_i\|} \tag{31}$$

$$tan(\alpha_i/2) = \frac{sin\ \alpha_i}{1 + cos\ \alpha_i} \tag{32}$$

where $w_i(x)$ is the weight function, and $\|x - x_i\|$ is the Eulerian distance between points M and M_i (in Figure 4). Point M is selected as the geometric center of the polygon, and n represents the number of vertices on the polygon, that is, the number of edges of the polygon. The interpolation function expressed in Equations (30)–(32) can be used for both convex and concave polygons.

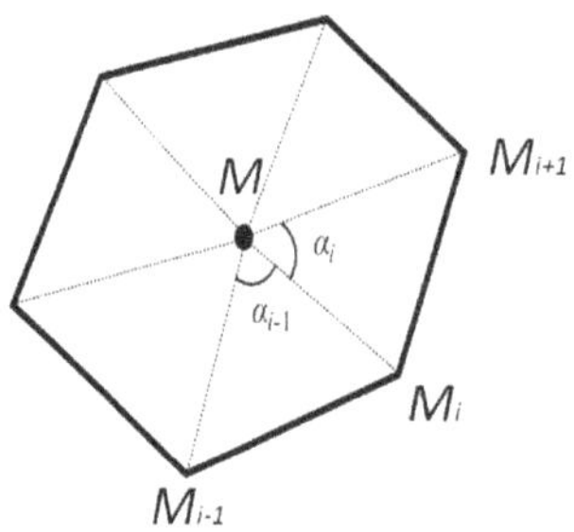

Figure 4. Illustration of mean-value coordinates.

Generally, the shape function is expressed in terms of the Cartesian coordinate system $(X_1, X_2) \in W_e$. In order to simplify the integration of the element matrix, it is necessary to construct a conforming approximation of the polygon using mean-value shape functions. Similar to the isoparametric element in FEM, the mean shape function is defined on a standard element in the local coordinate system $(\xi_1, \xi_2) \in W_0$. Four standard elements in the local coordinate system are illustrated as an example in Figure 5: a regular triangle, quadrilateral, pentagon, and hexagon. The vertices of each standard element are placed on the same circumscribed circle with a radius of 1.0, and the geometric center of each element coincides with the center of the circumscribed circle. Therefore, any point in a standard polygon element can be directly connected with each vertex without being occluded. The coordinates of the polygon vertices on the unit circle are $(cos\ 2\pi/n, sin\ 2\pi/n)$, $(cos\ 4\pi/n,$ $sin\ 4\pi/n), \ldots$, and $(1, 0)$, where n is the number of vertices. In this way, the shape function can be defined and used in the local coordinate system as shown in Figure 5, where only four kinds of polygons are shown, and any polygon in the global Cartesian coordinates can be transformed into a standard element using the corresponding local coordinate system through the polygon isoparametric mapping shape function **F**. An example of the mapping process for a pentagon is illustrated in Figure 6.

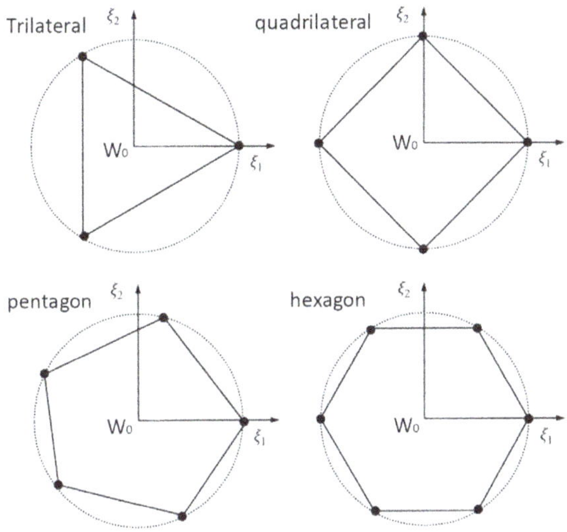

Figure 5. Four kinds of regularized polygon elements.

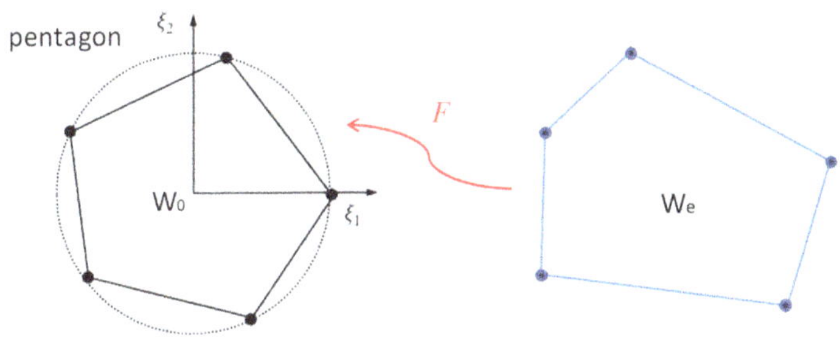

Figure 6. Illustration of mean-value coordinates.

After the mapping to the local coordinate system is completed (Figure 7), the standard polygon element is divided into triangular subunits with the center of the circle serving as a common vertex. Then, the element coefficient matrices, such as $[E^0]$, $[E^1]$, $[E^2]$, $[M^1]$, and so on, are computed by integrating over the triangular subunits using the standard orthogonal criterion. The triangular subunits in Figure 7 are used only for numerical integration. A detailed discussion of the integration methods can be found in references [29,52].

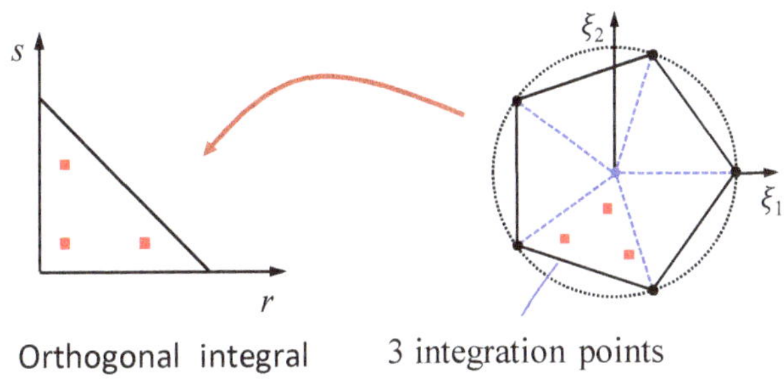

Figure 7. Standard quadrature rules.

2.2.2. Polyhedral Fluid Elements

As shown in Figure 8, for the dam upstream face with arbitrary convex polygon (triangle, quadrilateral, pentagon, hexagon, heptagon, octagon, etc.) grids, the reservoir model is discretized by the polyhedron SBFEM with polygonal semi-infinite prismatic elements, which are polyhedral fluid elements. A polygonal element (grid) is transformed into a standard unit in the local coordinate system by means of the polygon isoparametric mapping function F. Subsequently, the matrices ($[E^0]$, $[E^1]$, $[E^2]$, $[M^1]$, and so on) for the polyhedral element are integrated and calculated. In the end, the total matrices of all reservoir elements are integrated, and the hydrodynamic pressure can be directly solved as mentioned in Section 2.1.

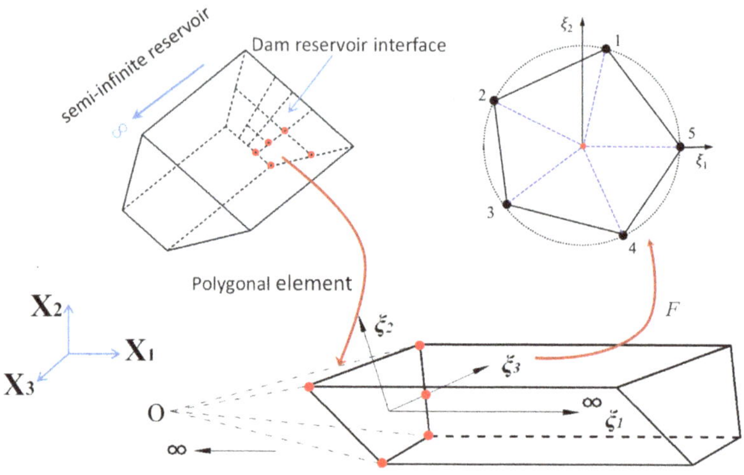

Figure 8. Polygonal semi-infinite prismatic reservoir element.

3. A Nonlinear Dynamic Coupling Method for Cross-Scale Dam-Reservoir Systems Based on the Polyhedron SBFEM

The establishment of the dynamic coupling method consists of two main phases: the procedure implementation for the hydrodynamic pressure computation method and analysis implementation for the coupling calculation equation of dam-reservoir systems.

3.1. Polyhedron SBFEM Procedure for Fluid

The polygonal semi-infinite prismatic fluid elements (i.e., polyhedral elements) are implemented based on the Windows programs GEODYNA [54], which was developed by the fifth author using object-oriented programming in Visual C++. Multicore parallel technology of the CPU coupled with the GPU is realized in the GEODYNA program, by which the computational capacity of solving a large-scale elasto-plastic analysis with millions of DOFs is provided. The GEODYNA program has been applied to the dynamic analysis of nonlinear structures [26–28,45,46,48,55–62].

3.2. Nonlinear Dynamic Coupling Method for Cross-Scale CFRD-Reservoir Systems

The nonlinear dam, including the foundation, is modeled by the polyhedron SBFEM with cross-scale mesh, and the reservoir is modeled by the polyhedron SBFEM for fluid. In this way, the equation for the dynamic coupling analysis between the CFRD and reservoir can be expressed as follows:

$$[M_s]\{\ddot{u}_r(t)\} + [C_s]\{\dot{u}_r(t)\} + [K_s]\{u_r(t)\} = -[M_s]\{\ddot{u}_g(t)\} - (1/\rho)\left[L_1\right]^T[M^1]^T\{p(\xi = 0)\} \tag{33}$$

where $[M_s]$, $[C_s]$, and $[K_s]$ are, respectively, the mass, damping, and stiffness matrices of the dam and foundation. $\{\ddot{u}_r(t)\}$, $\{\dot{u}_r(t)\}$, and $\{u_r(t)\}$ are, respectively, the relative acceleration, velocity, and displacement. $\{\ddot{u}_g(t)\}$ is the input earthquake acceleration from bedrock. $[L_1]$ is the conversion matrix between global coordinates and the local coordinates of the dam surface.

Substituting Equation (27) into Equation (33), the dynamic fluid–solid coupling analysis equations are derived further as:

$$([M_s]+[M_p])\{\ddot{u}_r(t)\}+[C_s]\{\dot{u}_r(t)\}+[K_s]\{u_r(t)\}=-([M_s]+[M_p])\{\ddot{u}_g(t)\} \quad (34)$$

$$\left[M_p\right]=-(1/\rho)[L_1]^T[M^1]^T\left\{[\Phi_{12}]\left[\Phi_{22}\right]^{-1}[M^1][L_1]+([\Phi_{12}]\left[\Phi_{22}\right]^{-1}[B_1]-[B_2])\rho\left[C^0\right][L_2]\right\} \quad (35)$$

$$[L_1]\{\ddot{u}_g(t)+\ddot{u}_r(t)\}=\{\ddot{u}_n\}, \qquad [L_2]\{\ddot{u}_g(t)+\ddot{u}_r(t)\}=\{\ddot{v}_n\} \quad (36)$$

where $[M_p]$ is the additional mass matrix of hydrodynamic pressure. $[L_2]$ is the conversion matrix between global coordinates and the local coordinates of the river valley surrounding the reservoir.

As shown in Equation (34), the dynamic interaction analysis between the dam and the reservoir water can be realized directly by superimposing the additional mass matrix of the hydrodynamic pressure $[M_p]$ into the mass matrix of the dam $[M_s]$. Then, a strong coupling method for a nonlinear cross-scale dam-reservoir system is established based on the polyhedron SBFEM.

4. Numerical Examples of Rigid Dams and River Valley

The following numerical examples of rigid dams, which have analytical solutions, have been selected for analysis and validation of the accuracy of the presented method for hydrodynamic pressure. The computed hydrodynamic pressure distribution due to earthquakes in the upstream–downstream, vertical, and dam axial directions have been compared with analytical solutions [63–66].

4.1. Dams with Polygonal Mesh on Upstream Face

The first group of examples are shown in Figure 9. An inclined dam face in a rectangular valley with a height and width of 180 m was selected. The water depth of the reservoir was 180 m (i.e., full reservoir condition), and the dam upstream face inclination angles were 30°, 45°, and 60°, respectively.

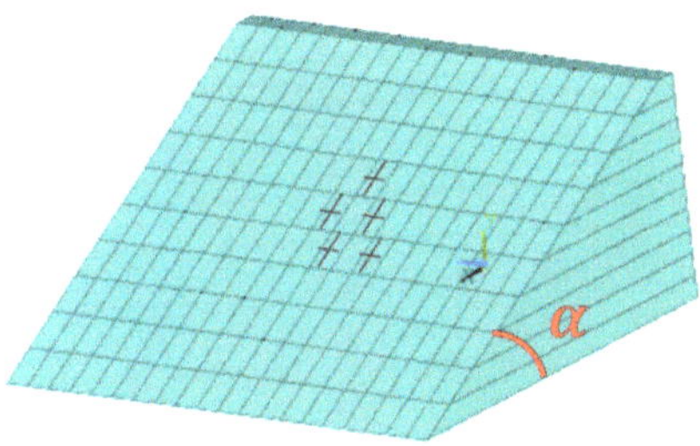

Figure 9. Inclined dam face in rectangular valley.

The second group of examples are shown in Figure 10. A 120-m high vertical dam face in an isosceles right triangular valley was selected. The water depth of the reservoir was 120 m (i.e., full reservoir condition).

The third group of examples are shown in Figure 11. A vertical dam face in the semi-circular valley with a radius of 100 m was selected. The water depth of the reservoir was 100 m (i.e., full reservoir condition).

The 2D dam upstream face meshes of the above three groups of calculation examples, which are plotted in Figures 9–11, were just used as the SBFEM meshes of the reservoir water. Furthermore it can be observed that the dam upstream faces contained pentagon, hexagon, heptagon, and octagon polygon grids, which often appear when polyhedron elements based on the polyhedron SBFEM are used to model dams. The peak acceleration of seismic excitation in different directions is expressed as **a**, the density of water is expressed as ρ, and the water depth is expressed as **H**.

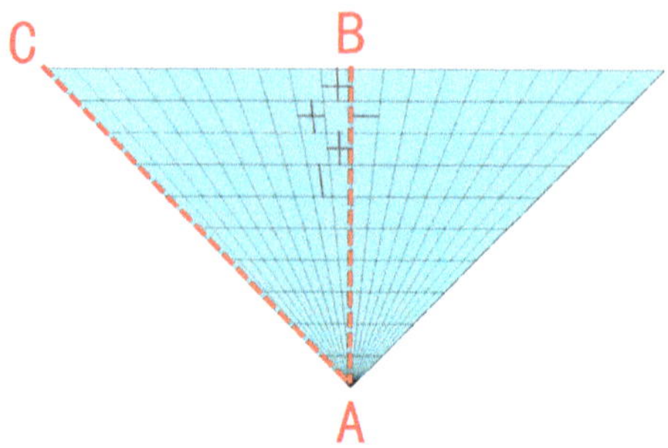

Figure 10. Vertical dam face in a triangular valley.

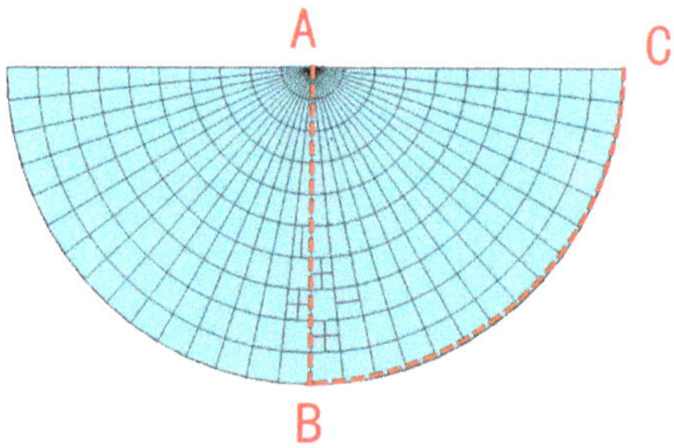

Figure 11. Vertical dam face in semi-circular valley.

4.2. Results and Discussion

Figure 12 shows the distribution of hydrodynamic pressure for the condition in which the dam face is inclined in a rectangular valley. Figures 13–15 provide the results for the case in the triangular valley. Figures 16–18 show the distribution of hydrodynamic pressure on the dam face in the semi-circular valley. From the above figures (Figures 12–18), it can be seen that the proposed hydrodynamic pressure calculation method based on polyhedron SBFEM could precisely compute the hydrodynamic pressure on the dam face induced by an earthquake in different directions and could accurately consider factors such as the inclination of the dam face and the complex shape of the river valley.

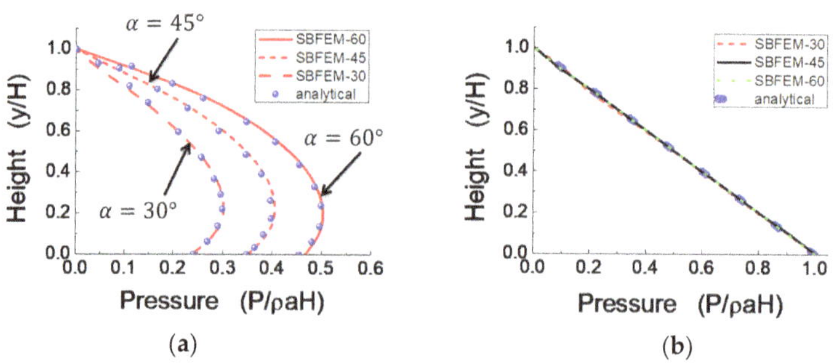

Figure 12. Rectangular valley: (**a**) upstream–downstream excitation and (**b**) vertical excitation.

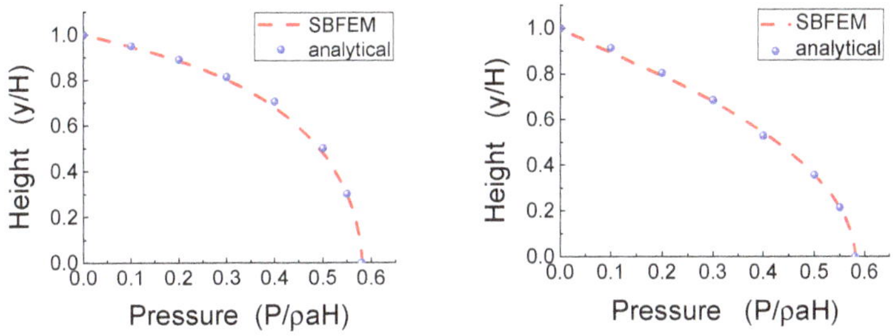

(a) (b)

Figure 13. Triangular valley with upstream–downstream excitation: (**a**) along line AB and (**b**) along line AC.

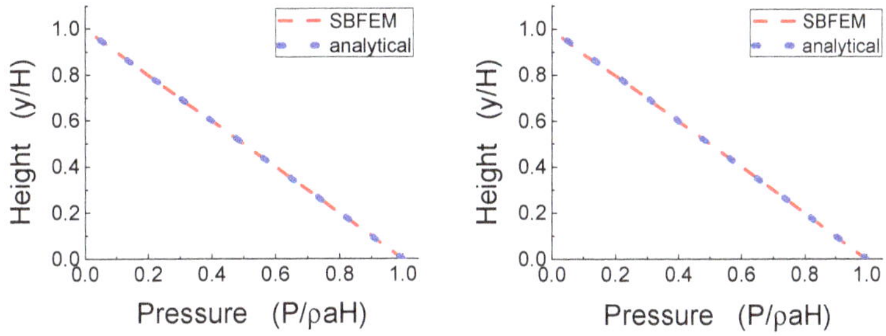

(a) (b)

Figure 14. Triangular valley with vertical excitation: (**a**) along line AB and (**b**) along line AC.

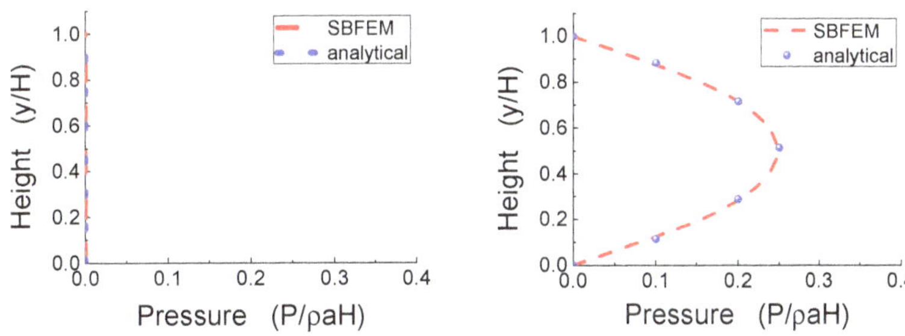

(a) (b)

Figure 15. Triangular valley with transverse excitation: (**a**) along line AB and (**b**) along line AC.

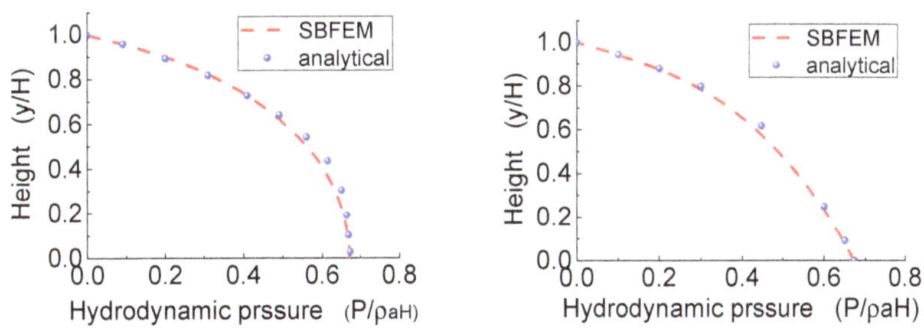

(a) (b)

Figure 16. Semi-circular valley with upstream–downstream excitation: (**a**) along line AB and (**b**) along line BC.

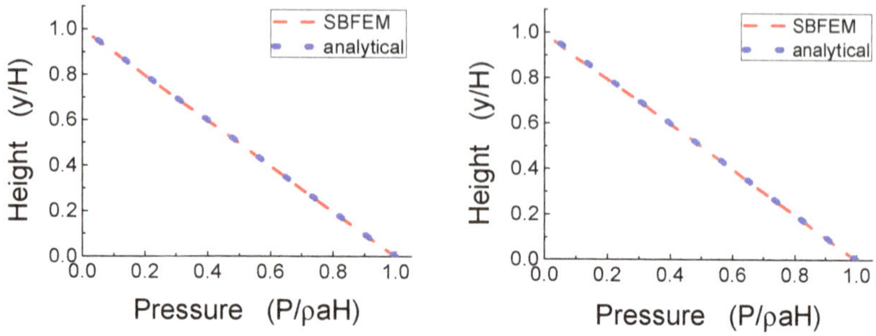

(a) (b)

Figure 17. Semicircular valley with vertical excitation: (**a**) along line AB and (**b**) along line BC.

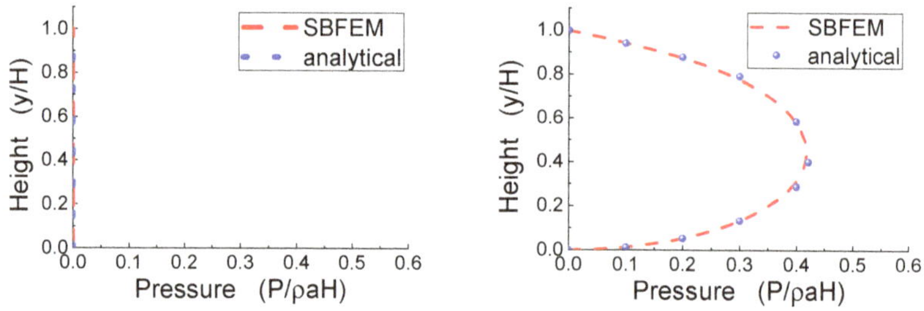

(a) (b)

Figure 18. Semicircular valley with transverse excitation: (**a**) along line AB and (**b**) along line BC.

Moreover, the proposed method has a high efficiency, because only 2D discretization was required to simulate the 3D semi-infinite reservoir water. Therefore, the proposed hydrodynamic pressure method can serve as a convenient analysis tool for further establishing a dynamic coupling analysis method for the dam–reservoir systems when polyhedral solid elements are utilized to build cross-scale models of dams.

5. Dynamic Coupling Analysis of Nonlinear Cross-Scale CFRD and Reservoir Systems

With the polyhedron SBFEM for solid and fluid being used to discretize the CFRD and reservoir, the elasto-plastic cross-scale dynamic coupling analysis was conducted to investigate the effect of hydrodynamic pressure on the dynamic response of CFRD. The results of the CFRD under static loading, including the filling and impounding processes, were introduced into the seismic analysis as the initial state.

5.1. Cross-Scale Model of the CFRD and Reservoir

The 3D polyhedron cross-scale mesh of a 100-m high dam and rock foundation is shown in Figure 19. The concrete slab grid was dense, and the rockfill grid was relatively coarse. In order to achieve a rapid cross-scale meshing from the slab to the rockfill, some polyhedral meshes appeared on upstream face of the slab inevitably. The upstream slope of the dam was 1 V:1.4 H, and the downstream slope was 1 V:1.6 H. The valley was a prismatic valley with a trapezoidal section with a bottom width of 70 m and a slope ratio of 1 V:1 H on both banks. The mesh of the dam and foundation had a total of 39,145 elements, of which the slab and interface both had 1272 elements. The slab and rockfill were simulated by polyhedral elements based on SBFEM [27]. Furthermore, the interface between the slab and cushion, slab joints, and peripheral joints were simulated by polygonal Goodman elements [27]. The displacement of the bottom boundary of the massless bedrock in the three directions of the global coordinate system was constrained, and the normal displacement of the lateral boundary was constrained.

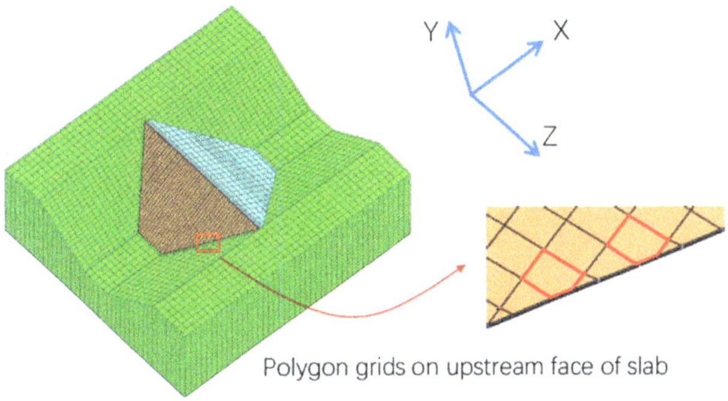

Figure 19. Cross-scale model of the CFRD and foundation.

The depth of the reservoir in front of the dam was 90 m. When the proposed method was used to calculate the hydrodynamic pressure, the mesh of the semi-infinite reservoir water could be directly generated using the 2D polygonal grid on the upstream face of the concrete slab, as indicated in Figure 3. Therefore, the pre-processing of the reservoir model was simplified to a large extent, and the overall analysis efficiency was improved.

5.2. Material Parameters, Damping Methods, and Ground Motion

A linear elastic model was adopted for both the concrete face slab (density ρ = 2.40 g/cm^3, elasticity modulus E = 25 GPa, Poisson's ratios ν = 0.167) and bedrock (density ρ = 2.50 g/cm^3, elasticity modulus E = 20 GPa, Poisson's ratios ν = 0.2). An improved P–Z generalized plastic model was used for rockfills [59,67], of which the 17 material parameters were calibrated by the results of the triaxial tests and listed in the Table 1. Furthermore, an ideal elasto-plastic model was used to model the interface between the face slab and rockfills, of which the parameters are listed in Table 2. The compression stiffness of the slab and peripheral joints was 25,000 MPa/m, and the shear stiffness was 1 MPa/m. The Rayleigh damping method was employed for the various material and mechanical models of the CFRD.

Table 1. Parameters of the rockfill.

G_0	K_0	M_g	M_f	α_f	α_g	H_0	H_{U0}	m_s
2400	2500	1.75	1.5	0.45	0.45	2900	2900	0.2
m_v	m_t	m_u	r_d	γ_{DM}	γ_U	β_0	β_1	
0.28	0.2	0.25	105	70	7	50	0.023	

Table 2. Parameters of the interface.

k_1	k_2	n	φ/°	c/Pa
300	1×10^{10}	0.8	41.5	0

Seismic waves were the input in the upstream–downstream direction for the dynamic coupling analysis. The time history of the seismic acceleration measured from a real earthquake in Figure 20 was selected. The peak ground acceleration (PGA) was 1.5 m/s^2. The results of the two conditions, which were considering the hydrodynamic pressure based on the polyhedron SBFEM condition and neglecting the hydrodynamic pressure condition, are compared and analyzed in the following section.

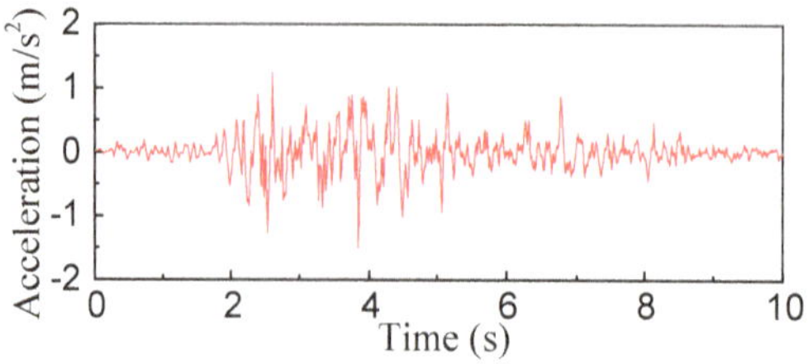

Figure 20. Input ground motion.

5.3. Results and Discussion

Due to the existence of reservoir water in front of the dam, the dam would definitely be affected by hydrodynamic pressure under an earthquake condition. The proposed hydrodynamic pressure calculation method based on the polyhedron SBFEM, which was proved to be accurate in Section 4.2, is adequate for analyzing the effect of hydrodynamic pressure on the dynamic response of the rockfill in CFRD. For the condition of ignoring hydrodynamic pressure, the additional mass matrix of hydrodynamic pressure $[M_p]$ in Equation (34) was a null matrix, which meant that errors would occur because it was inconsistent with the actual situation. When calculating the error of the dynamic response of rockfill and concrete face slabs in the CFRD caused by ignoring the hydrodynamic pressure, the results under the condition of considering the hydrodynamic pressure were used as a standard. The compressive stress of the face slabs was positive.

5.3.1. Rockfill

Table 3 lists the maximum absolute values of the dynamic acceleration and displacement for the rockfill under an earthquake and the corresponding errors caused by neglecting the hydrodynamic pressure. Compared to the results from the condition of calculating hydrodynamic pressure by the polyhedron SBFEM, the maximum errors of acceleration and displacement were 15.4% and 12.7%, respectively, when the hydrodynamic pressure was ignored. Figures 21 and 22 show the maximum distribution of the dynamic acceleration along the upstream–downstream direction (a_x) and the dynamic displacement along the vertical direction (d_y). As can be seen from Figures 21 and 22 and Table 3, the distribution rules of the dynamic response for the rockfill were consistent under the two conditions of hydrodynamic pressure, but the differences of the maximum value and the corresponding size of the area with a large response were obvious. The dynamic acceleration and displacement of the rockfill were smaller when hydrodynamic pressure was considered.

Table 3. Maximum dynamic response of a rockfill under an earthquake.

Hydrodynamic Pressure	Acceleration (m/s^2)		Displacement (m)	
	a_x	a_y	d_x	d_y
Polyhedron SBFEM	4.61	2.72	0.062	0.055
Neglecting	5.32	2.87	0.071	0.059
Error	15.4%	5.5%	12.7%	7.3%

In summary, when the hydrodynamic pressure is ignored, the dynamic acceleration and displacement response of a rockfill may be overestimated obviously, which is not conducive to the accurate and reasonable safety evaluation of CFRD under an earthquake.

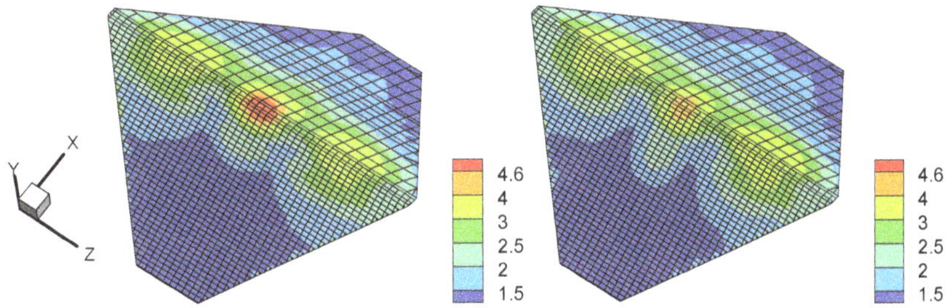

(a) (b)

Figure 21. Maximum distribution of rockfill acceleration along the upstream–downstream direction (m/s^2): (**a**) Neglecting hydrodynamic pressure and (**b**) the Polyhedron SBFEM.

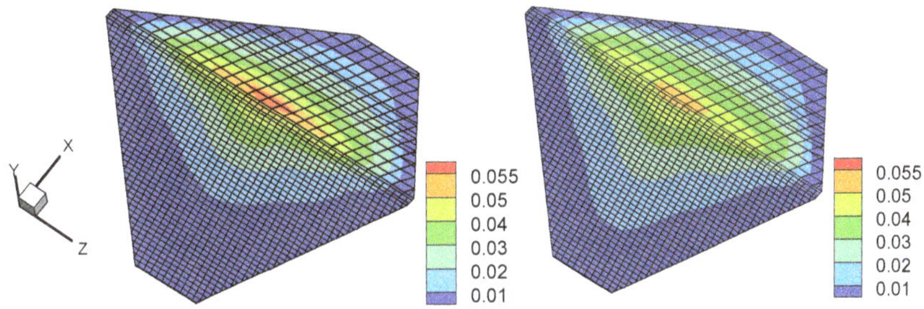

(a) (b)

Figure 22. Maximum distribution of rockfill displacement along the vertical direction (m): (**a**) Neglecting hydrodynamic pressure and (**b**) the Polyhedron SBFEM.

5.3.2. Concrete Face Slabs

Table 4 lists the maximum dynamic concrete face slabs' stresses under the two hydrodynamic pressure conditions and the relative errors of stresses caused by neglecting the hydrodynamic pressure. Compared with the results by the polyhedron SBFEM, the maximum errors in the dynamic face slabs' stresses along the slope direction and the dam axial direction for neglecting the hydrodynamic pressure condition were 22.5% and 24.9%, respectively. It can be observed from Table 4 that the errors of the maximum face slabs' stress along the slope direction and the dam axial direction caused by neglecting the hydrodynamic pressure were obvious, and the hydrodynamic pressure cannot be ignored in the dynamic stress analysis of face slabs. Figures 23 and 24 plot the distributions of the maximum dynamic tensile stress in face slabs along the slope direction and compressive stresses in face slabs along the dam axial direction. Figures 23 and 24 show that the distributions of the extreme stresses along the slope direction and the dam axial direction for both hydrodynamic pressure conditions were similar, but the extent of the high stress regions was significantly different.

Table 4. Maximum dynamic stress of face slabs under an earthquake.

Hydrodynamic Pressure	Slope Direction (MPa)		Dam Axial Direction (MPa)	
	Tensile	**Compressive**	**Tensile**	**Compressive**
Polyhedron SBFEM	−3.83	3.16	−2.05	3.38
Neglecting	−4.69	3.48	−1.88	2.54
Error	22.5%	9.2%	8.3%	24.9%

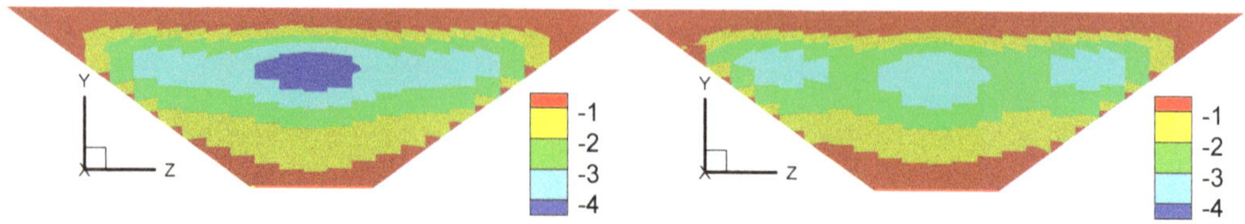

(**a**) (**b**)

Figure 23. Maximum distribution of concrete face slabs' tensile stresses along the slope direction (MPa): (**a**) Neglecting hydrodynamic pressure and (**b**) the Polyhedron SBFEM.

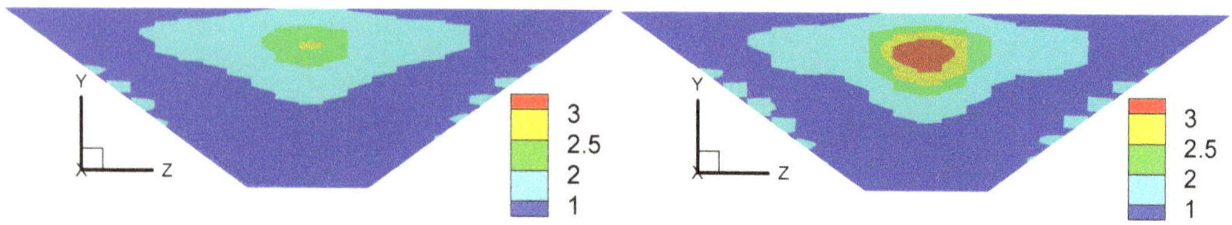

(**a**) (**b**)

Figure 24. Maximum distribution of concrete face slabs' compressive stresses along the dam axial direction (MPa): (**a**) Neglecting hydrodynamic pressure and (**b**) the Polyhedron SBFEM.

As can be concluded from the above Figures 23 and 24 and Table 4, for a high CFRD located in front of a reservoir, the hydrodynamic pressure has a big influence on the dynamic face slabs' stresses along the slope direction and the dam axial direction. It is possible to significantly overestimate or underestimate the dynamic concrete face slabs' stress when hydrodynamic pressure is ignored in the seismic analysis of a CFRD.

6. Conclusions

In this paper, polyhedral scaled boundary finite elements of fluid were constructed using the polygon mean-value shape function to compute the hydrodynamic pressure of a reservoir on a dam, which is applicable for not only a polygon mesh but also triangular and quadrilateral meshes on the upstream face of a dam. Moreover, a nonlinear cross-scale dynamic analysis of CFRD-reservoir systems under a seismic condition was carried out, and the effect of hydrodynamic pressure on the dynamic response of a rockfill in CFRD was investigated. The following primary conclusions are summarized:

(1) A 3D hydrodynamic pressure calculation method based on the polyhedron SBFEM was proposed, in which the reservoir in front of a dam was simulated with polygonal semi-infinite prismatic fluid elements. The pre-processing of the reservoir model was simplified to a large extent, as the 3D mesh of the reservoir could be generated automatically from the 2D grid of the upstream face of dam. A high efficiency was achieved also by reducing the one-dimensional discretization. The proposed method has a high accuracy and provides a convenient numerical tool for a dynamic coupling analysis of a dam–reservoir system, when the cross-scale dam is modeled by the polyhedron SBFEM.

(2) With an elastic–plastic CFRD being simulated by the polyhedron SBFEM and the hydrodynamic pressure of the reservoir being computed by the proposed polyhedron SBFEM for fluid, respectively, a nonlinear dynamic coupling method for cross-scale CFRD-reservoir systems based on the polyhedron SBFEM was developed. The results of a further numerical analysis showed that neglecting hydrodynamic pressure may produce obvious errors and lead to overestimation of the dynamic acceleration and displacement response of the rockfill, which is not conducive to an accurate and reasonable safety evaluation of a CFRD under an earthquake. Moreover, the hydrodynamic pressure had a big influence on the dynamic face slabs' stresses, and the hydrodynamic pressure cannot be ignored in the dynamic stress analysis of face slabs.

The hydrodynamic pressure calculation method proposed in this paper can also be applied to a dynamic linear or nonlinear analysis of cross-scale arch dams or gravity dams simulated by the polyhedron SBFEM for seismic safety evaluation.

Author Contributions: Conceptualization, H.X. and D.Z.; methodology, H.X. and J.X.; software, H.X. and K.C.; validation, H.X. and D.Y.; resources, H.X., K.C. and D.Z.; writing—original draft preparation, J.X. and H.X.; writing—review and editing, D.Y. and K.C.; supervision, J.X., H.X., D.Y., K.C. and D.Z. All authors have read and agreed to the published version of the manuscript.

Funding: This work was supported by the Project funded by the China Postdoctoral Science Foundation (2021M690998) and the National Natural Science Foundation of China (Grant Nos. U1965206, 52192674).

Institutional Review Board Statement: Not applicable.

Informed Consent Statement: Not applicable.

Data Availability Statement: Not applicable.

Conflicts of Interest: The authors declare no conflict of interest.

Nomenclature

$\bigtriangledown^2$	Laplacian operator
p	Hydrodynamic pressure
ρ	Fluid density
$\ddot{u}_n$	Normal accelerations of the dam–reservoir interface
$\ddot{\upsilon}_n$	Normal accelerations of the river –valley interface
$[N(\xi_2, \xi_3)]$	Polygon mean-value shape function
$[J]$	Jacobian matrix
w	weight function
$[E^0]$, $[E^1]$, $[E^2]$, $[C^0]$, $[M^1]$	Coefficient matrices
$\{q(\xi_1)\}$	Nodal force
$[Z]$	Hamilton coefficient matrix
$[\Lambda]$	Eigenvalue matrix
$[\Phi]$	Eigenvector matrix
$[A]$	The inverse of eigenvector matrix $[\Phi]$
$N_i(x)$	Interpolation function in mean-value coordinate system
$w_i(x)$	Weight function of mean-value coordinate system
$\|\|x - x_i\|\|$	Eulerian distance between points
W_e	Cartesian coordinate system
W_0	Local coordinate system
$[M_s]$, $[C_s]$, $[K_s]$	Mass, damping and stiffness matrices
$\{\ddot{u}_r(t)\}$, $\{\dot{u}_r(t)\}$, $\{u_r(t)\}$	Relative acceleration, velocity, and displacement
$\{\ddot{u}_g(t)\}$	Input earthquake acceleration
$[M_p]$	Additional mass matrix of hydrodynamic pressure
$[L_1]$, $[L_2]$	Conversion matrix
(x_1, x_2, x_3)	Global coordinates
(ξ_1, ξ_2, ξ_3)	Local scaled boundary coordinates
E	Elasticity modulus
ν	Poisson's ratios
SBFEM	Scaled boundary finite element method
CFRD	Concrete faced rockfill dam
3D	Three-dimensional
2D	Two-dimensional
DOF	Degrees of freedom
FEM	Finite element method
BEM	Boundary element method
PSBFEM	Polyhedron SBFEM
PGA	Peak ground acceleration

References

1. Song, C.M. *The Scaled Boundary Finite Element Method: Introduction to Theory and Implementation*; John Wiley & Sons Ltd.: Hoboken, NJ, USA, 2018.
2. Wolf, J.P.; Song, C. The scaled boundary finite-element method—A primer: Derivations. *Comput. Struct.* **2000**, *78*, 191–210. [CrossRef]
3. Ooi, E.T.; Song, C.M.; Natarajan, S. A scaled boundary finite element formulation with bubble functions for elasto-static analyses of functionally graded materials. *Comput. Mech.* **2017**, *60*, 943–967. [CrossRef]
4. Xue, B.; Lin, G.; Hu, Z. Scaled boundary isogeometric analysis for electrostatic problems. *Eng. Anal. Bound. Elem.* **2017**, *85*, 20–29. [CrossRef]
5. Zhang, X.; Wegner, J.L.; Haddow, J.B. Three-dimensional dynamic soil-structure interaction analysis in the time domain. *Earthq. Eng. Struct. Dyn.* **1999**, *28*, 1501–1524. [CrossRef]
6. Wegner, J.L.; Zhang, X. Free-vibration analysis of a three-dimensional soil-structure system. *Earthq. Eng. Struct. Dyn.* **2001**, *30*, 43–57. [CrossRef]
7. Genes, M.C.; Aslmand, M.; Kani, M. Efficient Dynamic Analysis of Foundation Via a Coupled Axisymmetric SBFEM-3D FEM. *Teknik Dergi* **2019**, *30*, 9327–9352. [CrossRef]
8. Schauer, M.; Rios Rodriguez, G. A coupled FEM-SBFEM approach for soil-structure-interaction analysis using non-matching meshes at the near-field far-field interface. *Soil Dyn. Earthq. Eng.* **2019**, *121*, 466–479. [CrossRef]
9. Yaseri, A. 2.5D coupled FEM-SBFEM analysis of ground vibrations induced by train movement. *Soil Dyn. Earthq. Eng.* **2017**, *104*, 307–318. [CrossRef]
10. Chen, D.; Dai, S. Dynamic fracture analysis of the soil-structure interaction system using the scaled boundary finite element method. *Eng. Anal. Bound. Elem.* **2017**, *77*, 26–35. [CrossRef]
11. Song, C.; Wolf, J.P. Semi-analytical representation of stress singularities as occurring in cracks in anisotropic multi-materials with the scaled boundary finite-element method. *Comput. Struct.* **2002**, *80*, 183–197. [CrossRef]
12. Zhang, P.; Du, C.; Tian, X.; Jiang, S. A scaled boundary finite element method for modelling crack face contact problems. *Comput. Methods Appl. Mech. Eng.* **2018**, *328*, 431–451. [CrossRef]
13. Yao, F.; Yang, Z.J.; Hu, Y.J. An SBFEM-Based Model for Hydraulic Fracturing in Quasi-Brittle Materials. *Acta Mech. Solida Sin.* **2018**, *31*, 416–432. [CrossRef]
14. Adak, D.; Pramod, A.; Ooi, E.T.; Natarajan, S. A combined virtual element method and the scaled boundary finite element method for linear elastic fracture mechanics. *Eng. Anal. Bound. Elem.* **2020**, *113*, 9–16. [CrossRef]
15. Chen, K.; Zou, D.; Tang, H.; Liu, J.; Zhuo, Y. Scaled boundary polygon formula for Cosserat continuum and its verification. *Eng. Anal. Bound. Elem.* **2021**, *126*, 136–150. [CrossRef]
16. Qu, Y.; Zou, D.; Kong, X.; Yu, X.; Chen, K. Seismic cracking evolution for anti-seepage face slabs in concrete faced rockfill dams based on cohesive zone model in explicit SBFEM-FEM frame. *Soil Dyn. Earthq. Eng.* **2020**, *133*, 106106. [CrossRef]
17. Jiang, S.; Du, C.; Ooi, E.T. Modelling strong and weak discontinuities with the scaled boundary finite element method through enrichment. *Eng. Fract. Mech.* **2019**, *222*, 106734. [CrossRef]
18. Liu, J.; Hao, C.; Ye, W.; Yang, F.; Lin, G. Free vibration and transient dynamic response of functionally graded sandwich plates with power-law nonhomogeneity by the scaled boundary finite element method. *Comput. Methods Appl. Mech. Eng.* **2021**, *376*, 113665. [CrossRef]
19. Zhang, P.; Du, C.; Tian, X.; Jiang, S. Buckling analysis of three-dimensional functionally graded sandwich plates using two-dimensional scaled boundary finite element method. *Mech. Adv. Mater. Struct.* **2021**, *3*, 431–451.
20. He, Y.; Guo, J.; Yang, H. Image-based numerical prediction for effective thermal conductivity of heterogeneous materials: A quadtree based scaled boundary finite element method. *Int. J. Heat Mass Transf.* **2019**, *128*, 335–343. [CrossRef]
21. Liu, L.; Zhang, J.; Song, C.; He, K.; Saputra, A.A.; Gao, W. Automatic scaled boundary finite element method for three-dimensional elastoplastic analysis. *Int. J. Mech. Sci.* **2020**, *171*, 105374. [CrossRef]
22. Wijesinghe, D.R.; Dyson, A.; You, G.; Khandelwal, M.; Song, C.; Ooi, E.T. Development of the scaled boundary finite element method for image-based slope stability analysis. *Comput. Geotech.* **2022**, *143*, 104586. [CrossRef]
23. Li, J.; Shi, Z.; Liu, L.; Song, C. An efficient scaled boundary finite element method for transient vibro-acoustic analysis of plates and shells. *Comput. Struct.* **2020**, *231*, 106211. [CrossRef]
24. Lin, G.; Xue, B.; Hu, Z. A mortar contact formulation using scaled boundary isogeometric analysis. *Sci. China Phys. Mech. Astron.* **2018**, *61*, 74621. [CrossRef]
25. Liu, J.; Zhang, P.; Lin, G.; Wang, W.; Lu, S. Solutions for the magneto-electro-elastic plate using the scaled boundary finite element method. *Eng. Anal. Bound. Elem.* **2016**, *68*, 103–114. [CrossRef]
26. Chen, K.; Zou, D.; Kong, X. A nonlinear approach for the three-dimensional polyhedron scaled boundary finite element method and its verification using Koyna gravity dam. *Soil Dyn. Earthq. Eng.* **2017**, *96*, 1–12. [CrossRef]
27. Chen, K.; Zou, D.; Kong, X.; Zhou, Y. Global concurrent cross-scale nonlinear analysis approach of complex CFRD systems considering dynamic impervious panel-rockfill material-foundation interactions. *Soil Dyn. Earthq. Eng.* **2018**, *114*, 51–68. [CrossRef]
28. Chen, K.; Zou, D.; Kong, X.; Yu, X. An efficient nonlinear octree SBFEM and its application to complicated geotechnical structures. *Comput. Geotech.* **2018**, *96*, 226–245. [CrossRef]

29. Zou, D.; Chen, K.; Kong, X.; Liu, J. An enhanced octree polyhedral scaled boundary finite element method and its applications in structure analysis. *Eng. Anal. Bound. Elem.* **2017**, *84*, 87–107. [CrossRef]
30. Vaghefi, M.; Behroozi, A.M. Radial basis function differential quadrature for hydrodynamic pressure on dams with arbitrary reservoir and face shapes affected by earthquake. *J. Appl. Fluid Mech.* **2020**, *13*, 1759–1768.
31. Pasbani Khiavi, M.; Sari, A. Evaluation of hydrodynamic pressure distribution in reservoir of concrete gravity dam under vertical vibration using an analytical solution. *Math. Probl. Eng.* **2021**, *2021*, 6669366. [CrossRef]
32. Gorai, S.; Maity, D. Seismic Performance Evaluation of Concrete Gravity Dams in Finite-Element Framework. *Pract. Period. Struct. Des. Constr.* **2022**, *27*, 04021072. [CrossRef]
33. Babaee, R.; Khaji, N. Decoupled scaled boundary finite element method for analysing dam–reservoir dynamic interaction. *Int. J. Comput. Math.* **2020**, *97*, 1725–1743. [CrossRef]
34. Liu, J.; Lin, G.; Li, J. Short-crested waves interaction with a concentric cylindrical structure with double-layered perforated walls. *Ocean Eng.* **2012**, *40*, 76–90. [CrossRef]
35. Song, H.; Tao, L. An efficient scaled boundary FEM model for wave interaction with a nonuniform porous cylinder. *Int. J. Numer. Methods Fluids* **2010**, *63*, 96–118. [CrossRef]
36. Nguyen, L.V.; Tornyeviadzi, H.M.; Bui, D.T.; Seidu, R. Predicting Discharges in Sewer Pipes Using an Integrated Long Short-Term Memory and Entropy A-TOPSIS Modeling Framework. *Water* **2022**, *14*, 300. [CrossRef]
37. Teng, B.; Zhao, M.; He, G.H. Scaled boundary finite element analysis of the water sloshing in 2D containers. *Int. J. Numer. Methods Fluids* **2006**, *52*, 659–678. [CrossRef]
38. Wang, Y.; Hu, Z.Q.; Guo, W.D. Hydrodynamic pressures on arch dam faces with irregular reservoir geometry. *J. Vibrat. Control* **2019**, *25*, 627–638. [CrossRef]
39. Zeinizadeh, A.; Mirzabozorg, H.; Noorzad, A.; Amirpour, A. Hydrodynamic pressures in contraction joints including waterstops on seismic response of high arch dams. *Structures* **2018**, *14*, 1–14. [CrossRef]
40. Wang, J.T.; Chopra, A.K. Linear analysis of concrete arch dams including dam-water-foundation rock interaction considering spatially varying ground motions. *Earthq. Eng. Struct. Dyn.* **2010**, *39*, 731–750. [CrossRef]
41. Gao, Y.; Jin, F.; Wang, X.; Wang, J. Finite Element Analysis of Dam-Reservoir Interaction Using High-Order Doubly Asymptotic Open Boundary. *Math. Probl. Eng.* **2011**, *2011*, 668–680. [CrossRef]
42. Cheng, H.; Zhang, G.; Jiang, C.; Liao, J.; Zhou, Q. Analysis of seismic damage process of high concrete dam-foundation system. *IOP Conf. Ser. Earth Environ. Sci.* **2019**, *304*, 042068. [CrossRef]
43. Wang, C.; Zhang, H.; Zhang, Y.; Guo, L.; Wang, Y.; Thira Htun, T.T. Influences on the seismic response of a gravity dam with different foundation and reservoir modeling assumptions. *Water* **2021**, *13*, 3072. [CrossRef]
44. Feng, S.J.; Chen, Z.L.; Chen, H.X. Effects of multilayered porous sediment on earthquake-induced hydrodynamic response in reservoir. *Soil Dyn. Earthq. Eng.* **2017**, *94*, 47–59. [CrossRef]
45. Xu, H.; Zou, D.; Kong, X.; Hu, Z. Study on the effects of hydrodynamic pressure on the dynamic stresses in slabs of high CFRD based on the scaled boundary finite-element method. *Soil Dyn. Earthq. Eng.* **2016**, *88*, 223–236. [CrossRef]
46. Xu, H.; Zou, D.; Kong, X.; Su, X. Error study of Westergaard's approximation in seismic analysis of high concrete-faced rockfill dams based on SBFEM. *Soil Dyn. Earthq. Eng.* **2017**, *94*, 88–91. [CrossRef]
47. Fu, Z.Z.; Chen, S.S.; Li, G.Y. Hydrodynamic pressure on concrete face rockfill dams subjected to earthquakes. *J. Hydrodyn.* **2019**, *32*, 152–168. [CrossRef]
48. Xu, H.; Zou, D.; Kong, X.; Hu, Z.; Su, X. A nonlinear analysis of dynamic interactions of CFRD-compressible reservoir system based on FEM-SBFEM. *Soil Dyn. Earthq. Eng.* **2018**, *112*, 24–34. [CrossRef]
49. Karalar, M.; Çavuşli, M. Assessing 3D seismic damage performance of a CFR dam considering various reservoir heights. *Earthq. Struct.* **2019**, *16*, 221–234.
50. Lin, G.; Wang, Y.; Hu, Z. An efficient approach for frequency-domain and time-domain hydrodynamic analysis of dam-reservoir systems. *Earthq. Eng. Struct. Dyn.* **2012**, *41*, 1725–1749. [CrossRef]
51. Floater, M.S. Mean value coordinates. *Comput. Aided Geom. Des.* **2003**, *20*, 19–27. [CrossRef]
52. Sukumar, N.; Tabarraei, A. Conforming polygonal finite elements. *Int. J. Numer. Methods Eng.* **2004**, *61*, 2045–2066. [CrossRef]
53. Jiang, W.; Xu, J.; Zheng, H.; Wang, Y.; Sun, G.; Yang, Y. Novel displacement function for discontinuous deformation analysis based on mean value coordinates. *Int. J. Numer. Methods Eng.* **2020**, *121*, 4768–4792. [CrossRef]
54. Degao, Z.; Xianjing, K.; Bin, X. *User Manual for Geotechnical Dynamic Nonlinear Analysis*; Institute of Earthquake Engineering, Dalian University of Technology: Dalian, China, 2005.
55. Pang, R.; Xu, B.; Zhou, Y.; Song, L. Seismic time-history response and system reliability analysis of slopes considering uncertainty of multi-parameters and earthquake excitations. *Comput. Geotech.* **2021**, *136*, 104245. [CrossRef]
56. Pang, R.; Xu, B.; Zhou, Y.; Zhang, X.; Wang, X. Fragility analysis of high CFRDs subjected to mainshock-aftershock sequences based on plastic failure. *Eng. Struct.* **2020**, *206*, 110152. [CrossRef]
57. Xu, B.; Pang, R.; Zhou, Y. Verification of stochastic seismic analysis method and seismic performance evaluation based on multi-indices for high CFRDs. *Eng. Geol.* **2020**, *264*, 105412. [CrossRef]
58. Li, Y.; Pang, R.; Xu, B.; Wang, X.; Fan, Q.; Jiang, F. GPDEM-based stochastic seismic response analysis of high concrete-faced rockfill dam with spatial variability of rockfill properties based on plastic deformation. *Comput. Geotech.* **2021**, *139*, 104416. [CrossRef]

59. Qu, Y.; Zou, D.; Kong, X.; Liu, J.; Zhang, Y.; Yu, X. Seismic damage performance of the steel fiber reinforced face slab in the concrete-faced rockfill dam. *Soil Dyn. Earthq. Eng.* **2019**, *119*, 320–330. [CrossRef]
60. Zou, D.; Sui, Y.; Chen, K. Plastic damage analysis of pile foundation of nuclear power plants under beyond-design basis earthquake excitation. *Soil Dyn. Earthq. Eng.* **2020**, *136*, 106179. [CrossRef]
61. Zou, D.; Xu, B.; Kong, X.; Liu, H.; Zhou, Y. Numerical simulation of the seismic response of the Zipingpu concrete face rockfill dam during the Wenchuan earthquake based on a generalized plasticity model. *Comput. Geotech.* **2013**, *49*, 111–122. [CrossRef]
62. Zou, D.; Han, H.; Liu, J.; Yang, D.; Kong, X. Seismic failure analysis for a high concrete face rockfill dam subjected to near-fault pulse-like ground motions. *Soil Dyn. Earthq. Eng.* **2017**, *98*, 235–243. [CrossRef]
63. Chwang, A.T. Hydrodynamic pressures on sloping dams during earthquakes. Part 2. Exact theory. *J. Fluid Mech.* **1978**, *87*, 343–348. [CrossRef]
64. Zhencheng, C. Dynamic water pressure on inclined dam face due to earthquake. *Acta Mech. Sin.* **1964**, *7*, 48–62. (In Chinese)
65. Werner, P.; Sundquist, K. On hydrodynamic earthquake effects. *Eos Trans. Am. Geophys. Union* **1949**, *30*, 636–657. [CrossRef]
66. Siao, T.-T.; Zhou, G.-M. Effects of shapes of valley cross-section on earthquake hydrodynamic pressure. *J. Hydraul. Eng.* **1965**, *1*, 1–15. (In Chinese)
67. Xu, B.; Zou, D.; Liu, H. Three-dimensional simulation of the construction process of the Zipingpu concrete face rockfill dam based on a generalized plasticity model. *Comput. Geotech.* **2012**, *43*, 143–154. [CrossRef]

water

MDPI

Article

Coordination Characteristics Analysis of Deformation between Polymer Anti-Seepage Wall and Earth Dam under Traffic Load

Hongyuan Fang [1], Hong Zhang [1], Binghan Xue [1,*], Jianglin Gao [2], Yan Li [2], Xinlei Gao [3] and Aiping Tian [4]

1 School of Water Conservancy Engineering, Zhengzhou University, Zhengzhou 450000, China; fanghongyuan1982@163.com (H.F.); hongzhang0053@163.com (H.Z.)
2 Jiangxi Academy of Water Science and Engineering, Nanchang 330000, China; gaojianglin@163.com (J.G.); liyan6609@163.com (Y.L.)
3 Harbin Institute of Technology National Engineering Research Center of Urban Water Resources, Harbin 150006, China; lgeg2018@163.com
4 Shanghai Dike and Pump Gate Construction and Operation Centre, Shanghai 200000, China; tianfxq@163.com
* Correspondence: xuebinghan@zzu.edu.cn

Abstract: Polymer anti-seepage walls have been widely used in the anti-seepage reinforcement projects of earth dams. Dam crest is always supposed to meet the requirements of traffic load which has significant influence on the dam body and anti-seepage wall. In order to analyze the coordination characteristics of the deformation between polymer anti-seepage wall and dam under traffic loads, a 3D finite element model of an earth dam that considers the coupling effect of seepage field and stress field was established. Besides, the influence of load amplitude, vehicle speed and driving position on the stress and deformation characteristics of polymer anti-seepage wall and dam was analyzed, with the displacement difference between dam and wall, wall Mises stress and wall subsidence as indicators. The results show that, compared with vehicle speed, the load amplitude and vehicle speed of traffic load exerted a greater impact on the coordination characteristics of the deformation of the dam. The variation range of the displacement difference caused by axial load change reached 87.1%, while that resulted from driving position change reached 90.3%. That is, when the passing vehicle has a light axle load and passes quickly over the anti-seepage wall, it has less impact on the dam.

Keywords: earth dam; polymer anti-seepage wall; traffic load; deformation coordination

Citation: Fang, H.; Zhang, H.; Xue, B.; Gao, J.; Li, Y.; Gao, X.; Tian, A. Coordination Characteristics Analysis of Deformation between Polymer Anti-Seepage Wall and Earth Dam under Traffic Load. *Water* **2022**, *14*, 1442. https://doi.org/10.3390/w14091442

Academic Editor: Giuseppe Oliveto

Received: 25 March 2022
Accepted: 27 April 2022
Published: 30 April 2022

1. Introduction

Dams bring huge economic, social and environmental benefits in flood control, water supply, irrigation, shipping and other aspects, playing an important role in economic and social progress. Seepage is one of the key issues of dam safety, and building anti-seepage walls are one of the common measures for dam seepage prevention [1]. With the appearance of various emerging materials, the materials of anti-seepage walls have also been improved, which are developed from the original ordinary concrete to reinforced concrete, plastic concrete, clay concrete, polymer anti-seepage walls, and so on. Non-aqueous reactive polyurethane, as an anti-seepage reinforcement material for dams, is featured with light weight, large expansion force, non-pollution, good anti-seepage performance, excellent durability and mechanical properties, etc. Compared with rigid concrete anti-seepage walls with large elasticity modulus, polymer anti-seepage walls make up for these deficiencies [2,3].

At present, some studies have been conducted on the properties of polymer grouting materials, grouting technology and construction effects [3–11], achieving significant progress and innovation in the construction technology of polymer anti-seepage walls. Li Jia et al. [12–14] studied the seismic response characteristics of earth dam structure of polymer anti-seepage walls by means of a physical centrifuge model, and compared the seismic response characteristics of earth dams with polymer anti-seepage walls and concrete core walls. The non-aqueous reactive two-component polyurethane polymer grouting material used in the technology of the polymer anti-seepage wall of the earth dam belongs to polyurethane materials. Although the anti-seepage wall exerts significant anti-seepage effect on the dam, the material interface under load is often the "weak link" of the structure [15,16] due to the different material properties between the soil body and the anti-seepage wall. Additionally, some progresses have been made on the study of material interface in recent years [17–19]. Deformation difference between the two material regions will inevitably occur during the construction period and water storage period. The interface between traditional concrete anti-seepage wall and dam often has certain defects and is not an ideal bonding state. As a result, such deformation differences may cause cracks in the material interface, the function of the dam is seriously affected. Therefore, it is necessary to study the coordination characteristics of the dam deformation. The evaluation indicators that can be used as deformation coordination properties include major principal stress, wall stress level, the maximum subsidence of dam, the change rate of wall subsidence, as well as vibration compaction time [20].

The dam crest is always supposed to meet the requirements of traffic load, and the traffic load exerts great influence on the dam body and leads to irreversible deformation of the dam body. With the increase of traffic flow and load capacity, the traffic load gradually has greater impact on the dam body. Sarkar examined the flexible pavement dynamic response under single, tandem and tridem axles at different speeds. Using two different hot-mix asphalt (HMA) layer thicknesses, the dynamic effects of moving axles were investigated on critical responses. These responses include the tensile strain at the bottom of asphalt layer, compressive strain on the top of subgrade and tensile and compressive strain on the surface layer [21]. Zhao et al. presented a numerical investigation into the plastic/dynamic characteristics of a saturated porous medium (capping and subgrade) subjected to moving axle loads. A detailed numerical analysis is described, whereby a coupled fluid-dynamic framework is developed for the saturated porous medium in conjunction with a generalized plasticity model, in order to examine the cyclic loading response of a soft subgrade soil. A relationship between the train speed, track settlement and drainage capacity of a sub-ballast (capping) layer is established [22]. Guo et al. comprehensively analyzed the dynamic response of full-depth asphalt pavement under moving load, a three-dimensional model of pavement structure and dynamic load moving zone are established based on ABAQUS finite element software. Based on the time history curves of different structures, the stress–strain states at the bottom of each structural layer in different structures under moving load are analyzed [23]. Liu et al. conducted hollow torsional shear tests on fiber-reinforced aeolian soil involving varying fiber contents, cyclic deviator stresses, cyclic shear stresses and consolidation confining pressures using the Small-Strain Hollow Cylinder Apparatus. This enabled an investigation of the deformation characteristics and noncoaxial angle changes of fiber-reinforced aeolian soil under a heart-shaped stress path [24]. Qian et al. performed a series of cyclic torsional shear tests to investigate the effect of principal stress rotation (PSR) on the stress–strain behaviors of saturated soft clay. The traffic-load-induced shear stress path was used in the cyclic test and the investigation mainly concerned the influence of PSR on the shear stiffness and non-coaxiality [25]. Currently, the analysis on traffic load mainly focuses on the field of roads, while few studies are conducted on the impact of traffic load on dams with polymer anti-seepage walls. Therefore, analyzing the influence of traffic load on dam safety is of important theoretical and practical value.

In order to analyze the stress and deformation characteristics of polymer anti-seepage walls and dams under traffic load, a three-dimensional finite element model of polymer anti-seepage dam wall that considers the coupling effect of seepage field and stress field was established, the influence of axle load, vehicle speed and driving position on the coordination characteristics of dam deformation was simulated, and the law of stress and deformation of dam under traffic load was revealed, which provides a basis for the design and safety evaluation of polymer anti-seepage walls in dam reinforcement engineering.

The remaining sections are arranged as follows: the indexes for evaluating the deformation coordination characteristics are introduced in Section 2; an overview of the model and load settings is presented Section 3; the calculation results are analyzed in Section 4; and the summary and conclusion are made in Section 5.

2. Deformation Coordination Index

Uneven subsidence will occur in the interface of the dam body and the anti-seepage wall due to the inconsistency of mechanical properties. The cross section of the dam body will dislocate in a nearly vertical direction because of the uneven subsidence. Meanwhile, the development of the uneven longitudinal displacement of the dam body will cause the dam body to separate on both sides of the failure surface, thus forming cracks with a certain width. In terms of the general dam body, the failure surface caused by the subsidence difference shows cracks of different widths. Thus, some indexes for evaluating the coordination characteristics of the deformation were proposed [26], thus that the judgment and analysis on the deformation difference of the dam could be carried out. The displacement difference was selected as the index to evaluate the deformation coordination of the polymer anti-seepage wall and the dam, and its definition is as follows:

The formula for the vertical displacement difference between the anti-seepage wall and the dam body is as follows:

$$\mathrm{H_y} = D_{xi}\big|_j - d_{xi}\big|_j \tag{1}$$

where $d_{xi}\big|_j$ means the vertical displacement of point i of the anti-seepage wall at time j and $D_{xi}\big|_j$ refers to the vertical displacement of the corresponding point i of the dam body at time j.

The formula for the horizontal displacement difference between the anti-seepage wall and the dam body is as follows:

$$\mathrm{H_x} = D_{fi}\Big|_j - d_{fi}\Big|_j \tag{2}$$

where $d_{fi}\Big|_j$ indicates the deflection of point i of the anti-seepage wall at time j and $D_{fi}\Big|_j$ stands for the lateral displacement of the corresponding dam body i at time j.

3. Calculation Model

3.1. Overview of Model

According to the information on the Yellow River dam section which is crossed by the Zhengjiao inter-city railroad bridge, a 3D finite element model as shown in Figure 1 was established by using ABAQUS software, in which the height of the dam body is 15.0 m, and that of the dam foundation is 10.0 m, the overall length along the X-axis is 124.5 m and the width along the Z-axis direction is 2.1 m. The height of the anti-seepage wall is 17.0 m, and its lower part is embedded in the dam foundation for 2.0 m. The C3D8P element with pore pressure degree of freedom was used for the solution and the Mohr–Coulomb elastic-plastic constitutive model was adopted for the soil, assuming that the foundation is homogeneous, the literatures [27,28] carried out studies on the randomness of parameters, and the linear elastic constitutive model was employed for the wall. The total number of elements is 30,024 and that of nodes is 35,087, in which the number of the units and nodes of the anti-seepage wall are 1584 and 2210, respectively. The material parameters are shown in Table 1. Traffic load was applied at the top of the dam, and the seepage boundary was set within the range from the bottom of the dam near water to the height of 13.5 m. Goodman element without thickness was adopted to simulate the contact surface between the anti-seepage wall and the dam [29]; the setting parameters are shown in Tables 2 and 3.

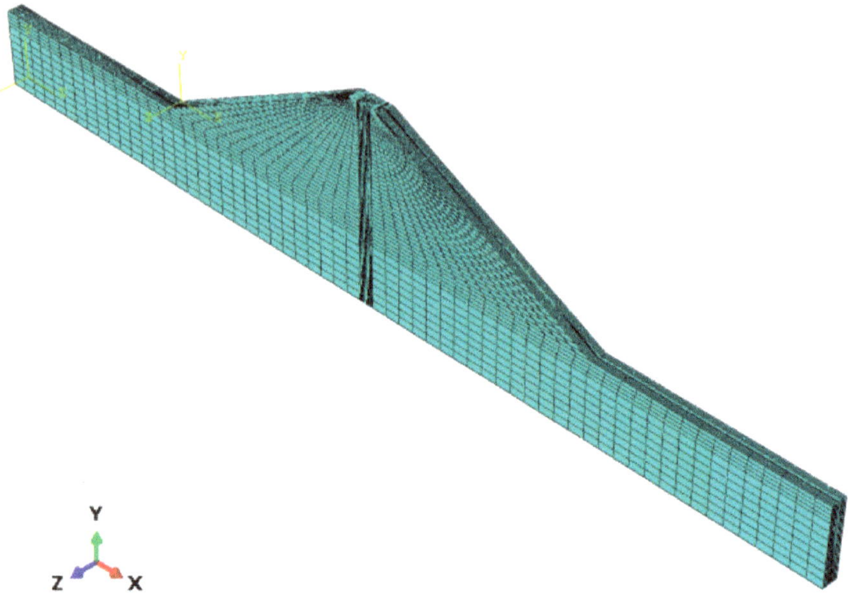

Figure 1. Three-dimensional finite element model.

Table 1. Material parameters of anti-seepage wall and soil body.

Material	Density (kg/m^3)	Young's Modulus (kPa)	Poisson's Ratio	Permeating Coefficient (cm/s)	Cohesion (kPa)	Internal Friction Angle	Dilatancy Angle
Dam	1930	100,000	0.29	5×10^{-7}	12.38	20	0
Polymer	300	100,176	0.258	1×10^{-10}	—	—	—
Rigid concrete	2400	20,000,000	0.3	1×8^{-10}	—	—	—
Plastic concrete	2400	500,000	0.25	1×8^{-10}	—	—	—

Table 2. Parameters of Goodman element between the polymer and the dam body.

S/N	Parameter	Symbol	Value
1	Tangential stiffness coefficient	K_1	300
2	Normal stiffness coefficient	K_2	300
3	Test constant	n	0.34
4	Damage ratio	R_f	0.95
5	Interfacial friction angle	δ	11.3
6	Water bulk density	R_w	100
7	Atmospheric pressure	P_a	1000

Table 3. Parameters of Goodman element between the concrete and the dam body.

S/N	Parameter	Symbol	Value
1	Tangential stiffness coefficient	K_1	1400
2	Normal stiffness coefficient	K_2	1400
3	Test constant	n	0.35
4	Damage ratio	R_f	0.75
5	Interfacial friction angle	δ	15
6	Water bulk density	R_w	100
7	Atmospheric pressure	P_a	1000

In order to compare the deformation coordination characteristics of the anti-seepage walls with different materials and the dam body, models of polymer, rigid concrete and plastic concrete walls were established. The thickness of the polymer anti-seepage wall is 0.03 m, and that of the concrete anti-seepage wall is 0.24 m. The total number of model elements is 25,100, and that of nodes is 29,799, in which the number of the units and nodes of the anti-seepage wall are 1400 and 1980, respectively. The intersection of the central axis of the anti-seepage wall and the dam foundation was taken as the origin point for reference. Additionally, it was supposed that the X-axis was positive along the boundary line between the dam foundation and the dam body to the far river, and the Y-axis was positive along the central line of the anti-seepage wall to the dam crest. The schematic diagram of coordinate system and details of the model dimensions are shown in Figure 2.

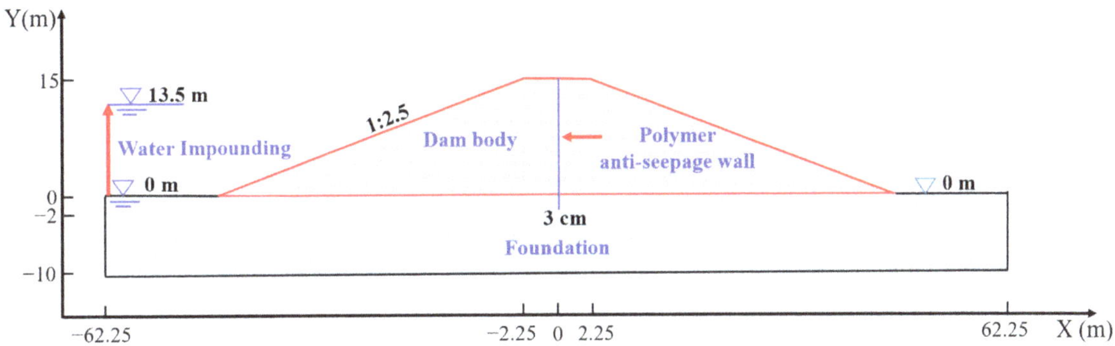

Figure 2. Schematic diagram of coordinate system and model dimensions.

3.2. Traffic Load

In order to simulate the actual road bearing dynamic load, the traffic load was described by the means of a single half-sine function as shown in Figure 3, and the process of approaching and leaving with standard axle loads was stimulated by adopting the form of impact load. According to The Specification for Design of Highway Asphalt Pavement, trucks with single-axle and double-wheel rear wheels are regarded as standard vehicles, with the standard axle load being 100 kN. Additionally, the traffic load expression is as follows:

$$p(t) = 0.11737 * sin(10\pi t) * 0.7,\ 0 \leq t \leq 0.1s \quad (3)$$

where t refers to the calculation time and $p(t)$ means the traffic load at time t. In this paper, the single-circle load was utilized to simulate the rear wheel and double-wheel set of vehicles. The contact surface with ground is shown in Figure 4. The diameter of circle load is 300 mm, the double spacing of wheels is 1.8 m and the grounding area is simplified to two rectangular areas of 0.2 m × 0.3 m at the top of the dam.

During the calculation, ground stress balance was conducted for the calculation model, the water level in front of the dam was set to 13.5 m. The settings of the traffic load amplitude, vehicle speed and driving position are shown in Table 4, and the driving position is represented by the X coordinates of the central line of the axle as shown in Figure 5.

Table 4. Condition setting table.

S/N	Load Amplitude (MPa)	Vehicle Speed (km/h)	Driving Position (m)
1	0.7	20	1
2	0.9	40	0.5
3	1.1	60	0
4	1.3	80	−0.5
5	1.5	100	−1

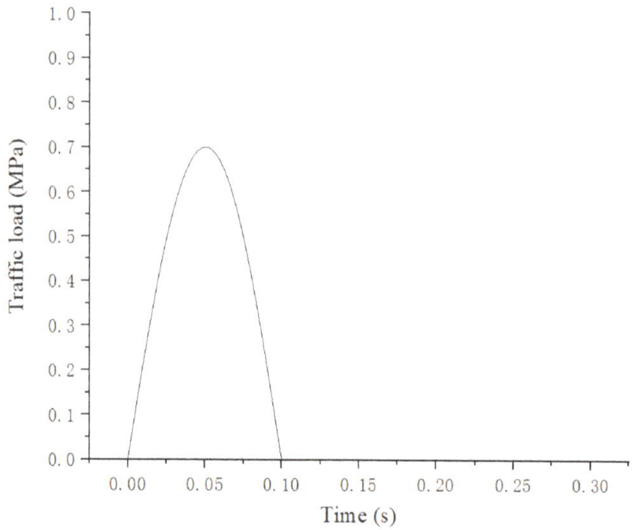

Figure 3. Contact surface between tire and ground.

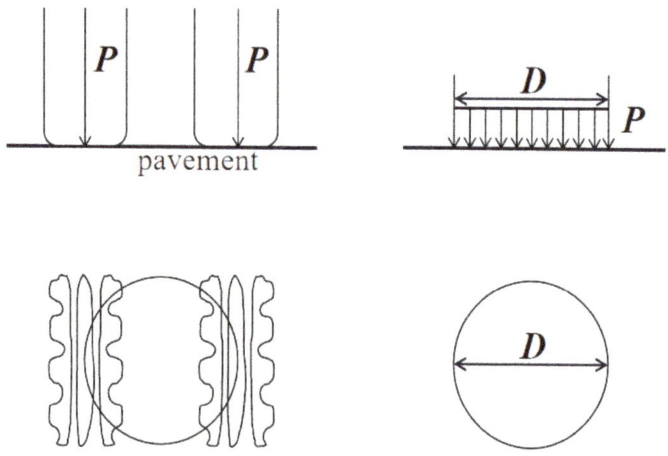

Figure 4. Contact surface between tire and ground.

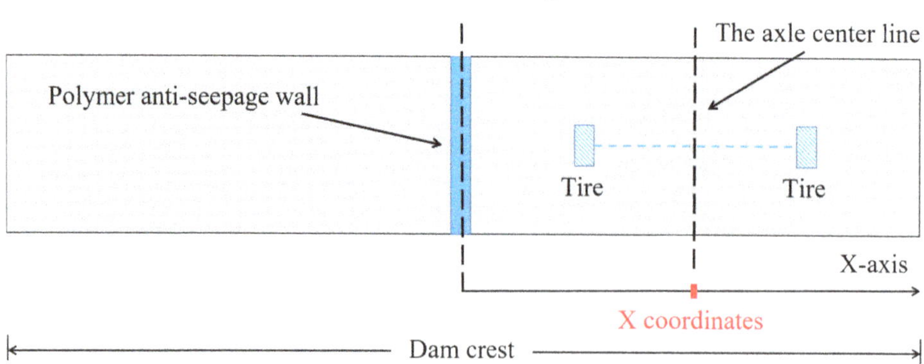

Figure 5. Driving position diagram.

3.3. Seepage-Stress Coupling Analysis

As a water retaining structure, the stability of the dam is greatly affected by seepage, and the stress deformation of the dam body and seepage is a typical seepage-stress coupling problem. Therefore, the interaction between the seepage wall and the dam body, which is an important part of the dam body, also has seepage-stress coupling problems. This section introduces the principles and methods related to the analysis of seepage stress coupling in dam engineering by combining the seepage-stress coupling function of ABAQUS finite element calculation software.

The dam soil is a porous material, and the pores contain two parts of fluid: liquid and gas. The volume of the soil body includes both soil particles and pores. Since the tensile stress is positive and the compressive stress is negative in the ABAQUS calculation, but both gas pressure and liquid pressure are positive in terms of pressure, the effective stress principle expression is:

$$\bar{s} = s + (\mu u_w + (1-\mu)\mu_a) \tag{4}$$

where $\bar{s}$ is the effective stress, s is the total stress, μ is the effective stress parameter, u_w is the liquid pressure and u_a is the gas pressure, when the soil is fully saturated, $\mu = 1$ and when the soil is completely dry, $\mu = 0$.

The effective stress of the material in ABAQUS is the basis of the stress–strain relationship and calculation results. The pore water pressure in the coupled seepage-stress analysis can be chosen based on the total pore pressure or super pore pressure, and in this paper, the gravity applied load is used, which means the analysis is based on the total pore pressure.

The permeability law used in ABAQUS is the Forchheimer equation, and the permeability coefficient is expressed as:

$$\bar{k} = \frac{k_s}{(1+\beta\sqrt{v_w v_w})}k \tag{5}$$

where k is the seepage coefficient of saturated soil; v_w is the flow speed, β is the parameter reflecting the effect of flow speed on the seepage coefficient, when $\beta = 0$; Equation (5) is simplified to Darcy formula; k_s is the saturation coefficient, when the soil is saturated, $k_s = 1$, and the coefficient k_s reflects the relationship between the permeability coefficient of non-saturated soil and saturated soil.

In addition, the permeability coefficient k can also be a function related to the pore ratio, characterizing the effect of soil skeleton changes on the permeability coefficient, and the effect of the stress field on the seepage field can be calculated by setting the relationship function between the permeability coefficient and the pore ratio.

According to Equation (5), the effect of saturation on permeability coefficient can be reflected by k_s. In ABAQUS, when saturation $S_r < 1$, $k_s = {S_r}^3$, and $k_s = 1$ when $S_r \geq 1$. For the coupling effect of seepage and stress in unsaturated soil, it is realized by the relationship between negative pore pressure and permeability coefficient, and in saturated soil, it is realized by setting the function between permeability coefficient and pore ratio. Due to the hygroscopic and dewatering characteristics of unsaturated soils, it is necessary to perform unsaturated soil seepage and stress calculations to determine the relationship between hygroscopicity and dewatering and the permeability coefficient

Based on the above analysis, it can be seen that the matrix suction in unsaturated soil is related to the permeability coefficient, realizing the interaction between seepage flow and stress in unsaturated soil.

The coupled seepage seepage-stress analysis needs to unify the saturated and unsaturated seepage calculations. To simplify the calculation, it is assumed in this paper that both saturated and unsaturated soils obey the law, ignore the effect of flow velocity, and distinguish by permeability coefficient and pore water pressure, and the C3D8P element with pore pressure degree of freedom was used for the solution. The water pressure in the pores is shown in Figure 6.

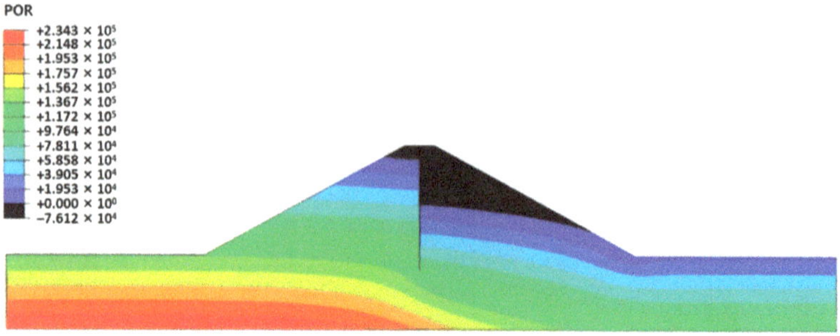

Figure 6. The water pressure in the pores.

4. Calculation Results

4.1. Comparison of Deformation Coordination Characteristics between the Anti-Seepage Wall with Different Materials and the Dam Body

In order to study the deformation coordination characteristics between the wall with different materials and the dam, the vehicle speed of 20 km/h was taken as an example. Specifically, when vehicles drive above the anti-seepage wall, traffic load with an amplitude of 0.7 MPa was applied, and the anti-seepage walls with different materials were adopted. As shown in Figure 7, the variation trend of the displacement difference between the concrete anti-seepage wall and the dam body is basically the same along the wall height, that is, it gradually decreases from dam crest downward. While, the displacement difference between the dam body and the rigid concrete anti-seepage wall at each height is much larger than that between the dam body and the plastic concrete anti-seepage wall, the maximum values of the displacement difference appear at the dam crest, which are 1.6 cm and 0.58 cm, respectively; the displacement difference between the polymer anti-seepage wall and the dam body gradually increases along the dam crest downward, reaching a maximum value of 0.18 cm at a wall height of 5 m, and gradually decreases. From the perspective of the maximum displacement difference, the rigid concrete anti-seepage wall is larger, and the polymer anti-seepage wall is smaller. Hence, it can be seen that the deformation morphological characteristics of polymer anti-seepage wall and the dam body are better than that of the rigid and plastic concrete anti-seepage walls.

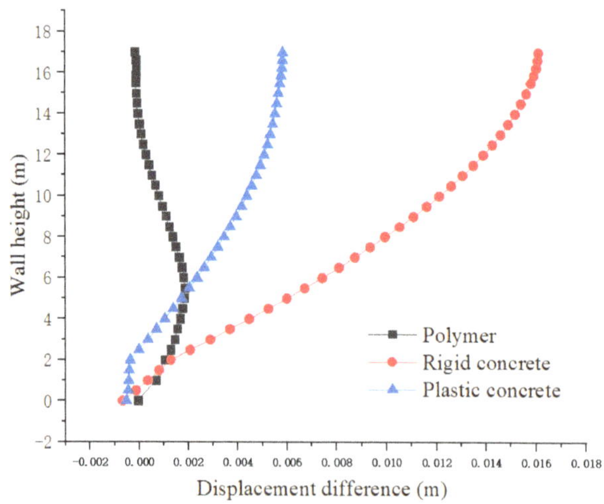

Figure 7. Comparison of displacement difference of anti-seepage walls with different materials along the wall.

4.2. Deformation Coordination Characteristics between Polymer Anti-Seepage Wall and Dam Body

4.2.1. Influence of Amplitude

In order to determine the influence depth of traffic load, the vehicle speed of 20 km/h was taken as an example. In detail, the subsidence, Mises stress and displacement difference distribution of the dam seepage wall along the wall height under different load amplitude conditions when vehicles drive above the polymer wall are shown in Figures 8–10.

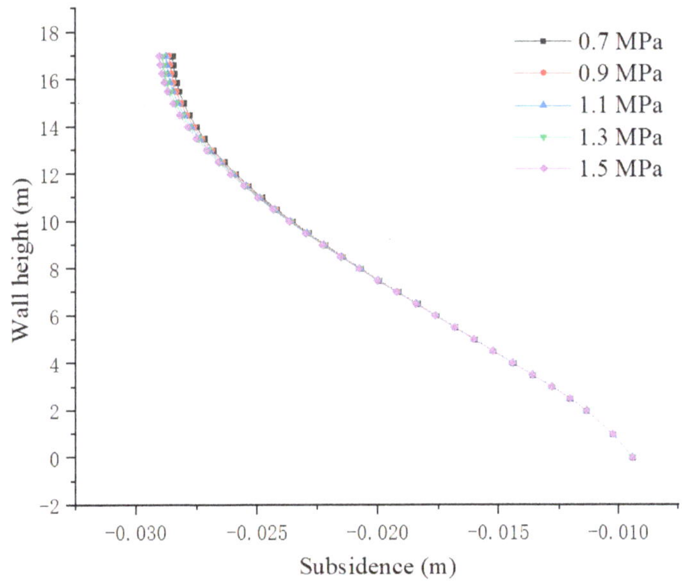

Figure 8. Distribution of subsidence along the anti-seepage wall under the action of different load amplitudes.

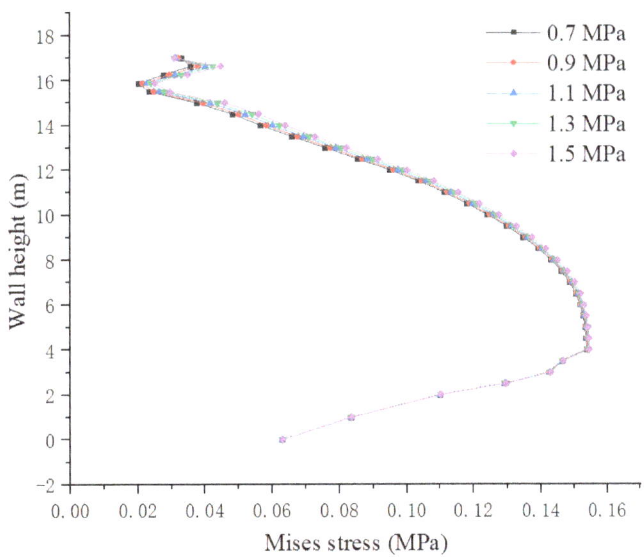

Figure 9. Distribution of Mises stress along the anti-seepage wall under different load amplitudes.

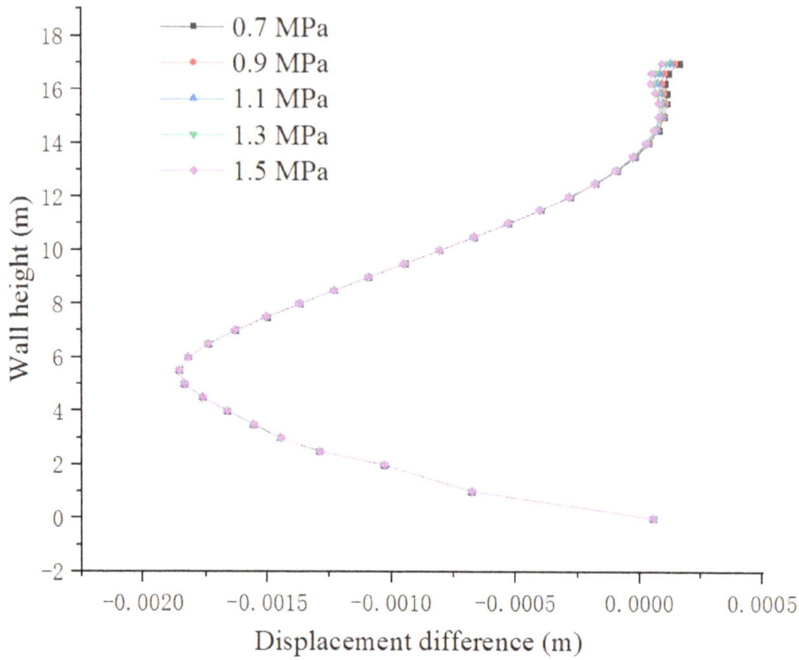

Figure 10. Distribution of the displacement difference along the anti-seepage wall under different load amplitudes.

As shown in Figures 8 and 10, the maximum position of the change amplitude of the subsidence and the displacement difference are at the wall height of 16.625 m. Figure 9 shows that in the range of 0–4 m, the Mises stress gradually increases with the height of the wall. Additionally, in the range of 4–15 m, it gradually decreases with the increase of the wall height. In the range of 15–17 m, a certain fluctuation occurs, the greatest variation appearing at the wall height of 16.625 m. To sum up, under the traffic load and pore pressure, the most responsive part of the dam is in the middle dam, but the part that is more sensitive to traffic load is at the dam crest. The influence depth of traffic load on the dam strain is about 5 m. The height of 16.625 m should be chosen, that is, the dam height of 14.625 m is the research section.

In order to analyze the variation law of dam subsidence with the amplitude of traffic load, when vehicles drive above the polymer wall, the vehicle speed of 20 km/h was taken as an example—Figure 11 shows the subsidence of the dam body and the anti-seepage wall under different load amplitudes. It can be seen that under traffic load, the subsidence curve presents "saddle-shaped double peak", an obvious drop in subsidence at the junction occurs, and the maximum subsidence of the dam body reaches about 2.98 cm. Under the same conditions, with the increase of the load amplitude, the subsidence of the dam body and the wall gradually increases. The subsidence difference of the dam body at the left wheel under the load amplitudes from 0.7 to 1.5 MPa, reaching 1.1 mm, with the amplitude being 3.7%. The cloud diagrams of wall subsidence and Mises stress under different load amplitudes are as follows.

The subsidence curves of the wall at the dam height of 14.625 m under different load amplitudes when the vehicle travels above the polymer wall at a speed of 20 km/h are shown in Figure 12. It can be seen that, under the same conditions, the subsidence of the wall increases with the increase of amplitude, and the subsidence difference of wall under the load amplitudes from 0.7 to 1.5 MPa is 0.55 mm, with an amplitude of variation of 1.7%.

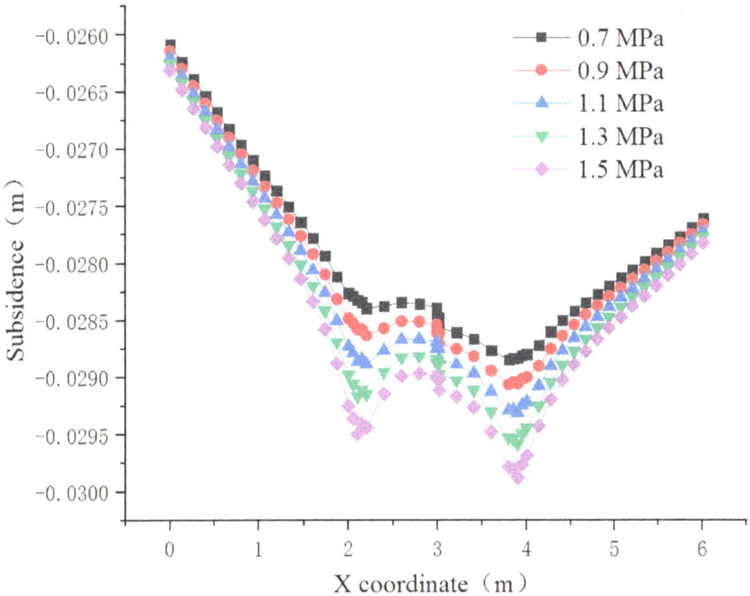

Figure 11. Distribution of dam subsidence at 14.625 m under different load amplitudes.

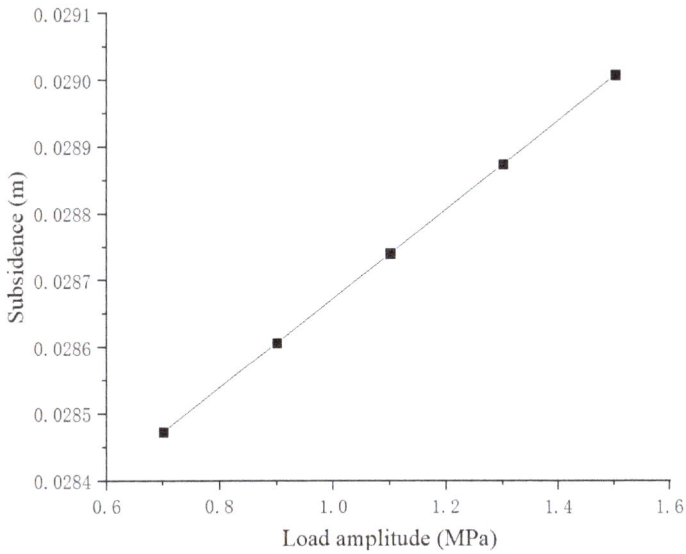

Figure 12. Variation of wall subsidence at 14.625 m under different load amplitudes.

In order to analyze the variation law of Mises stress of the dam with the amplitude of the traffic load, the vehicle speed of 20 km/h was taken as an example. In detail, the Mises stress of the dam body and the anti-seepage wall under different load amplitude conditions when vehicles drive above the polymer wall are shown in Figure 13. It can be seen that the Mises stress of the wall has an obvious peak value at the wheel, and the maximum Mises stress reaches 0.23 MPa. Under the same conditions, with the increase of load amplitude, the Mises stress of the dam and the wall gradually increases, with considerable change. Additionally, the difference of the Mises stress of dam at the wheel point under the load amplitudes from 0.7 to 1.5 MPa reaches 0.063 MPa, with the amplitude of variation of 27.8%, indicating that the traffic load amplitude has a significant impact on Mises stress of the dam.

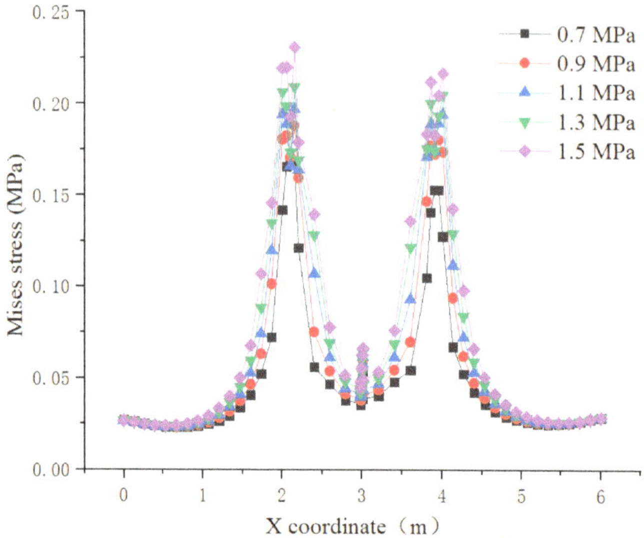

Figure 13. Distribution of dam Mises stress at 14.625 m under different load amplitudes.

The Mises stress curves of the wall at the dam height of 14.625 m under different load amplitudes when the vehicle travels above the polymer wall at a speed of 20 km/h are shown in Figure 14. It can be seen that, under the same conditions, the Mises stress of the wall grows with the increase of amplitude. The difference of the Mises stress of the wall under the load amplitudes from 0.7 to 1.5 MPa is 8996 Pa, with the amplitude of variation of 25%, indicating that the traffic load amplitude exerts a significant effect on the Mises stress of the wall.

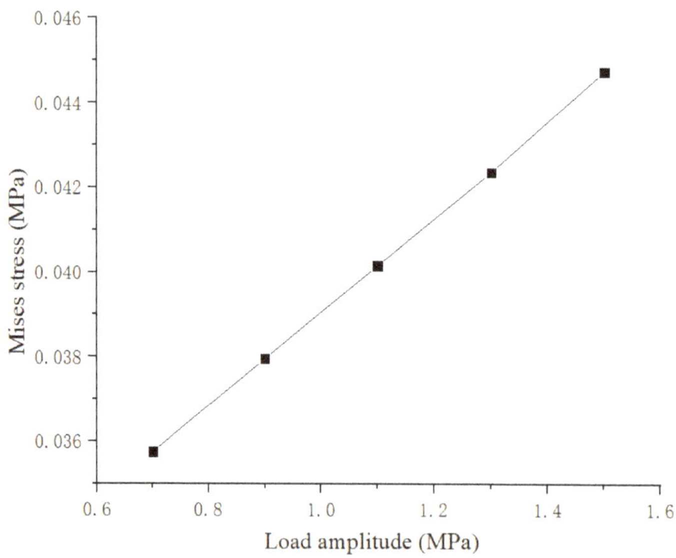

Figure 14. Variation of wall Mises stress at 14.625 m under different load amplitudes.

In order to analyze the variation law of dam displacement difference with traffic load amplitude, vehicle speed of 20 km/h was taken as an example. The displacement difference of the dam and the anti-seepage wall under different load amplitudes when vehicles drive above the polymer wall are shown in Figure 15. It can be seen that the displacement difference on both sides of the dam is symmetrical on the central line of the dam, as there is a wheel load equal to the wall on both sides of dam. Additionally, the displacement difference of the wall shows obvious peak value under the four edges of wheel action, with the maximum displacement difference of 0.37 mm. Under the same conditions, with the rise of the load amplitude, the displacement difference between the dam and the wall also gradually increases, with more obvious change among them. The displacement difference of the dam at the edge of the left wheel under the load amplitudes from 0.7 to 1.5 MPa is 0.23 mm, with the amplitude of variation of 62.2%, indicating that the amplitude has a significant impact on the displacement difference of the dam.

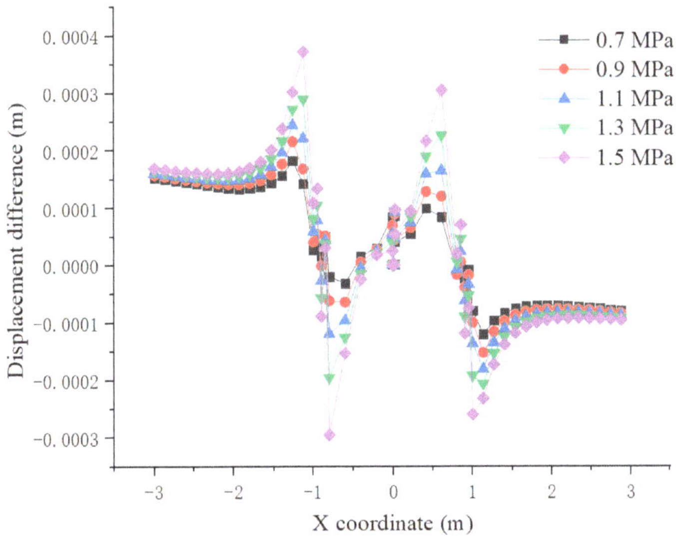

Figure 15. Distribution of displacement difference at 14.625 m under different load amplitudes.

When the vehicle travels above the polymer wall at a speed of 20 km/h, the displacement difference curve of the wall at the dam height of 14.625 m under different load amplitudes is shown in Figure 16. It can be observed that under the same conditions, the displacement difference enhances with the increase of the amplitude, and the displacement difference under the load amplitudes from 0.7 to 1.5 MPa is 0.061 mm, with the amplitude of variation of 87.1%.

It shows that the amplitude of traffic load exerts a vital impact on the deformation coordination between the polymer anti-seepage wall and the dam. The vertical subsidence, Mises stress and displacement difference of the dam rises with the increase of traffic load amplitude. Therefore, in terms of the dam with traffic load, the lower the traffic load amplitude is, the more conducive it is to the coordinated deformation between the dam and the polymer anti-seepage wall. In order to protect the dam, the vehicles crossing the dam should be strictly limited in weight, and overweight vehicles should be prohibited from boarding the dam.

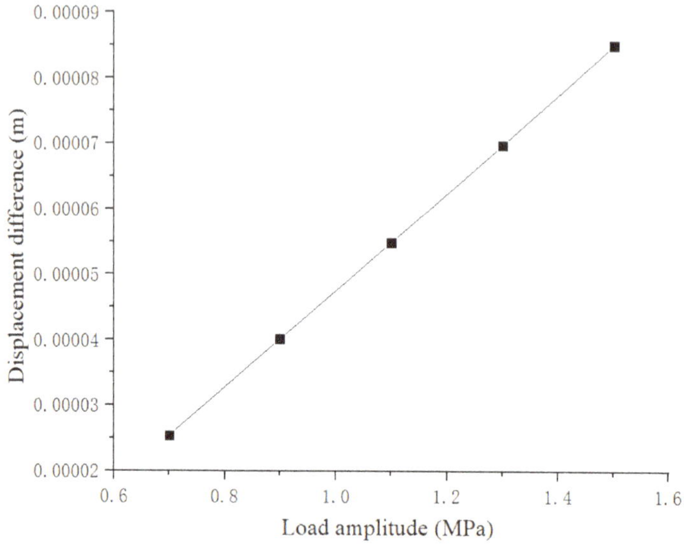

Figure 16. Variation of displacement difference at 14.625 m under different load amplitudes.

4.2.2. Influence of Vehicle Speed

In order to analyze the variation law of dam subsidence stress with the vehicle speed, when vehicles drive above the polymer wall, the vehicle load amplitude of 0.7 MPa was taken as an example. Figure 17 shows the response of the dam body and the anti-seepage wall under different vehicle speeds. It can be observed that under traffic load, the subsidence curve presents "saddle-shaped double peak", and an obvious drop in subsidence at the junction occurs, with the maximum subsidence of the dam body reaching about 2.85 cm. Under the same conditions, with the acceleration of the vehicle speed, the subsidence of the dam and the wall gradually decreases. The subsidence difference of the dam at the left wheel under the speed from 20 to 100 km/h is 0.5 mm, with the amplitude of variation of 1.8%.

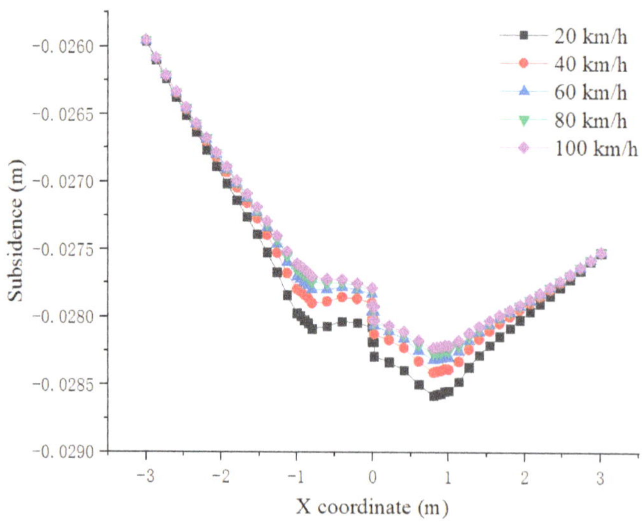

Figure 17. Distribution of dam subsidence at 14.625 m under different vehicle speeds.

The subsidence curve of the wall at the dam height of 14.625 m under different vehicle speeds when the vehicle with load amplitude of 0.7 MPa drives above the polymer wall is shown in Figure 18. It shows that under the same conditions, the subsidence of the wall decreases with the increase of the vehicle speed. The subsidence difference of the wall under the speed from 20 to 100 km/h is 0.27 mm, with the amplitude of variation of 0.89%.

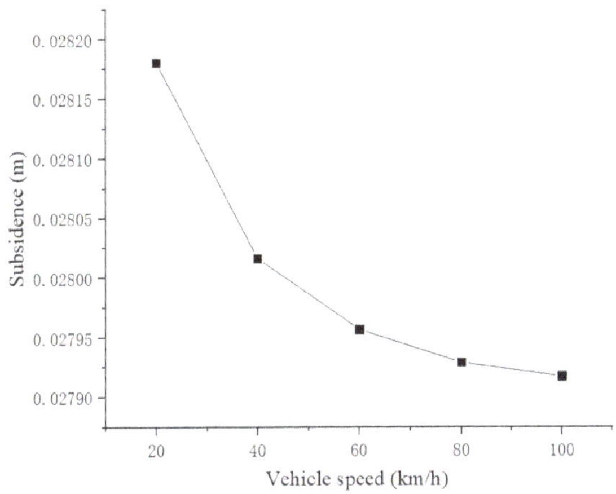

Figure 18. Variation of wall subsidence at 14.625 m under different vehicle speeds.

In order to analyze the variation law of Mises stress of the dam with the vehicle speed, the amplitude of vehicle load of 0.7 MPa/h was taken as an example. The Mises stress of the dam body and the anti-seepage wall under different vehicle speeds when vehicles drove above the polymer wall is shown in Figure 19. It can be seen that the subsidence of the wall has an obvious peak value at the wheel, with the maximum Mises stress reaching 0.16 MPa. Under the same conditions, with the acceleration of the vehicle speed, the Mises stress of the dam and the wall gradually decreases, with considerable change. The difference of Mises stress of the dam body at the left wheel under the vehicle speed from 20 to 100 km/h is 0.072 MPa, with the amplitude of variation of 45.9%.

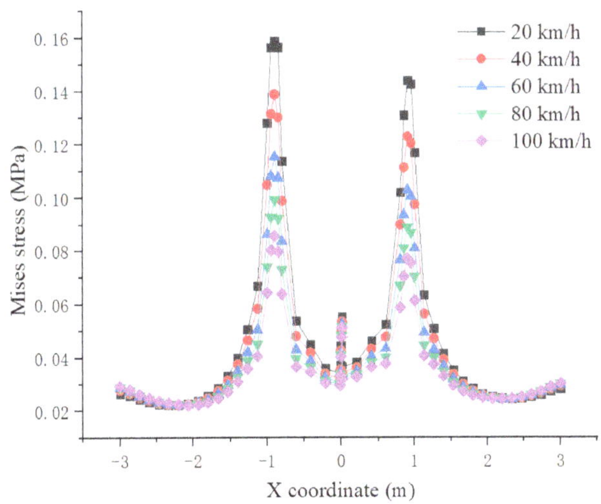

Figure 19. Distribution of Mises stress of the dam at 14.625 m under different vehicle speeds.

The Mises stress curve of the wall at the dam height of 14.625 m under different vehicle speeds when the vehicles drove above the polymer wall with the load amplitude of 0.7 MPa is shown in Figure 20. It shows that under the same conditions, the Mises stress of the wall decreases with the acceleration of the vehicle speed. The difference of the Mises stress of the wall under the vehicle speed from 20 to 100 km/h is 4001 Pa, with the amplitude of variation of 11.9%.

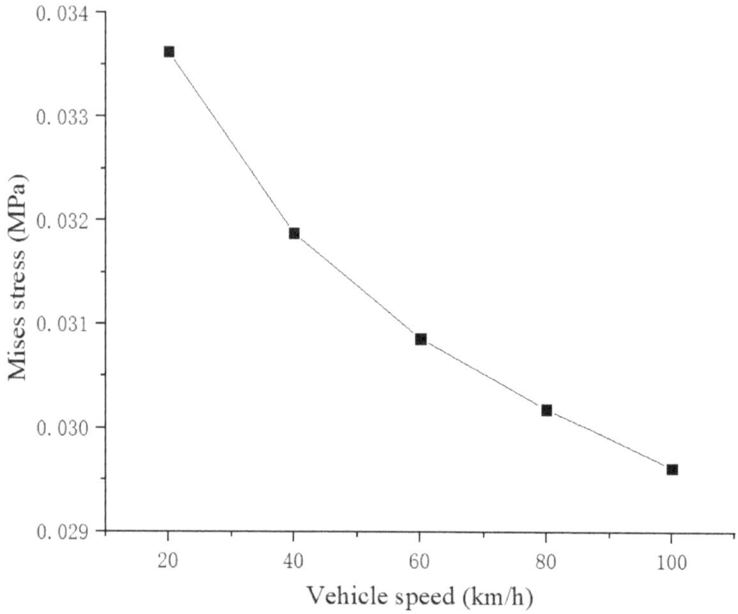

Figure 20. Variation of Mises stress of the wall at 14.625 m under different vehicle speeds.

The displacement difference of the dam and the anti-seepage wall under different vehicle speeds when vehicles drive above the polymer wall with the vehicle load amplitude of 0.7 MPa is shown in Figure 21. It can be observed that the displacement difference on both sides of the dam is symmetrical on the central line of the dam, as there is a wheel load equal to the wall on both sides of the dam. The displacement difference of the wall shows obvious peak value under the four edges of wheel action, with the maximum displacement difference of 0.37 mm. Under the same conditions, with the acceleration of the vehicle speed, the displacement difference of the dam gradually decreases. The displacement difference of the dam at the left wheel edge under the speed from 20 to 100 km/h is 0.04 mm, with the amplitude of variation of 38.5%.

The displacement difference curve of the wall at the dam height of 14.625 m under different load amplitudes when the vehicles drove above the polymer wall with the load amplitude of 0.7 MPa is shown in Figure 22. It can be discovered that under the same conditions, the subsidence of the wall rises with the increase of the vehicle speed. The subsidence difference of the wall under the speed from 20 to 100 km/h is 0.19 mm, with the amplitude of variation of 17.8%.

It can be seen from the above analysis that under the same conditions, the deformation coordination between the polymer anti-seepage wall and the dam is affected greatly by different vehicle speeds. The faster the vehicles drive, the smaller the vertical subsidence and Mises stress of the wall are. Additionally, the displacement difference between the wall and the dam grows with the increase of the vehicle speed. In consideration of the small value of the displacement difference, for the dam bearing traffic load, without considering the speed limit, the faster the vehicles drive, the more conducive it is to the coordination between the dam and the polymer anti-seepage wall.

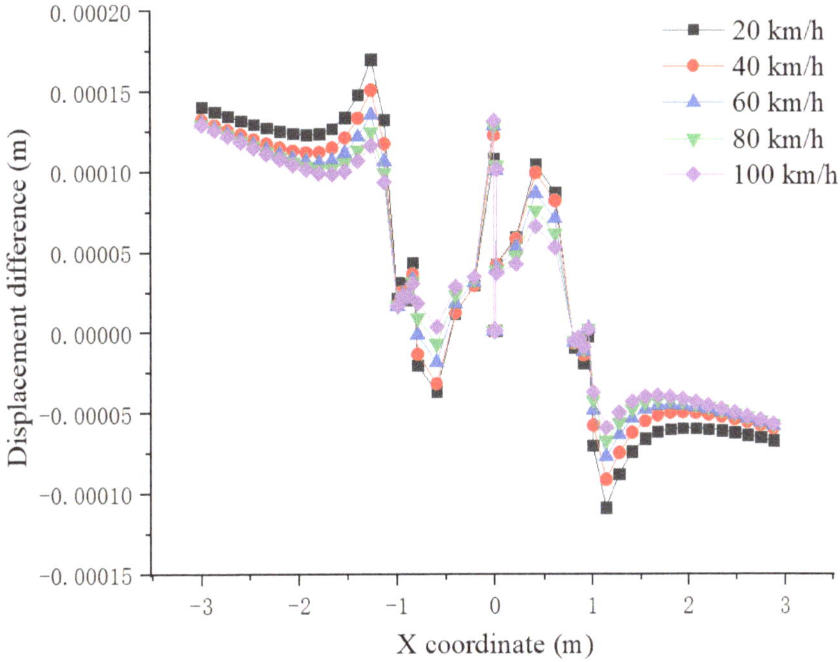

Figure 21. Distribution of displacement difference at 14.625 m under different vehicle speeds.

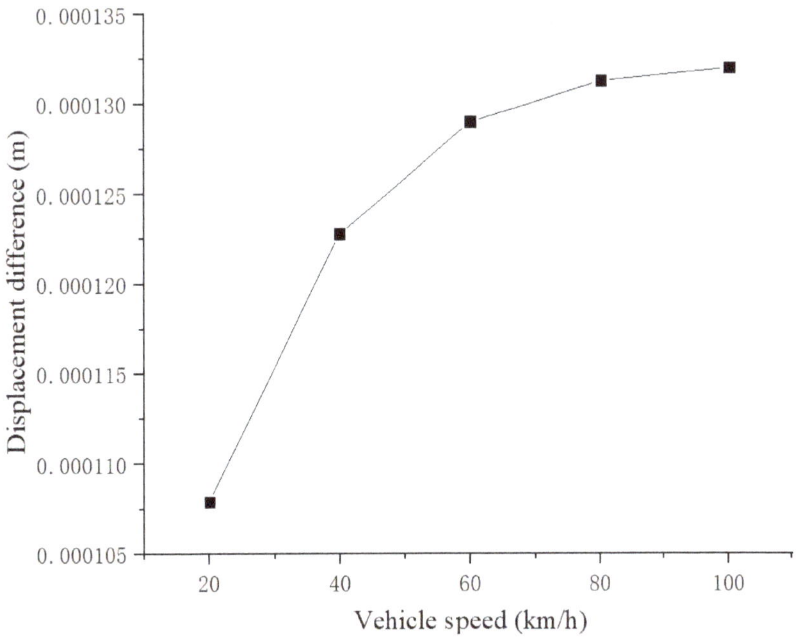

Figure 22. Variation of displacement difference at 14.625 m under different vehicle speeds.

4.2.3. Influence of Driving Position

In order to analyze the variation law of dam subsidence with the driving position, the vehicle speed of 20 km/h was taken as an example. The subsidence of the dam and the anti-seepage wall at different driving positions when the amplitude reaches 0.7 MPa is shown in Figure 23. It can be seen that there is a significant difference in the subsidence at the junction of the dam and the wall. The influence trend of the driving position on the subsidence of the dam has a certain symmetry on both sides of the dam. Vehicles on the left side of the dam exert more significant effects than these on the right side with the same distance. This results from the higher pore water pressure on the left side and the coupling of seepage stress. When the vehicle acts on both sides of the dam, the maximum value of the subsidence of the dam appears at the junction of the dam and the anti-seepage wall. When the vehicle acts on the middle of the dam, the peak value of subsidence appears at the wheel, and the peak value at this time is the minimum value of 2.86 cm among the peaks of each working condition. The maximum peak value appears in the 0.5 m wheelbase group, with the maximum value of 2.91 cm, and the variation range of 1.7%

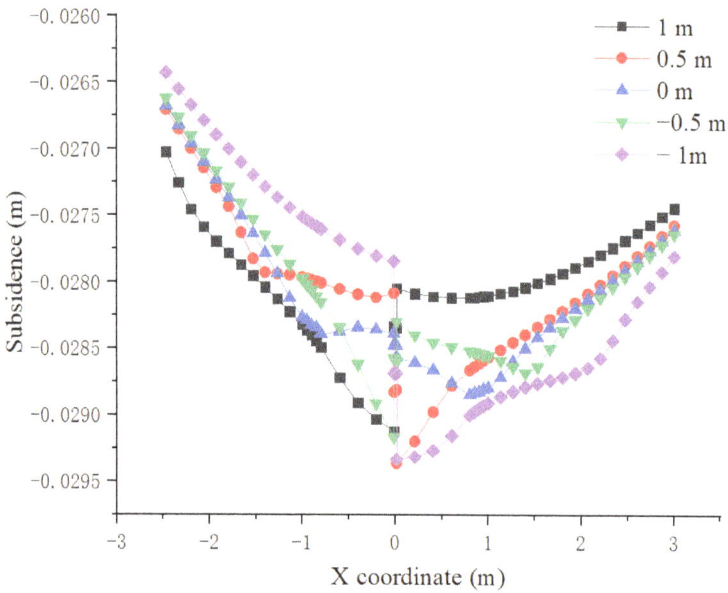

Figure 23. Distribution of dam subsidence at 14.625 m under different load positions.

The subsidence curve of the wall at dam height of 14.625 m at different driving positions when the vehicles drove with the load amplitude of 0.7 MPa and the vehicle speed of 20 km/s is shown in Figure 24. It can be seen that under the same conditions, from the far right to the far left of the dam, the influence of driving load gradually decreases, and a peak value of 2.88 cm appears at the wheelbase of 0.5 m, with an amplitude of variation of 1.6%. For the subsidence of wall, the most unfavorable driving position is near the wheelbase of 0.5 m in the middle of the dam.

In order to analyze the variation law of the Mises stress of the dam with the driving position, the vehicle speed of 20 km/h was taken as an example. The Mises stress of the dam and the anti-seepage wall at different driving positions when the amplitude is 0.7 MPa, is shown in Figure 25. It can be seen that the influence trend of the driving position on the Mises stress of the dam has a certain symmetry on both sides of the dam. The maximum peak value is 0.19 MPa at the wheel on the right side. Additionally, when the vehicle acts on the dam, the peak value of the Mises stress is the smallest, with the minimum peak value of 0.062 MPa.

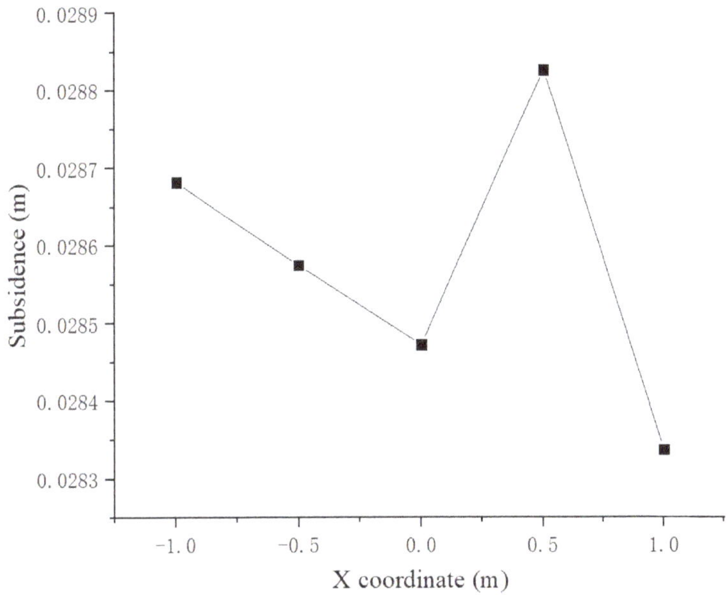

Figure 24. Variation of wall subsidence at 14.625 m under different load positions.

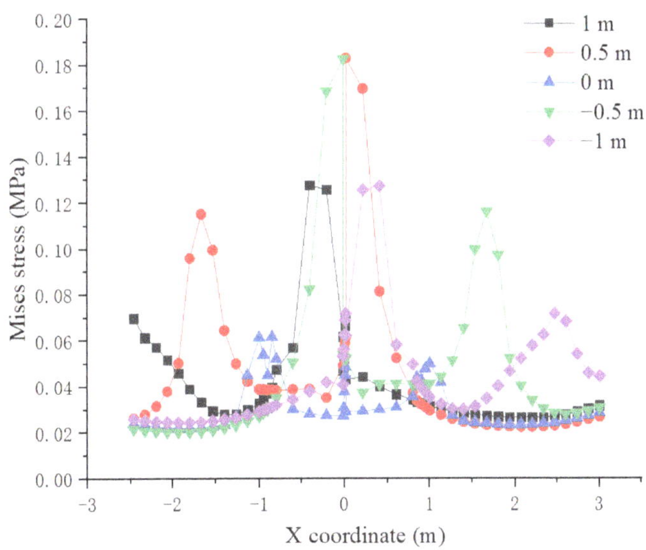

Figure 25. Mises stress distribution of the dam at the height of 14.625 m under different driving positions.

The Mises stress curve of the wall at the dam height of 14.625 m at different driving positions when the vehicles drove with the load amplitude of 0.7 MPa and the vehicle speed of 20 km/s is shown in Figure 26. It can be seen that under the same conditions, from the far right to the far left of the dam, the influence of traffic load remains within a certain range. However, the influence is significantly reduced when the load is on the center of the dam. Meanwhile, the Mises stress is 0.27 MPa, and the maximum value appears at the wheelbase of 0.5 m, the Mises stress at this time is 0.54 MPa, with the variation range of 50%. Therefore, with regard to the Mises stress of the wall, the best driving position is near the middle of the dam crest, and the most unfavorable driving position is near the 0.5 m near the water side of the dam.

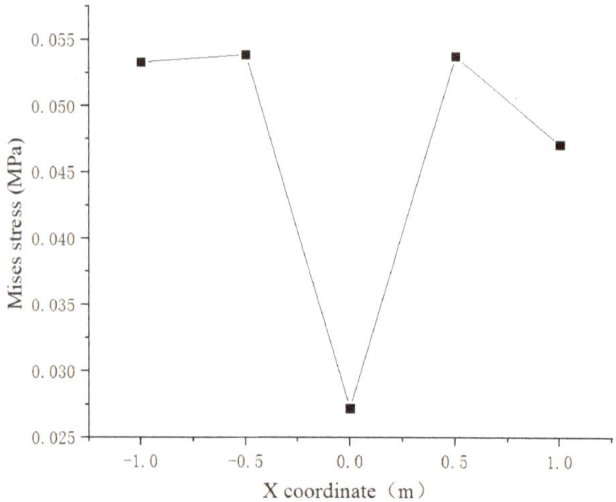

Figure 26. Variation of wall Mises stress at 14.625 m under different load positions.

In order to analyze the variation law of displacement difference stress of the dam with the driving position, the vehicle speed of 20 km/h was taken as an example. The displacement difference of the dam and the anti-seepage wall at different driving positions when the amplitude is 0.7 MPa is shown in Figure 27. It can be seen that the influence trend of the driving position on the subsidence of the dam has a certain symmetry on both sides of the dam, with significant fluctuations. When the vehicle acts on the middle of the dam crest, the peak value of the displacement difference is 0.085 mm, which is the minimum value among the peak values of various working conditions. The maximum peak appears in the −1 m wheelbase group, namely, 0.84 mm, with the variation range of 89.9%.

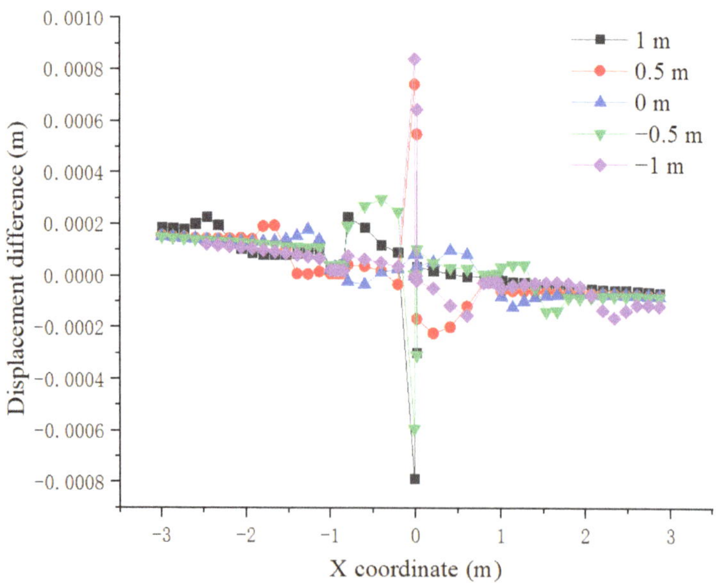

Figure 27. Distribution of displacement difference at 14.625 m under different load positions.

The displacement difference curve of the wall at the dam height of 14.625 m at different driving positions when the vehicles drove with the load amplitude of 0.7 MPa and the vehicle speed of 20 km/s is shown in Figure 28. It can be seen that under the same conditions, from the far right to the far left of the dam, the influence of the driving load on the displacement difference between the wall and the dam remains in a certain range, which is significantly reduced when the load acts on the dam, and the displacement difference is 0.085 mm at this time. The maximum value appears when the wheelbase is −1 m, with the displacement difference of 0.84 mm, and change range between the maximum value and the minimum value is 90.3%. Thus, it can be seen that for the Mises stress of the wall, the driving position with less influence is in the middle of the dam crest, and that with greater influence is near 1 m on the far water side of the dam.

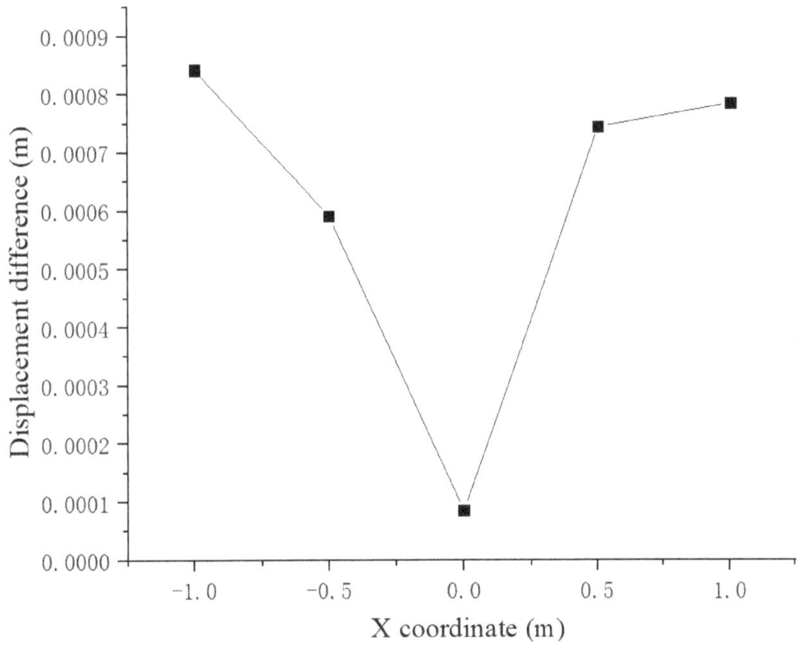

Figure 28. Variation of displacement difference at 14.625 m under different load positions.

According to the above analysis, it is known that the driving position has a significant influence on the deformation coordination characteristics between the polymer anti-seepage wall and the dam. When vehicles drive above the anti-seepage wall, the subsidence of the dam, the Mises stress and the deformation difference are all at the lowest or lower level. Regarding the dam body bearing traffic load, when vehicles drive above the anti-seepage wall, it is more conducive to the coordinated deformation between the dam and the polymer anti-seepage wall. Therefore, when designing the driveway at the top of the dam, it should be designed directly above the anti-seepage wall.

5. Conclusions

A 3D finite element model of polymer anti-seepage dam wall considering the coupling effect of seepage field and stress field was established. Compared with the 2D model, the 3D model is more similar to the actual situation of the dam, and the 3D model can better reflect the complex working conditions, especially those applied to the top of the dam, such as traffic loads. Based on the 3D finite element model, the influence of axle load, vehicle speed and driving position on the stress-deformation characteristics of the dam and polymer anti-seepage wall was analyzed, with the displacement difference of the dam and anti-seepage wall, the Mises stress of the wall and the subsidence of the wall as indicators. The main conclusions are as follows:

Under the traffic load and pore pressure, the response of the middle dam is larger, but the top of the dam is more sensitive to the traffic load, and the influence depth of the traffic load on the dam strain is about 5 m.

Compared with a traditional concrete wall, the capacity of coordinative deformation between the polymer anti-seepage wall and dam is better, and the ratio of the displacement difference between the polymer and rigid concrete at the dam crest with the dam is about 1:96.

The subsidence and Mises stress of the dam are positively correlated with traffic load amplitude, and negatively correlated with vehicle speed. The displacement difference between the wall and the dam rises with the increase of vehicle speed and amplitude. When the amplitude rises from 0.7 to 1.5 MPa, the displacement difference increases by 87.1%. Additionally, when the vehicle speed grows from 20 to 100 km/h, the displacement difference increases by 17.8%.

When the driving position approaches the wall, the displacement difference between the wall and the dam gradually decreases, and the Mises stress of the wall first increases slowly and then decreases sharply, both of which reach their minimum values when the vehicle drives just above the anti-seepage wall.

The most influential factors of vehicle load on the coordination characteristics of dam deformation include the size of vehicle axle load and the driving position, while the vehicle speed has little effect on the coordination characteristics of dam deformation.

From the perspective of the deformation coordination between the polymer anti-seepage wall and the dam, a vehicle with low axle load passes directly above the anti-seepage wall as quickly as possible, which exerts less influence on the dam.

Author Contributions: Conceptualization, H.F. and B.X.; Data curation, A.T.; Supervision, H.F., J.G., Y.L. and X.G.; Writing—original draft, H.Z.; Writing—review & editing, B.X. All authors have read and agreed to the published version of the manuscript.

Funding: This research was funded by the National Natural Science Foundation of China grant number 52109169, the Natural Science Foundation of Henan grant number 212300410279 and the China Postdoctoral Science Foundation grant number 2021M702951.

Institutional Review Board Statement: Not applicable.

Informed Consent Statement: Informed consent was obtained from all subjects involved in the study.

Data Availability Statement: The data that support the findings of this study are available from the corresponding author upon reasonable request.

Conflicts of Interest: The authors declare no conflict of interest.

References

1. Guo, C.C.; Wang, F.M. Mechanism study on the construction of ultra-thin anti-seepage wall by polymer injection. *J. Mater. Civ. Eng.* **2012**, *24*, 1183–1192. [CrossRef]
2. Yu, X.; Wang, Y.K.; Wang, G.; Xue, B.H.; Zhao, X.H.; Du, X.M. Study on Working Behaviors and Improvement Strategies of Concrete Cutoff Wall with Slurry Cake in Thick Soil Foundation. *Int. J. Geomech.* **2022**, *22*, 04022075. [CrossRef]
3. Fang, H.Y.; Zhao, P.; Zhang, C.; Pan, W.; Yu, Z.S.; Cai, K.; Wang, C.X.; Wang, J.; Du, M.R.; He, W.; et al. A cleaner polyurethane elastomer grouting material with high hardening strain for the fundamental rehabilitation: The comprehensive mechanical properties study. *Constr. Build. Mater.* **2022**, *318*, 125951. [CrossRef]
4. Hao, M.M.; Wang, F.M.; Li, X.L.; Zhang, B.; Zhong, Y.H. Numerical and experimental studies on diffusion law of grouting with expansible polymer. *J. Mater. Civ. Eng.* **2018**, *30*, 04017290. [CrossRef]
5. Wang, Z.Y.; Du, M.R.; Fang, H.Y.; Zhang, C.; Li, M.J.; Shi, M.S. Influence of different corrosion environments on mechanical properties of a roadbed rehabilitation polyurethane grouting material under uniaxial compression. *Constr. Build. Mater.* **2021**, *301*, 124092. [CrossRef]
6. Li, M.J.; Fang, H.Y.; Du, M.R.; Zhang, C.; Su, Z.; Wang, F.M. The behavior of polymer-bentonite interface under shear stress. *Constr. Build. Mater.* **2020**, *248*, 118680. [CrossRef]

7. Li, M.J.; Du, M.R.; Wang, F.M.; Xue, B.H.; Zhang, C.; Fang, H.Y. Study on the mechanical properties of polyurethane (PU) grouting material of different geometric sizes under uniaxial compression. *Constr. Build. Mater.* **2020**, *259*, 119797. [CrossRef]
8. Fang, H.; Zheng, Q.; Du, M.; Liu, J.; Zhang, C.; Wang, Z.; Wang, F. Study of the Mechanical Properties and Constitutive Model of the Roadbed Rehabilitation Polyurethane Grouting Materials Under Uniaxial Compression. *J. Appl. Polym. Sci.* **2022**, *139*, 52276. [CrossRef]
9. Hao, M.M.; Li, X.L.; Wang, X.L.; Zhong, Y.H.; Zhang, B.; Wang, F.M.; Zhang, Y.L. Experimental Study On Viscosity Characteristics of Expanding Polymer Grout. *J. Wuhan Univ. Technol. Mater. Sci.* **2021**, *36*, 297–302. [CrossRef]
10. Li, M.J.; Zhang, C.; Fang, H.Y.; Du, M.R.; Su, Z.; Wang, F.M. Effects of Water Content On Shear Properties of Bentonite-Polymer Composite Structure. *Eng. Geol.* **2021**, *287*, 106098. [CrossRef]
11. Zhang, S.; Xue, B.H.; Wang, J.; Gao, J.L.; Wang, C.J.; Li, W.Z. Preliminary Study of Nondestructive Testing of the Polymer Cutoff Wall Based on Vibration Theory. *Adv. Mater. Sci. Eng.* **2021**, *2021*, 4444684. [CrossRef]
12. Li, J.; Zhang, J.W.; Xu, J.G.; Wang, F.M.; Wang, B. Dynamic behavior of polymer anti-seepage wall for earth dam by centrifuge test. *Int. J. Geomech.* **2018**, *18*, 04018179. [CrossRef]
13. Li, J.; Zhang, J.; Hu, H. Numerical analysis on dynamic responses of earth-rock dam with polymer anti-seepage wall under seismic load. In Proceedings of the 2016 3rd International Conference on Information Science and Control Engineering (ICISCE), Beijing, China, 8–10 July 2016; pp. 492–496. [CrossRef]
14. Li, J.; Zhang, J.W.; Wang, Y.K.; Wang, B. Seismic response of earth dam with innovative polymer anti-seepage wall. *Int. J. Geomech.* **2020**, *20*, 04020079. [CrossRef]
15. Zhang, G.X.; Liu, Y.; Zheng, C.Y.; Feng, F. Simulation of the influence of multi-defects on long-term working performance of high arch dam. *Sci. China Technol. Sci.* **2011**, *54*, 1–8. [CrossRef]
16. Luo, D.N.; Hu, Y.; Li, Q.B. An interfacial layer element for finite element analysis of arch dams. *Eng. Struct.* **2016**, *128*, 400–414. [CrossRef]
17. Sator, C.; Becker, W. Closed-form solutions for stress singularities at plane bi- and tri-material junctions. *Arch. Appl. Mech.* **2012**, *82*, 643–658. [CrossRef]
18. Carpinteri, A.; Paggi, M. Analytical study on the singularities arising at multi-material interfaces in 2D linear elastic problems. *Eng. Fract. Mech.* **2007**, *74*, 59–74. [CrossRef]
19. Liu, D.H.; Li, Z.L.; Liu, J.L. Experimental study on real-time control of roller compacted concrete dam compaction quality by using unit compaction energy indices. *Constr. Build. Mater.* **2015**, *96*, 567–575. [CrossRef]
20. Huangfu, Z.H.; Wu, Y.L.; Guo, W.L. Deformation compatibility analysis on clay core wall dams based on multi-objective optimization theory. *J. Hydroelectr. Eng.* **2020**, *39*, 99–108. (In Chinese) [CrossRef]
21. Sarkar, A. Numerical comparison of flexible pavement dynamic response under different axles. *Int. J. Pavement Eng.* **2016**, *17*, 377–387. [CrossRef]
22. Zhao, H.Y.; Indraratna, B.; Ngo, T. Numerical simulation of the effect of moving loads on saturated subgrade soil. *Comput. Geotech.* **2021**, *131*, 103930. [CrossRef]
23. Guo, M.; Hou, F.J.; Zhang, S.X.; Li, X.; Li, Y.L.; Bi, Y.F. Research and Evaluation on Dynamic Response Characteristics of Various Pavement Structures. *Adv. Mater. Sci. Eng.* **2022**, *2022*, 5302142. [CrossRef]
24. Liu, J.S.; Zhu, K.X.; Shen, Y.; Ren, Y.; Zuo, J.P.; Zhang, X.F. Experimental Investigation On the Deformation and Noncoaxial Characteristics of Fiber-Reinforced Aeolian Soil Under Traffic Load. *Int. J. Geomech.* **2022**, *5*, 04022054. [CrossRef]
25. Qian, J.G.; Du, Z.B.; Lu, X.L.; Gu, X.Q.; Huang, M.S. Effects of Principal Stress Rotation On Stress-Strain Behaviors of Saturated Clay Under Traffic-Load-Induced Stress Path. *Soils Found.* **2019**, *59*, 41–55. [CrossRef]
26. Li, N.H.; Wang, J.L.; Mi, Z.K.; Li, D.H. Connotation of deformation safety of high concrete face rockfill dams and its application. *Chin. J. Geotech. Eng.* **2012**, *32*, 193–201. (In Chinese)
27. Kyriakidis, P.; Gaganis, P. Efficient Simulation of (Log)Normal Random Fields for Hydrogeological Applications. *Math. Geosci.* **2013**, *45*, 531–556. [CrossRef]
28. Liu, Y.; Hu, J.; Wei, H.; Saw, A.L. A Direct Simulation Algorithm for a Class of Beta Random Fields in Modelling Material Properties. *Comput. Methods Appl. Mech. Eng.* **2017**, *326*, 642–655. [CrossRef]
29. Xue, B.H.; Du, X.M.; Wang, J.; Yu, X. A Scaled Boundary Finite-Element Method with B-Differentiable Equations for 3D Frictional Contact Problems. *Fractal Fract.* **2022**, *6*, 133. [CrossRef]

water

MDPI

Article

Numerical Limit Analysis of the Stability of Reinforced Retaining Walls with the Strength Reduction Method

Jinsheng Li [1], Xueqi Li [2], Mingyuan Jing [2] and Rui Pang [2,3,*]

[1] CCCC Investment Company Limited, Beijing 100088, China; jinsheng_369@126.com
[2] School of Hydraulic Engineering, Faculty of Infrastructure Engineering, Dalian University of Technology, Dalian 116024, China; 18235102685@163.com (X.L.); 22006113@mail.dlut.edu.cn (M.J.)
[3] State Key Laboratory of Coastal and Offshore Engineering, Dalian University of Technology, Dalian 116024, China
* Correspondence: pangrui@dlut.edu.cn

Abstract: The failure mechanism of MSE (mechanically stabilized earth) walls was studied via numerical analysis with the finite element strength reduction method, which was verified as an effective technique by simulating the experimental results reported in previous papers. The finite element program was applied to explore the effects of reinforcement, geometry, and seismic parameters on failure mechanism control at the design stage of MSE walls to avoid the unavoidable errors experienced in common numerical analysis caused by the assumptions of the failure mode and complex input parameters. The research parameters included the wall height, length, and spacing of the geogrid-reinforced retaining wall and seismic load. The results indicated that the wall height and reinforcement length play a major role in failure mode change. When the reinforcement length is less than 2 m, overturning failure could occur, which was unrelated to the other parameters in all cases studied in this paper. In this paper, the parametric study results were presented by evaluating the critical reinforcement length, generating the failure surface pattern, and summarizing design recommendation.

Keywords: numerical analysis; finite element strength reduction; parameter study; MSE walls; failure mechanism

Citation: Li, J.; Li, X.; Jing, M.; Pang, R. Numerical Limit Analysis of the Stability of Reinforced Retaining Walls with the Strength Reduction Method. *Water* **2022**, *14*, 2319. https://doi.org/10.3390/w14152319

Academic Editor: Giuseppe Oliveto

Received: 7 July 2022
Accepted: 20 July 2022
Published: 26 July 2022

1. Introduction

MSE walls represent a more economical alternative to traditional gravity-type walls. MSE walls are mainly applied in bridge abutments, wing walls, and areas where excavation and slope construction cannot be conducted. Under poor foundation conditions, MSE walls provide significant technical and cost advantages.

Over the past few decades, due to the contradiction between land restrictions and infrastructure development, an increasing number of MSE walls has been applied in slope construction and research [1–12]. Field experiments are an important way to study the failure mechanism of MSE walls, which can be divided into full-scale and proportional experiments [13–16]. The relationship among wall deflection, earth pressure behind the wall, wall height, and the secondary geogrid was obtained by measuring the wall pressure and strain in field experiments [13,14]. It was confirmed that secondary reinforcement played an important role in decreasing wall-facing deflection and generating a uniform, lateral earth-pressure distribution. Yazdandoust and Ghalandarzadeh [15] performed shaking table scaled model tests to obtain the failure pattern of reinforced walls, which reflects the influence of a nonuniform acceleration distribution on the value of the seismic coefficient for reinforced soil structures. Safaee et al. [17] measured values of the most critical dynamic parameters of single-layer and multi-layer walls subjected to different seismic loads. The behavior of wall stability was obtained from the comparison of single-layer and multi-layer walls.

MSE wall stability analysis theory is also an effective way to explore the wall failure mechanism. According to the limit equilibrium method, Bilgin [1] studied the effect of the reinforcement length on the MSE wall failure mode. In his research, it was concluded that the reinforcement length could be reduced to less than 70% of the wall height under the condition of perfect parameters. In addition, the type of reinforcement exerted an important impact on the wall stability [18]. However, simple theoretical analysis is limited by the inherent shortcomings of analysis theory.

The most common research topic in numerical parameter analysis is the limit equilibrium theory, which constitutes the theoretical basis of current design manuals [2,13,19,20]. The influence of the extension and the strengthening of the stability of reinforced soil wall was studied by using the finite element model [2]. Leshchinsky et al. [19] proposed a new limit analysis framework verified by parametric analysis. Their research results revealed that the proposed framework was reasonable, including the influence of facing blocks, seismicity, reinforcement length, and secondary reinforcement. Finite element analysis of MSE walls yielded more accurate results, but the calculation process is time-consuming [20]. By examining the overall stability evaluation of the finite element method, Razeghi et al. [16] provided suggestions for wall designers to quickly check the overall stability of retaining walls. Jiang et al. [13] indicated that secondary reinforcement resulted in a uniform, lateral earth-pressure distribution. However, these simulation experiments were based on theoretical analysis of the slope stability, which often produced a high safety factor for circular failure surfaces and a low safety factor for V-shaped failure surfaces. Liu et al. [21] proposed a novel finite element limit equilibrium method (FELEM) to enhance the applicability of slope stability of FELEM which was validated by five slope problems. The limit equilibrium method must assume a general form of the failure mechanism for calculation, which often leads to inaccurate calculation results [5]. The influence of geosynthetic reinforcement on the stability of the retaining structure was conducted by using the finite element limit analysis method. Hassen et al. [22] proposed a new calculation method (multiphase model) to numerically analyze the stability of reinforced soil structures which showed good performance and computing capabilities. Kazimierowicz-Frankowska and Kulczykowski [23] analyzed that the selected analysis method can accurately predict the deformation of reinforced soil structure under service load. By numerical analysis, Mirmoradi et al. [24] studied the factors affecting the foundation stability, including foundation stiffness and geometry, wall height, and reinforcement stiffness. The numerical model calculation carried out parameter analysis to investigate the influence of the reinforcement spacing, wall height, and foundation location, and reinforcement design on the stability of back-to-back reinforced soil-retaining walls [25]. The limit equilibrium method must assume a general form of the failure mechanism for calculation, which often leads to inaccurate calculation results. Finite element analysis of MSE walls yielded more accurate results, but the calculation process is time consuming.

The main objective of this paper is to investigate the stability of reinforced retaining walls using the lower- and upper-bound principles in the classical plasticity theory. The analyses are carried out by the software OptumG2 (Copenhagen, Denmark) [26], which is based on the methodology in Sloan [27], giving rigorous lower and upper bounds on the failure load. This is known as numerical limit analysis, which only requires soil strength parameters that are familiar to geotechnical engineers. In this paper, this method is used to examine the effects of wall geometry, reinforcement, and seismic parameters on the failure mechanism and factor of safety of geogrid reinforced retaining walls. As numerical analysis can account for a wider range of influential parameters, it is a useful complement to experimental studies (typically limited). The numerical analysis results can help engineers better understand the mechanism of the problem. The lower and upper bounds are invaluable in practice, which enable accurate failure loads to be obtained by error estimates and the adaptive meshing technique.

2. Numerical Limit Analysis

The classical limit analysis method was proposed by Drucker and Prager [28] and has been applied in many geotechnical engineering practices. Following previous work in limit analysis, many studies have been performed to upgrade limit analysis; e.g., Sloan [27] achieved great progress in regard to the FELA method, which was implemented in other research studies [29].

2.1. Theory

This paper analyzed the stability of MSE walls with the finite element strength reduction method, which was originally developed by Sloan and includes the theory of lower and upper bounds [30,31]. The adopted analysis software is OptumG2 [26], which is related to 2D modeling.

2.2. Lower-Bound Principle

The lower-bound theory involves an objective function that should be maximized when the structure is subjected to a collapse load under the equilibrium equality constraints expressed in Equation (1), the discontinuity equilibrium defined in Equations (2)–(5), and the yield condition described in Equations (6)–(10).

In the equilibrium state of each element, the constraint must achieve equilibrium in each element, as expressed in Equations (2)–(5), which is consistent with Equation (1).

$$\begin{aligned}\frac{\partial \sigma_x}{\partial x}+\frac{\partial \tau_{xy}}{\partial y}+b_x=0\\ \frac{\partial \sigma_y}{\partial y}+\frac{\partial \tau_{xy}}{\partial x}+b_y=0\end{aligned} \tag{1}$$

$$\left[A^e_{\text{equil}}\right]\{\sigma\}=\left\{b^e_{\text{equil}}\right\} \tag{2}$$

where:

$$\left[A^e_{\text{equil}}\right]=\frac{1}{2A^e}\left[T_1 \quad T_2 \quad T_3\right], T_i=\begin{bmatrix}\eta_i & 0 & \zeta_i\\ 0 & \zeta_i & \eta_i\end{bmatrix} \tag{3}$$

$$\{\sigma\}=\left\{\sigma_{x,1} \quad \sigma_{y,1} \quad \tau_{xy,1} \quad \sigma_{x,2} \quad \sigma_{y,2} \quad \tau_{xy,2} \quad \sigma_{x,3} \quad \sigma_{y,3} \quad \tau_{xy,3}\right\}^T \tag{4}$$

$$\left\{b^e_{\text{equil}}\right\}=\left\{0 \quad \gamma^e\right\}^T \tag{5}$$

where σ and τ are the stress in soil elements; A^e refers to stress boundary conditions of the area of the element; η_i and ζ_i are constants that depend on the nodal co-ordinates; γ^e is the soil unit weight.

In the limit analysis method, the element corresponds to an individual node, so under the condition of stress balance in the region, the normal and shear nodal stresses along the edge of an element must be equal. The constraints on the different regions at each edge can be described with Equation (3), where α is expressed with respect to the x-axis.

$$\left[A^d_{\text{equil}}\right]\{\sigma\}=\left\{b^d_{\text{equil}}\right\} \tag{6}$$

$$\left[A^d_{\text{equil}}\right]=\begin{bmatrix}T & -T & 0 & 0\\ 0 & 0 & T & -T\end{bmatrix}, T=\begin{bmatrix}sin^2\alpha & cos^2\alpha & -sin2\alpha\\ -0.5sin2\alpha & 0.5sin2\alpha & cos2\alpha\end{bmatrix} \tag{7}$$

$$\{\sigma\}=\left\{\sigma_{x,1} \quad \sigma_{y,1} \quad \tau_{xy,1} \quad \sigma_{x,2} \quad \sigma_{y,2} \quad \tau_{xy,2} \quad \sigma_{x,3} \quad \sigma_{y,3} \quad \tau_{xy,3} \quad \sigma_{x,4} \quad \sigma_{y,4} \quad \tau_{xy,4}\right\}^T \tag{8}$$

$$\left\{b^d_{\text{equil}}\right\}=\left\{0 \quad 0 \quad 0 \quad 0\right\}^T \tag{9}$$

The Mohr-Coulomb strength criterion provides additional yield condition constraints to ensure that no point stress exceeds the yield value (Sloan, 2013), which is defined as Equation (3). Function f contains the yield limit formed by all the above stresses.

$$f(\sigma_i) \leq 0 \tag{10}$$

2.3. Upper-Bound Principle

The objective function of the upper-bound theory should be minimized when the internal power dissipation rate decreases, which satisfies the continuum flow rule defined in Equation (11), the velocities in the discontinuities satisfy the flow rule expressed in Equation (12), and the stresses in the elements satisfy the yield condition described in Equation (13).

$$\begin{aligned} \dot{\varepsilon}^p_{xx} &= \dot{\lambda}\partial f/\partial\sigma_{xx} \\ \dot{\varepsilon}^p_{yy} &= \dot{\lambda}\partial f/\partial\sigma_{yy} \\ \dot{\gamma}^p_{xy} &= \dot{\lambda}\partial f/\partial\tau_{xy} \\ \dot{\lambda} &\geq 0, \dot{\lambda} f(\sigma^e) = 0 \end{aligned} \tag{11}$$

$$\begin{aligned} \Delta u_n &= \dot{\lambda}\partial f(\sigma_n,\tau)/\partial\sigma_n \\ \Delta u_s &= \dot{\lambda}\partial f(\sigma_n,\tau)/\partial\tau \\ \dot{\lambda} &\geq 0, \dot{\lambda} f(\sigma_n,\tau) = 0 \end{aligned} \tag{12}$$

$$f(\sigma^e) \leq 0 \tag{13}$$

where $\dot{\lambda}$ is the plastic multiplier and $f(\sigma^e)$ is the yield condition for each element. A more detailed introduction to the lower- and upper-bound principles was provided by Lyamin and Sloan [32]. The subscript in these equations indicates the direction of the stress/strain in three-dimensional co-ordinates. Subscript n indicates the normal direction.

2.4. Mesh Detail

A comparison of 10,000 and 20,000 adaptive, refined element meshes for this problem is shown in Figure 1. When the number of grid elements exceeded 10,000, there was a slight difference between the failure surface diagrams and safety factor values. The simulation process relied on an adaptive mesh, which could reduce the computational costs, while a refined mesh could closely capture the failure mechanism. In all cases in this study, the above lower- and upper-bound principles were applied to analyze the factor of safety (FS), and 10,000 upper- and lower-bound elements were considered with three adaptivity iterations.

(a) (b)

Figure 1. Comparison of the different numbers of elements. (**a**) Safety factor = 0.8752 (elements = 10,000). (**b**) Safety factor = 0.8781 (elements = 20,000).

3. Verification of the Numerical Model

3.1. Case 3-1

Two model walls in the critical state were analyzed according to the AASHTO design method with the FLAC model according to EHWA-RD-03-04 [33,34]. The purpose was to compare OptumG2 predictions to results obtained with an existing limit analysis method. The following two cases were selected to represent different failure mechanisms identified with OptumG2, as shown in Figure 2.

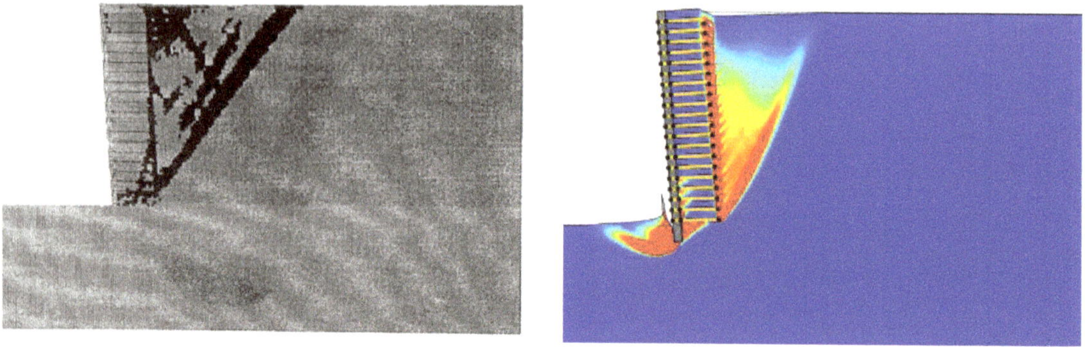

(**a**) Numerical model of FLAC (**b**) Numerical model of OptumG2

Figure 2. Comparison of the FLAC and OptumG2 results for case 3-1.

The data input for verification is listed in Table 1, in which the same wall geometry, rigid block facing, water-free foundation, and reinforcement were considered. Major FLAC and OptumG2 simulation results are provided in Table 1. Case 3-1 indicated that, according to FLAC analysis, the wall occurred in the failure state due to overturning failure (FS = 1.09). The OptumG2 results for Case 3-1 confirmed that under a reinforcement length of $l = 1.5$ m, the wall was at the verge of overturning failure (FS = 1.075), which coincided with the FLAC analysis results regarding the failure mechanism and safety factor. OptumG2 predicted the same overturning mode of failure and safety factor identified via FLAC calculations. The safety factor for the reinforcements obtained via FLAC and OptumG2 calculations is provided in Table 2.

Table 1. Input data for verification of the numerical analysis models.

Input Data	Case 3-1	Case 3-2
Wall height	8.2	12.1
Reinforcement spacing (m)	0.4	1.34
Reinforcement length (m)	1.5	7.5
Reinforcement soil unit weight (kN/m^3)	22	15.64
Reinforcement soil angle of friction (°)	45	39.5
Retaining soil unit weight (kN/m^3)	22	15.64
Retaining soil angle of friction (°)	45	39.5
Foundation soil unit weight (kN/m^3)	22	15.64
Foundation soil angle of friction (°)	45	39.5
Ultimate strength of geogrid reinforcement (kN/m)	9.0	10.0
Soil–geogrid angle of friction (°)	35	39

Table 2. Safety factor results for Case 1 with the numerical model.

Safety Factor	Lower Bound	Upper Bound	Average Value	FLAC Analysis
Case 3-1	1.039	1.110	1.075	1.09 (1)

3.2. Case 3-2

Another case [35] was selected to represent different failure mechanisms identified with both OptumG2 (Fs = 1.059) and FLAC (Fs = 1.07), as shown in Figure 3. The data input is listed in Table 2. The triangle marked in the image is the failure surface in which there is a slight difference between OptumG2 and FLAC. The failure mechanism results shown in Figure 3 indicate that the failure modes determined with the two simulation methods were consistent under the same input parameters (global failure).

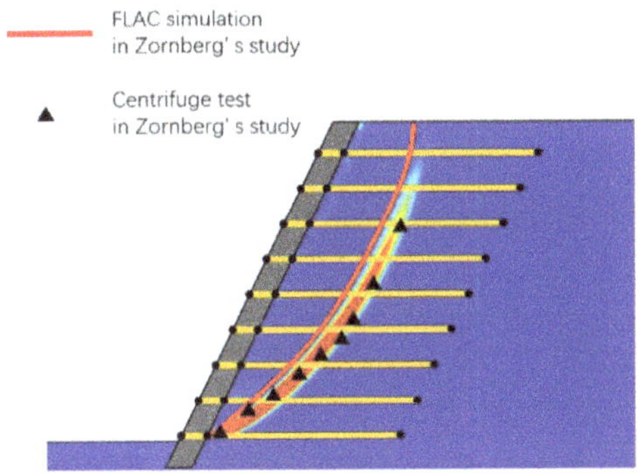

Figure 3. Comparison of the FLAC and OptumG2 results for case 3-2.

4. Parametric Study and Results

There are two major parts of the research results. The first part of the results includes the relationship between the identified failure mechanism and influence of the following parameters: (A) reinforcement parameter, (B) geometry of the wall, and (C) horizontal seismic load. The second part illustrates the influence of variables on the safety factor. In the Experimental Discussion section, the obtained conclusion was supported by calculating the minimum length of the reinforcement zone to maintain the failure mechanisms in each case. Each wall was simulated but the failure mechanism was altered by increasing the reinforcement length while maintaining the wall height at 10 m. Three common failure mechanisms are often considered: overturning failure, sliding failure, and global failure. In the simulations, the failure mechanism was defined by determining the two critical reinforcement lengths of sliding failure. The critical reinforcement lengths of sliding failure were identified as follows: (1) the failure surface was straight, (2) a slip surface was fully developed through the reinforced wall, and (3) there was a horizontal movement of the wall, as shown in Figure 4.

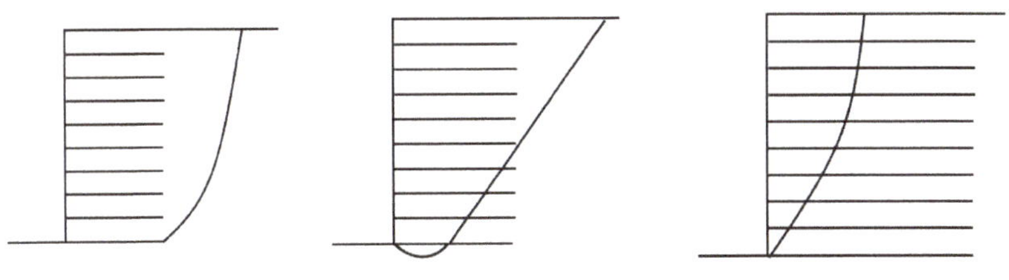

(**a**) overturning (**b**) sliding (**c**) global failure

Figure 4. Three different failure mechanisms in the cases.

4.1. Numerical Model

4.1.1. Parameter Value Ranges and Baseline Case

A basic model was defined, as shown in Figure 5. The block in front of the reinforcement soil in the base case comprised a stiff material, with a thickness of 500 mm and a height of 300 mm without a footing underneath the wall. The height of the reinforced wall is 10 m. In addition, the soil models in this paper, including reinforced and retained soil models, were all elastic to perfectly plastic models. The ranges considered in this paper referred to the parameters of most MSE walls in the field [36].

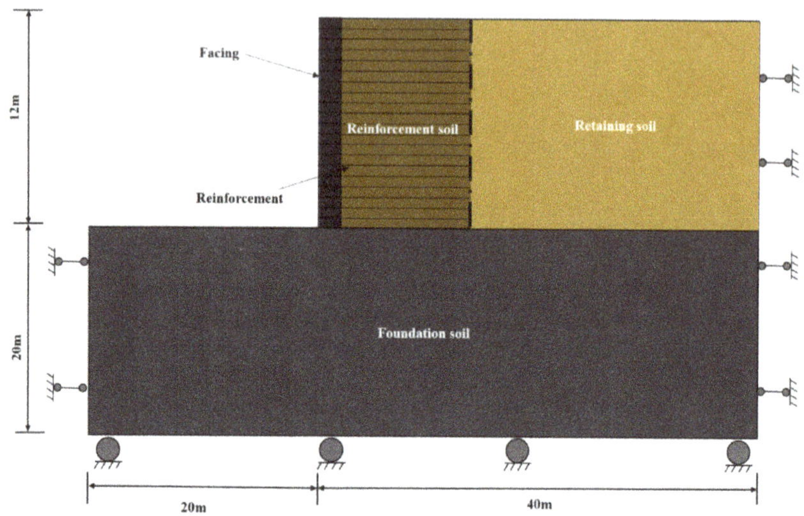

Figure 5. Geometry of the numerical model.

Some meaningful simulation results of the geogrid (reinforcement) length in the transformation process between the above three failure mechanisms are shown in this section. The influence of each parameter in the numerical model was identified by exploring its impact on the critical value of the geogrid length by varying a single parameter while the other parameters remained constant.

4.1.2. Geometry and Boundary Conditions

In the present study, the parameter range in the base case encompasses average values previously reported in the literature. The height and width of the foundation were 20 and 60 m, respectively, which were beyond the standard values to minimize boundary effects, as shown in Figure 5. The height and width of the wall facing were 12 m and 0.5 m, respectively. The width of the reinforcement soil zones was 10 m.

The deformation-limiting boundary conditions of the model in this paper are consistent with those in most numerical analysis experiments in the literature (Jiang et al., 2019). The bottom was constrained along the normal and tangential directions, and the sides were constrained only along the normal direction. The influence of groundwater was not considered in this study.

4.1.3. Soil Constitutive Models and Properties

The soil models in this paper, including reinforced and retained soil models, were all elastic–perfectly plastic models, as listed in Table 3. The foundation soil and block facing were simulated with linearly elastic models. In the literature, relevant experiments [37] with the MC model have been reported in terms of the establishment of a soil model for MSE wall simulation, which have demonstrated the feasibility of MSE simulation.

Table 3. Soil parameters in the numerical model.

Material	Reinforced Soil	Retained Soil	Foundation Soil	Block Facing
Constitutive mode	Mohr-Coulomb	Mohr-Coulomb	Linearly elastic	Linearly elastic
Unit weight (kN/m^3)	18	18	18	23
Young's modulus (MPa)	20	20	2000	-
Poisson's ratio	0.3	0.3	0.3	-
Cohesion (kPa)	0	0	0	-
Friction angle (°)	35	35	35	35
Dilation angle (°)	5	5	5	-

4.1.4. Reinforcement Properties

The type of reinforcement in the numerical simulations was a geogrid, which entailed a linearly elastic–perfectly plastic model allowing small deformations. The weightless geogrid cannot sustain uniaxial compression and offers no resistance to bending. The detailed parameters are listed in Table 4.

Table 4. Modeling of the reinforcement stiffness.

Materials	Secant Stiffness at 2%, $J_{2\%}$ (kN/m^2)	Tensile Strength (kN/m^2)
geogrid	400	20

4.1.5. Interface Properties

The numerical model considered two types of interfaces, as listed in Table 5. The shear stress of the interface surface is directly proportional to the displacement, which reflects linear elastic to perfectly plastic properties. The reduced strength of the interface in the numerical calculation was 0.85. The cohesion of the backfill-reinforcement interface was assumed to reach zero, and the dilation angle was 5°.

Table 5. Interface parameters.

Interface	Friction Angle (°)	Dilation Angle (°)	Cohesion (kPa)	Normal	Shear Stiffness
Backfill-reinforcement	35	5	0	-	-
Block-reinforcement	25	0	0	-	-

4.1.6. Critical Reinforcement Lengths

The significant impact of the geogrid length on the failure mechanism is shown in Figure 6. When the geogrid length was smaller than 4 m, the safety factor rapidly increased with increasing geogrid length, and the failure mechanism transitioned from overturning into sliding. However, when the geogrid length was larger than 10 m, the safety factor slowly increased. This suggests that when the geogrid length reaches the critical length, the overall materials in the reinforced retaining wall experience antifailure deformation, and the overall structure fully absorbs the failure energy. The two inflection points in Figure 6 denote the approximate values of the critical length. More details on the failure mechanism are described, considering the geometric parameters of the MSE wall.

4.2. Wall Height

The geometry of the reinforced retaining wall is a crucial factor influencing the evaluation results of MSE wall design stability. Figure 7 shows that FS significantly increased with increasing geogrid length and decreased with increasing wall height when the wall height was varied from 5 to 15 m. In addition, the increase in FS at a wall height of 5 m changed more obviously than that at a wall height of 15 m.

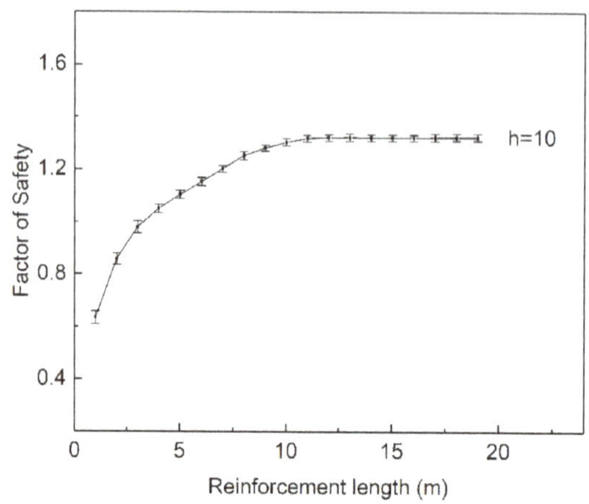

Figure 6. Effect of the reinforcement length on the factor of safety.

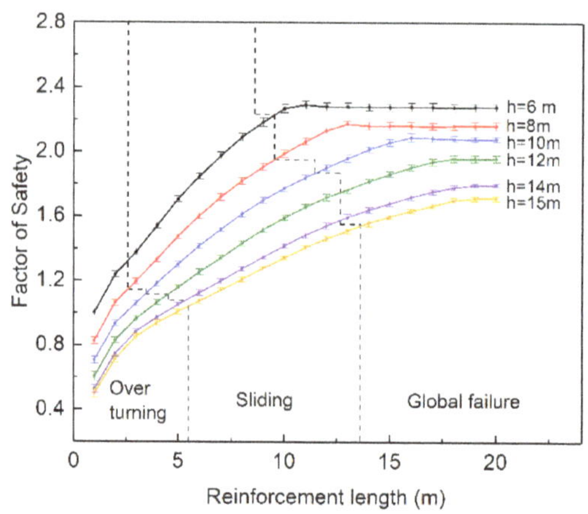

Figure 7. Safety factor with the wall height and critical reinforcement length.

Based on the three regions separated by the dotted line in Figure 7, it could be observed that the three geogrid lengths corresponded to distinct failure mechanisms. Both the geogrid length and wall height contributed to the observed change in the failure mechanism. Moreover, when the change in wall height did not exceed 9 m, the failure mechanism was mainly affected by the length of the geogrid, and an increase in wall height could not alter the overturning failure mechanism, as shown in Figure 8. Each study case is marked with black points in Figure 8, and the parameters and failure mechanism are summarized in Table 6. Three failure mechanisms in cases with geogrid lengths l = 2 m, 10 m, and 15 m are shown in Figure 9. When the range of reinforced soil is very limited, the soil failure surface transects the bottom, and the reinforced wall is directly overturned, with which an increase in the reinforced soil area imposes a significant influence on the safety factor, as shown in Figure 9a. When the geogrid length was further increased, the slope-sliding failure mechanism emerged, as shown in Figure 9b. However, when the length of the geogrid exceeded the critical length, due to its large scale and high strength, the failure surface did not penetrate the reinforced wall, as shown in Figure 9c. An increase in the wall height could yield many negative effects, including an increase in the active earth pressure, a decrease in the factor of safety, and an increase in the critical length, which could reduce the overall stability of the structure, and wall failure mode variation required greater reinforcement.

Table 6. Summary of the failure mechanism cases.

Case	Parameter in the Numerical Model				Failure Mechanism
	Wall Height (m)	Reinforcement Length (m)	Reinforcement Space (m)	*K* (g)	
1	6	2	0.6	0	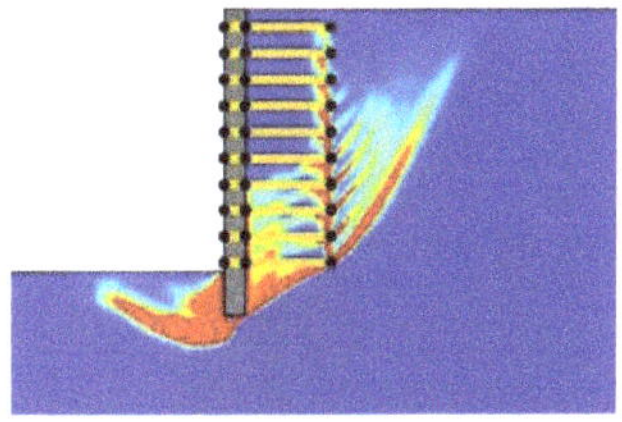
2	10	3	0.6	0	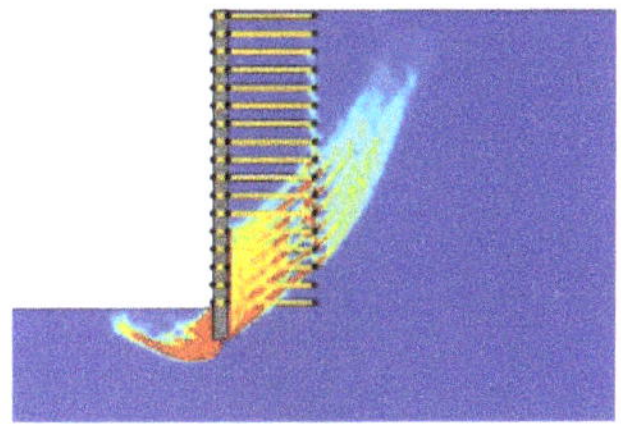
3	13	4	0.6	0	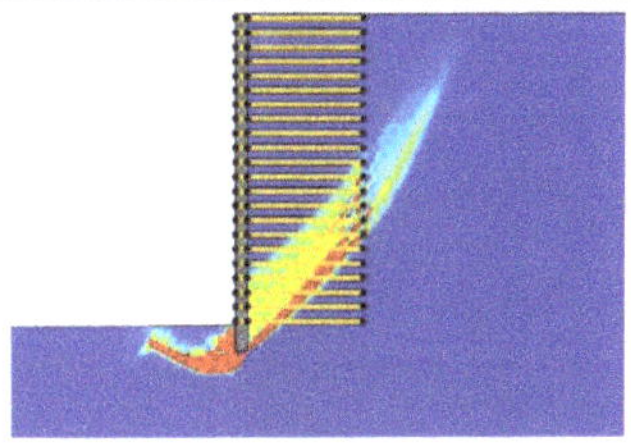
4	6	9	0.6	0	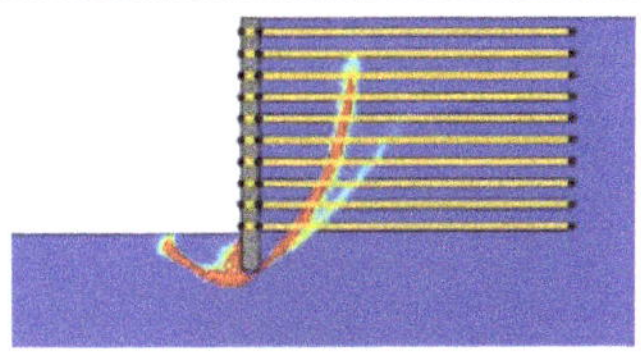
5	10	10	0.6	0	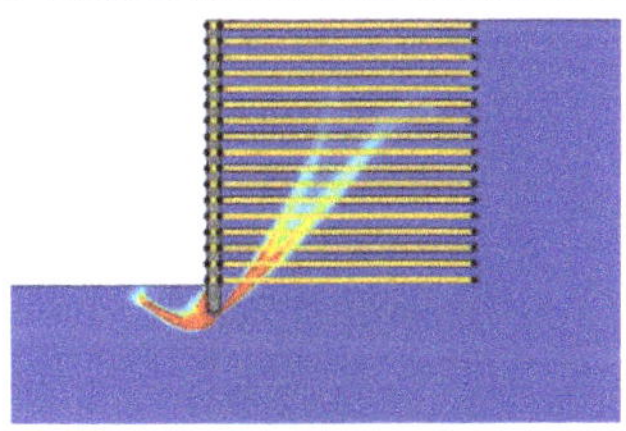
6	13	12	0.6	0	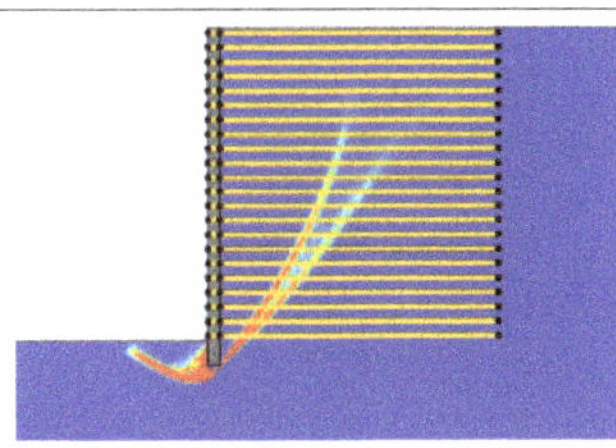

Table 6. *Cont.*

Case	Parameter in the Numerical Model				Failure Mechanism
	Wall Height (m)	Reinforcement Length (m)	Reinforcement Space (m)	K (g)	
7	10	2	0.4	0	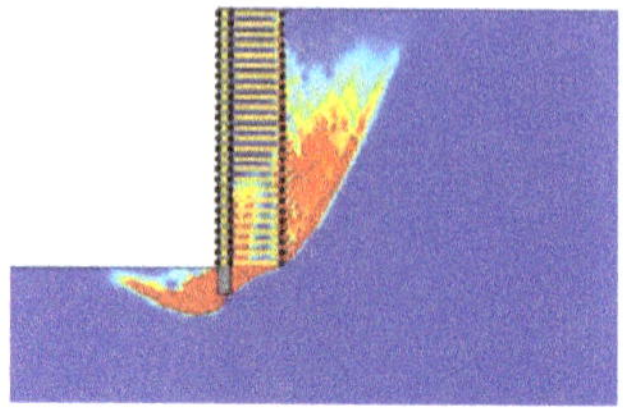
8	10	3	0.6	0	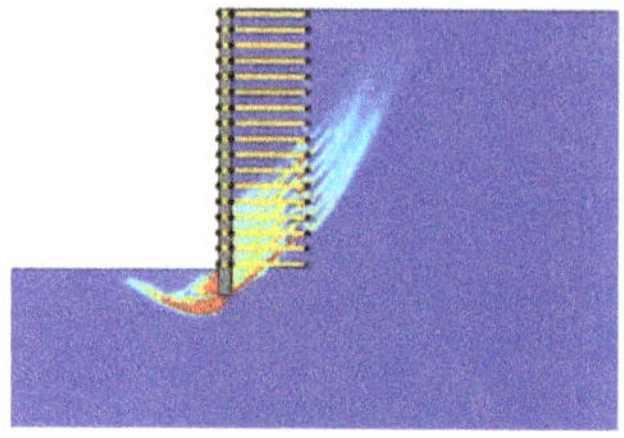
9	10	5	0.9	0	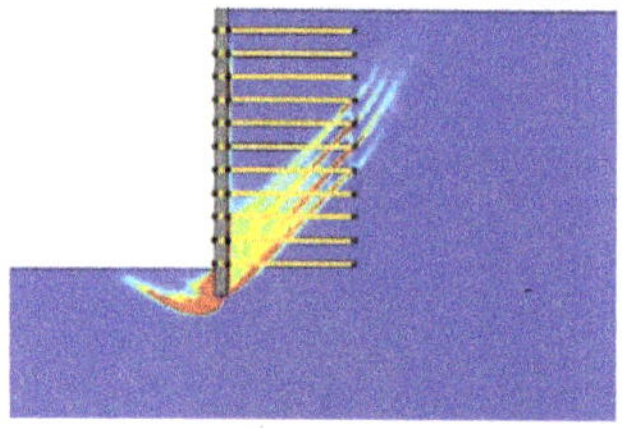
10	10	8	0.4	0	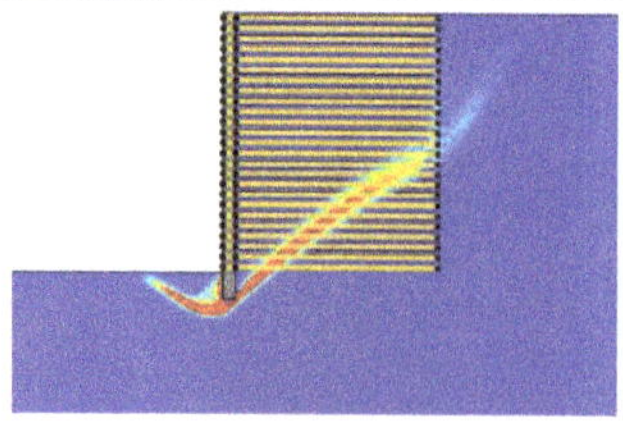
11	10	8	0.6	0	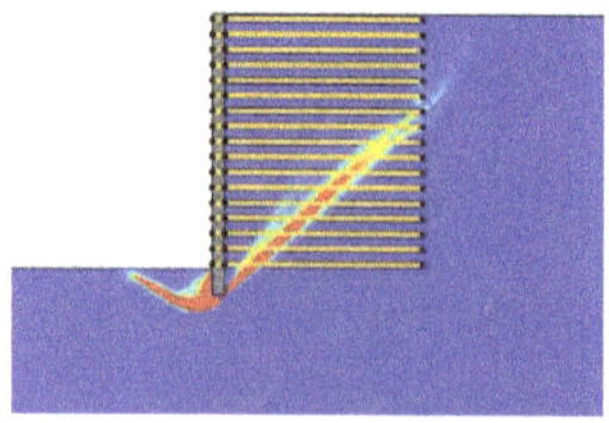
12	10	9	0.9	0	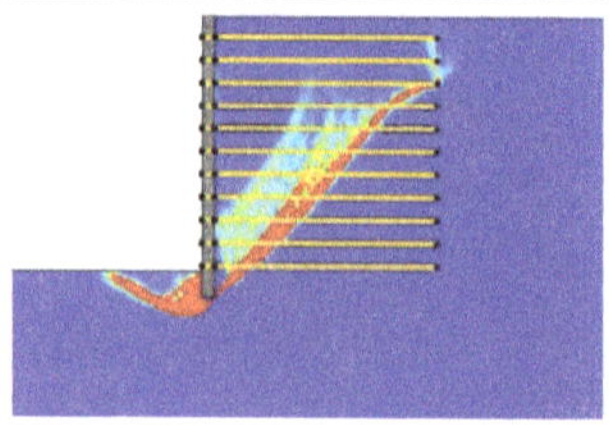

Table 6. *Cont.*

Case	Parameter in the Numerical Model				Failure Mechanism
	Wall Height (m)	Reinforcement Length (m)	Reinforcement Space (m)	*K* (g)	
13	10	5	0.6	0.1	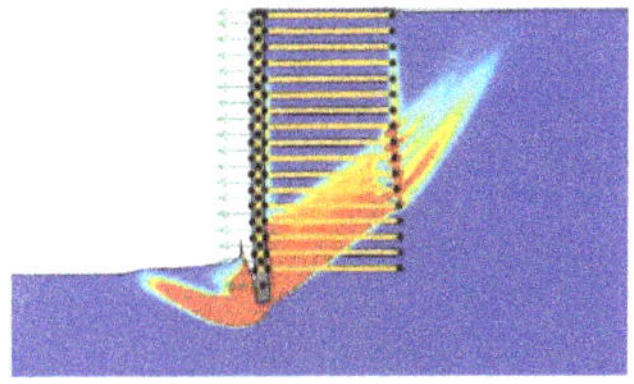
14	10	7	0.6	0.2	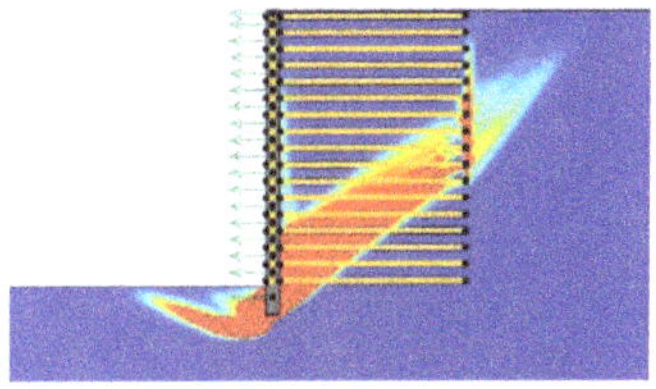
15	10	12	0.6	0.1	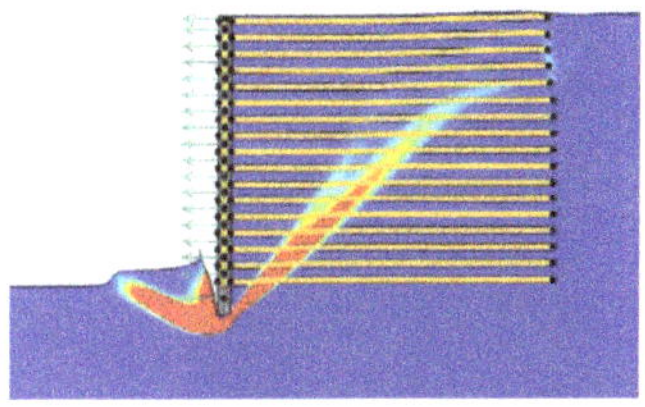
16	10	15	0.6	0.2	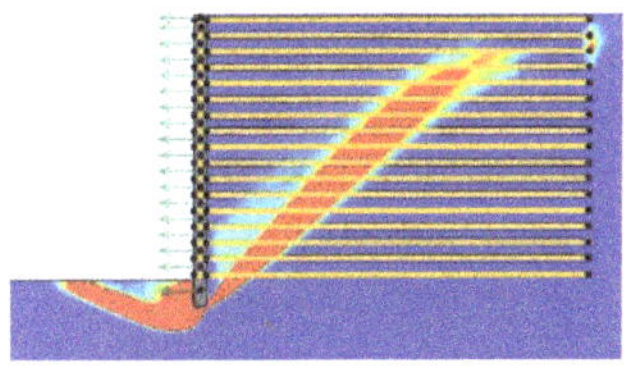

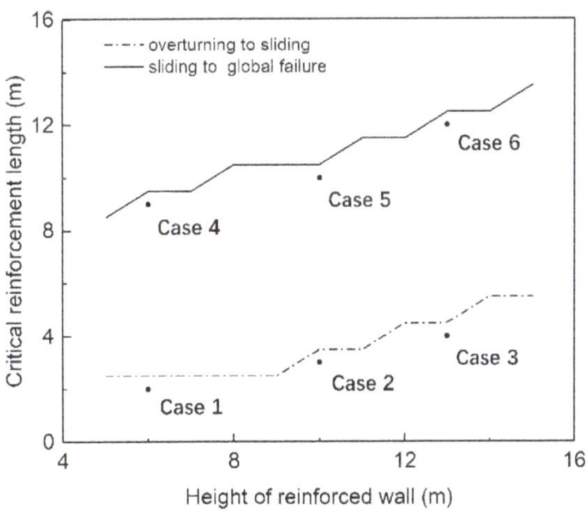

Figure 8. Effect of the reinforced wall height on the critical reinforcement length.

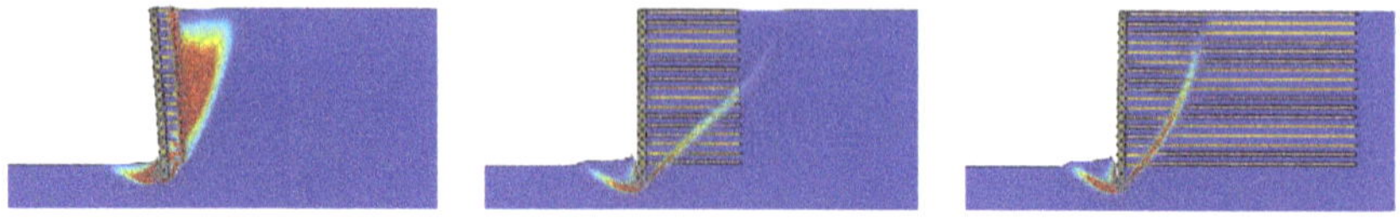

(a) overturning (b) sliding (c) global failure

Figure 9. Three failure mechanism cases.

Cases 1 to 6 in Table 6 represent the failure mechanism with the wall height. In Cases 1 to 3, although the transition state also hardly occurred, it was obvious that the change in wall height did not greatly influence the transition state of the failure mechanism similarly to Case 1. In Cases 4 to 6, the numerical results indicated the transition state from sliding failure to global failure. In these cases, a higher wall resulted in a longer duration of the transition state.

4.3. Reinforcement Spacing

The influence of the reinforcement spacing on the stability of MSE walls is shown in Figure 10. Since the AASHTO manual requires that the spacing should be smaller than 0.8 m, the variation range of the spacing considered in the parameter study is 0.3–0.9 m. The results are shown in Figure 10, in which the change in spacing did not affect the critical value of global failure occurrence but greatly impacted the overturning failure mode when the reinforcement vertical spacing was smaller than 0.5 m. In addition, when the reinforcement spacing exceeded 0.8 m, the failure mode mainly depended on the reinforcement length. This occurred because the low density of reinforcement reduced the bearing capacity of the reinforcement zone.

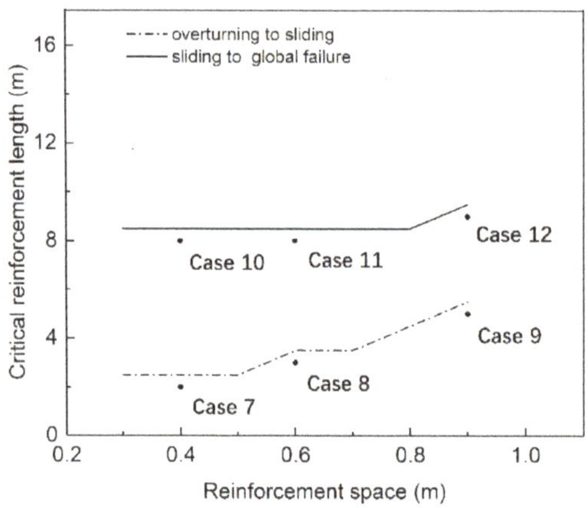

Figure 10. Effect of the reinforcement spacing on the critical reinforcement length.

A summary of all parametric study results is given in Table 6, in which Cases 7 to 12 represent the failure mechanism as a function of the spacing. In each case, numerical experiments below the critical reinforcement value were performed. In Cases 7 to 9, if the reinforcement length was larger than 1 m, the failure mode transitioned into sliding failure, which reflects the transition state change from overturning failure to sliding failure. In the cases with a small spacing, the transition state hardly occurred. Due to the large spacing of the reinforcement zone, the influence of the reinforcement length on the strength decreased. In Cases 10 to 12, the inclination angle of the sliding failure surface gradually increased. Because the strength of the reinforcement zone decreased with increasing spacing, the failure surface more easily penetrated the reinforcement zone. These results in Figure 10 are consistent with the safety factor change trend.

4.4. Horizontal Seismic Load Originating from Earthquake

Cases with seismic coefficient values of $K = g_h/g_v$, with $g_v = 9.8\ m/s^2$, were considered. The effect of the seismic load, ranging from 0.05 g to 0.2 g, on the critical reinforcement length is shown in Figure 11. An increase in seismic coefficient value required a larger critical length of the reinforcement to satisfy the stability requirements of the sliding and global failure modes.

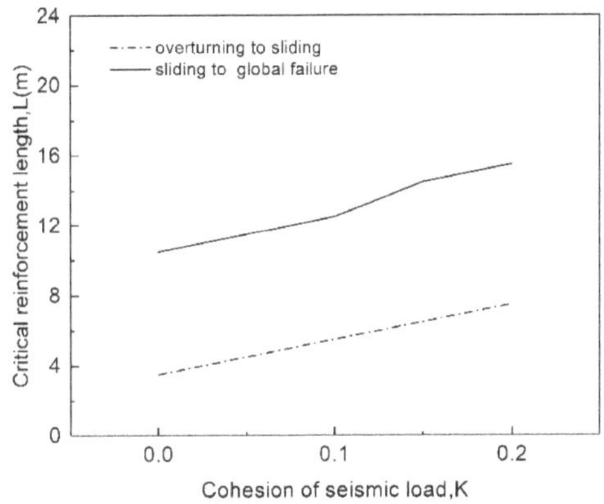

Figure 11. Effect of the seismic load on the critical reinforcement length.

Cases 13 to 16, as listed in Table 6, represent the base case failure mechanism under seismic loading. An increased seismic coefficient value resulted in a longer transition state and higher inclination angle of the failure surface.

5. Recommendations for Design

The existing design method involves internal and external stability analysis based on the limit state method. In this design method, the location of the damaged surface is often assumed, and assessment calculations are then carried out. This method generally incorporates experience-based knowledge, and the assumptions before calculation are often difficult to verify in practical applications. However, this design method does not consider the relationship between the parameters of MSE walls and failure mode. The equilibrium conditions of the analysis method proposed in this paper are applicable to the whole soil area, and the safety factor is defined in a very small range, so that engineers can meet different design requirements in the seismic design of structures according to the range. The research results of this paper provide engineers with rich references. Specific suggestions and contributions are as follows:

(1) Length of the reinforcement. A minimum reinforcement length of 0.7 H is recommended for MSE walls. In areas with poor foundation conditions and areas of a high seismic grade, larger lengths are required, as listed in Table 7.

(2) Spacing of the reinforcement. When the spacing of the reinforcement is smaller than 0.6 m, the position of the sliding surface could occur behind the reinforced area. In the simulation analysis experiments in this paper, if the overturning failure mode emerged, the length-to-height ratio of reinforcement varied between 0.23 and 0.4. When the ratio was higher than 0.9, the global failure mode emerged. However, under normal conditions, parametric analysis indicated that the wall stability was not only determined by the length-to-height ratio but also determined by the reinforcement length. When the foundation conditions were limited, the stability of the wall could be improved by increasing the reinforcement length and reducing the spacing, as listed in Table 8.

(3) Horizontal seismic load. The seismic load could significantly reduce the wall stability. Maintaining the wall in the global failure mode required higher wall design conditions. Under the baseline conditions, the reinforcement length and the length-to-

height ratio of reinforcement could be increased to improve the wall stability. When the reinforcement length was limited due to the construction environment, the wall stability could be improved by decreasing the spacing of the reinforcement.

Table 7. Minimum length of the reinforcement.

Case	*L/H*	Length (m)
Base conditions	0.8	12
Seismic loading	0.9	15

Table 8. Maximum spacing of the reinforcement.

Case	Spacing (m)
Base conditions	0.7
Seismic loading	0.5
Limited reinforcement length	0.5

6. Conclusions

The critical reinforcement values resulting in MSE wall failure mechanism transition under the effect of various parameters were studied under different conditions. The influence of the length of the reinforcement in different cases on the stability of MSE walls was studied. The research obtained rich and interesting results, provided design suggestions for engineers, and made contributions to the field of seismic design of retaining walls. According to this research, the following conclusions can be drawn:

(1) Both the reinforcement length and wall height greatly affected the change in failure mode of MSE walls, based on the parameter study in this paper. When the wall height was greater than 9 m, an increase in height could reduce the strengthening effect of the reinforcement, in which maintaining a favorable failure mode required a longer reinforcement length.

(2) With the properties involved in this paper, the critical length of the reinforcement was determined as approximately 0.4-H and 0.9-H, which divided the various failure modes into overturning failure, sliding failure, and global failure.

(3) The reinforcement spacing was an important factor influencing the failure mode of MSE walls. Increasing the reinforcement spacing from 0.3 to 0.9 m reduced the safety factor and altered the failure mode. In particular, when the reinforcement spacing was above 0.5 m, the critical length of the reinforcement increased from overturning failure to sliding failure, and when the reinforcement spacing was above 0.8 m, the critical length of the reinforcement decreased from sliding failure to global failure.

(4) The seismic coefficient obviously affected the stability of MSE walls. The required reinforcement length to maintain the wall stability in the case with a seismic coefficient value of 0.2 was almost 1.5 times larger than that in the case without a seismic load.

The assumption in this study is that elastic–perfectly plastic models cannot consider deformation. In stability analysis, there are numerous parameters in the elastic–plastic constitutive model, and inappropriate parameter selection could cause large errors. The following research direction will be to integrate the advantages of both models for parametric analysis.

Author Contributions: Conceptualization, J.L.; data analysis, X.L.; formal analysis, M.J.; project, R.P. All authors have read and agreed to the published version of the manuscript.

Funding: China National Natural Science Foundation (Grant Nos. 52009017, 51979026).

Institutional Review Board Statement: Institutional review board approval of our university was obtained for this study.

Informed Consent Statement: All study participants provided informed consent.

Data Availability Statement: Data openly available in a public repository.

Conflicts of Interest: The authors declare no conflict of interest.

References

1. Bilgin, Ö. Failure mechanisms governing reinforcement length of geogrid reinforced soil retaining walls. *Eng. Struct.* **2009**, *31*, 1967–1975. [CrossRef]
2. Chen, J.-F.; Liu, J.-X.; Xue, J.-F.; Shi, Z.-M. Stability analyses of a reinforced soil wall on soft soils using strength reduction method. *Eng. Geol.* **2014**, *177*, 83–92. [CrossRef]
3. Song, F.; Liu, H.; Ma, L.; Hu, H. Numerical analysis of geocell-reinforced retaining wall failure modes. *Geotext. Geomembr.* **2018**, *46*, 284–296. [CrossRef]
4. Pang, R.; Xu, B.; Zhou, Y.; Song, L. Seismic time-history response and system reliability analysis of slopes considering uncertainty of multi-parameters and earthquake excitations. *Comput. Geotech.* **2021**, *136*, 104245. [CrossRef]
5. Fathipour, H.; Payan, M.; Chenari, R.J. Limit analysis of lateral earth pressure on geosynthetic-reinforced retaining structures using finite element and second-order cone programming. *Comput. Geotech.* **2021**, *134*, 104119. [CrossRef]
6. Pang, R.; Xu, B.; Kong, X.J.; Zhou, Y.; Zou, D.G. Seismic performance evaluation of high CFRD slopes subjected to near-fault ground motions based on generalized probability density evolution method. *Eng. Geol.* **2018**, *246*, 391–401. [CrossRef]
7. Pang, R.; Xu, B.; Zou, D.G.; Kong, X.J. Stochastic seismic performance assessment of high CFRDs based on generalized probability density evolution method. *Comput. Geotech.* **2018**, *97*, 233–245. [CrossRef]
8. Pang, R.; Xu, B.; Kong, X.J.; Zou, D.G.; Zhou, Y. Seismic reliability assessment of earth-rockfill dam slopes considering strain-softening of rockfill based on generalized probability density evolution method. *Soil Dyn. Earthq. Eng.* **2018**, *107*, 96–107. [CrossRef]
9. Pang, R.; Xu, B.; Zhou, Y.; Zhang, X.; Wang, X.L. Fragility analysis of high CFRDs subjected to mainshock-aftershock sequences based on plastic failure. *Eng. Struct.* **2020**, *206*, 110152. [CrossRef]
10. Li, Y.; Pang, R.; Xu, B.; Wang, X.; Fan, Q.; Jiang, F. GPDEM-based stochastic seismic response analysis of high concrete-faced rockfill dam with spatial variability of rockfill properties based on plastic deformation. *Comput. Geotech.* **2021**, *139*, 104416. [CrossRef]
11. Xu, B.; Pang, R.; Zhou, Y. Verification of stochastic seismic analysis method and seismic performance evaluation based on multi-indices for high CFRDs. *Eng. Geol.* **2020**, *264*, 105412. [CrossRef]
12. Zai, D.; Pang, R.; Xu, B.; Fan, Q.; Jing, M. Slope system stability reliability analysis with multi-parameters using generalized probability density evolution method. *B. Eng. Geol. Environ.* **2021**, *80*, 8419–8431. [CrossRef]
13. Jiang, Y.; Han, J.; Zornberg, J.; Parsons, R.L.; Leshchinsky, D.; Tanyu, B. Numerical analysis of field geosynthetic-reinforced retaining walls with secondary reinforcement. *Géotechnique* **2019**, *69*, 122–132. [CrossRef]
14. Jiang, Y.; Han, J.; Parsons, R.L.; Brennan, J.J. Field Instrumentation and Evaluation of Modular-Block MSE Walls with Secondary Geogrid Layers. *J. Geotech. Geoenviron. Eng.* **2016**, *142*, 05016002. [CrossRef]
15. Yazdandoust, M.; Ghalandarzadeh, A. Pseudo-Static Coefficient in Reinforced Soil Structures. *Int. J. Phys. Model. Geotech.* **2020**, *20*, 320–337. [CrossRef]
16. Razeghi, H.R.; Viswanadham, B.; Mamaghanian, J. Centrifuge and numerical model studies on the behaviour of geogrid reinforced soil walls with marginal backfills with and without geocomposite layers. *Geotext. Geomembr.* **2019**, *47*, 671–684. [CrossRef]
17. Safaee, A.M.; Mahboubi, A.; Noorzad, A. Experimental investigation on the performance of multi-tiered geogrid mechanically stabilized earth (MSE) walls with wrap-around facing subjected to earthquake loading. *Geotext. Geomembr.* **2020**, *49*, 130–145. [CrossRef]
18. Bilgin, Ö.; Mansour, E. Effect of reinforcement type on the design reinforcement length of mechanically stabilized earth walls. *Eng. Struct.* **2014**, *59*, 663–673. [CrossRef]
19. Leshchinsky, D.; Kang, B.; Han, J.; Ling, H. Framework for Limit State Design of Geosynthetic-Reinforced Walls and Slopes. *Transp. Infrastruct. Geotechnol.* **2014**, *1*, 129–164. [CrossRef]
20. Reed, E.C.; Vandenberge, D.R. Comparison of FEA and analytical methods for determining stability of a RAP supported MSE wall. *DFI J. J. Deep Found. Inst.* **2018**, *12*, 122–129. [CrossRef]
21. Liu, S.; Su, Z.; Li, M.; Shao, L. Slope stability analysis using elastic finite element stress fields. *Eng. Geol.* **2020**, *273*, 105673. [CrossRef]
22. Hassen, G.; Donval, E.; De Buhan, P. Numerical stability analysis of reinforced soil structures using the multiphase model. *Comput. Geotech.* **2021**, *133*, 104035. [CrossRef]
23. Kazimierowicz-Frankowska, K.; Kulczykowski, M. Deformation of model reinforced soil structures: Comparison of theoretical and experimental results. *Geotext. Geomembr.* **2021**, *49*, 1176–1191. [CrossRef]
24. Mirmoradi, S.; Ehrlich, M.; Magalhães, L. Numerical evaluation of the effect of foundation on the behaviour of reinforced soil walls. *Geotext. Geomembr.* **2021**, *49*, 619–628. [CrossRef]
25. Xu, P.; Yang, G.; Li, T.; Hatami, K. Finite element limit analysis of bearing capacity of footing on back-to-back reinforced soil retaining walls. *Transp. Geotech.* **2021**, *30*, 100596. [CrossRef]
26. OptumCE. OptumG2 v. 2021. Available online: https://optumce.com/products/brochure-and-datasheet/ (accessed on 24 January 2021).

27. Sloan, S. Geotechnical stability analysis. *Géotechnique* **2013**, *63*, 531–571. [CrossRef]
28. Drucker, D.C.; Prager, W. Soil mechanics and plastic analysis or limit design. *Q. Appl. Math.* **1952**, *10*, 157–165. [CrossRef]
29. Schmüdderich, C.; Lavasan, A.A.; Tschuchnigg, F.; Wichtmann, T. Behavior of Nonidentical Differently Loaded Interfering Rough Footings. *J. Geotech. Geoenviron. Eng.* **2020**, *146*, 04020041. [CrossRef]
30. Lyamin, A.; Sloan, S.W. Upper bound limit analysis using linear finite elements and non-linear programming. *Int. J. Numer. Anal. Methods Géoméch.* **2002**, *26*, 181–216. [CrossRef]
31. Sloan, S.W. Lower bound limit analysis using finite elements and linear programming. *Int. J. Numer. Anal. Methods Géoméch.* **1988**, *12*, 61–77. [CrossRef]
32. Lyamin, A.V.; Sloan, S.W.; Krabbenhøft, K.; Hjiaj, M. Lower bound limit analysis with adaptive remeshing. *Int. J. Numer. Methods Eng.* **2005**, *63*, 1961–1974. [CrossRef]
33. Zornberg, J.G.; Sitar, N.; Mitchell, J.K. Limit Equilibrium as Basis for Design of Geosynthetic Reinforced Slopes. *J. Geotech. Geoenviron. Eng.* **1998**, *124*, 684–698. [CrossRef]
34. Vulova, C.; Leshchinsky, D. *Effects of Geosynthetic Reinforcement Spacing on the Performance of Mechanically Stabilized Earth Walls*; No. FFHWA-RD-03-048; The National Academies of Sciences, Engineering, and Medicine: Washington, DC, USA, 2003.
35. Berg, R.R.; Christopher, B.R.; Samtani, N.C. *Design of Mechanically Stabilized Earth Walls and Reinforced Soil Slopes–Volume I*; No. FHWA-NHI-10-024; United States, Department of Transportation, Federal Highway Administration: Washington, DC, USA, 2009.
36. Koerner, R.M.; Koerner, G.R. An extended data base and recommendations regarding 320 failed geosynthetic reinforced mechanically stabilized earth (MSE) walls. *Geotext. Geomembr.* **2018**, *46*, 904–912. [CrossRef]
37. Yu, Y.; Bathurst, R.J.; Allen, T.M. Numerical Modeling of the SR-18 Geogrid Reinforced Modular Block Retaining Walls. *J. Geotech. Geoenviron. Eng.* **2016**, *142*, 04016003. [CrossRef]

water

MDPI

Article

Risk Assessment of Dike Based on Risk Chain Model and Fuzzy Influence Diagram

Xiaobing Wang, Xiaozhou Xia, Renjie Teng, Xin Gu and Qing Zhang *

College of Mechanics and Materials, Hohai University, Nanjing 211100, China
* Correspondence: lxzhangqing@hhu.edu.cn

Abstract: For the risk assessment of flood defense, a comprehensive understanding of risk factors affecting dike failure is essential. Traditional risk assessment methods are mostly based on experts' experience and focus on just one type of failure mode of flood defensive structures. The risk resources, including the analytical factors and non-analytical factors, were summarized firstly according to the general experience of dikes. The uncertainty of the resources that affect dike safety can be quantified by membership degree. Hence, a fuzzy influence diagram based on fuzzy mathematics was proposed to assess the safety of the dikes. We evaluated the multi-failure modes at the same time by a fuzzy influence diagram. Taking a dike as an example, the expected value of the dike failure was 6.25%. Furthermore, the chance of damage to this dike was "very unlikely" according to the descriptive term of the Intergovernmental Panel on Climate Change (IPCC). The evaluation result was obtained as a probabilistic value, which enabled an intuitive perception of the safety of the dikes. Therefore, we provided some reasonable suggestions for project management and regular maintenance. Since the proposed method can account for uncertainties, it is well suited for the risk assessment of dikes with obvious uncertainties.

Keywords: dike failure; risk chain model; fuzzy influence diagram; risk assessment

Citation: Wang, X.; Xia, X.; Teng, R.; Gu, X.; Zhang, Q. Risk Assessment of Dike Based on Risk Chain Model and Fuzzy Influence Diagram. *Water* **2023**, *15*, 108. https://doi.org/10.3390/w15010108

Academic Editors: Rui Pang, Binghan Xue, Xiang Yu, Yantao Zhu and Xiaoying Lin

Received: 22 November 2022
Revised: 23 December 2022
Accepted: 26 December 2022
Published: 28 December 2022

1. Background

Dikes play an important role in the flood defense system of China. It was reported there were around 320,000 km of dikes to protect more than 640 million people and numerous cities [1]. Therefore, how to manage dikes using a reasonable method is an important task for reducing flood disaster. Because of some historical reasons, dikes in China were constructed in ancient times, and many dikes were made with local materials. As a result, the design standard and strength of the dikes located in remote areas of China are relatively low. Therefore, conducting the risk assessment of the dikes and discovering the hidden dangers in time are significant to protect the life and property of every citizen in the dike-protected areas.

In recent years, the traditional management of the dikes gradually changed to risk management. Developed countries, such as the UK, the Netherlands, and Japan, have taken the lead in introducing risk management methods into the risk assessment of dikes. For example, Mouri et al. [2] adopted the probability method to conduct the risk assessment of flooding considering natural and societal factors. Marijnissen et al. [3] used an extensive probabilistic method to re-evaluate the multifunctional dikes in the Netherlands considering the society and economy. Balistrocchi et al. [4] discussed the advantages of the two methods, i.e., numerical simulations and fragility curves, to study the failure probability analysis of levees. He proposed a model to calculate the failure probability of a mammal bioerosion levee considering the peak flow discharge and flood duration. Agbaje et al. [5] studied the influence of the effective friction angle on slope stability based on the Grag deposits in the East of England, UK. It pointed out that the total factor of safety was usually higher, leading to overestimation of the stability of a slope.

There are three main failure modes in China: overtopping, scouring, and seepage [6]. If the crest of the dike is below the run-up level, the waves will cause a flow of water over the top of the dike, across the crest, and onto the landside slope. This phenomenon is called overtopping. The pore pressure affects the stability of the levee body and/or foundation through infiltration and affects the back erosion piping or slope stability. This phenomenon is called seepage failure. Scouring failure is the damage caused by the water flow and waves, which usually leads to dangerous situations such as the inner slope collapse of the dike. Overtopping failure is considered a kind of hydrology failure, and the scouring and seepage failures are considered structural failures. However, most of the risk analyses focused on one of the above three failure mechanisms by only considering the analytical factors. For example, the risk analysis of overtopping was used to measure the overtopping failure [7–14]; the scouring analysis and slope stability analysis were conducted only for the dike body [15–19]; and most seepage analyses were also performed separately [20–22]. There were a few pieces of literature that assessed dike damage by considering failure modes with both analytical and non-analytical factors [23].

In another way, the traditional models, including the analytic hierarchy process, event tree method, and fault tree method, are mainly used to perform the system risk assessment. Song et al. [24] studied the risk assessment of tunnel construction based on the fuzzy analytic hierarchy process. Wang et al. [25] used a fuzzy analytic hierarchy process to identify the safety mode of the section of Jingnan dike that is located in Jingzhou City, China. Gu et al. [26] proposed an experts' weighting model for the safety evaluation of a dike using fuzzy mathematics and the dynamic clustering method. The analytic hierarchy process mostly relies on the subjective experts' experience and evaluates only using the appearance, which often leads to deviation in the evaluation results [27]. Fault tree and event tree analysis methods are widely used; however, the accurate probability value of basic events needs to be known in the calculation process [28].

The influence diagram method is a risk assessment method proposed by Howard [29]. The influence diagram is composed of two parts, the risk node and the risk relationship. It can represent the directed graph of the mutual influence between the risk elements and can provide relatively accurate calculation results for risk assessment. Moreover, it is also suitable for the risk assessment of system engineering. A fuzzy influence diagram is a combination of the fuzzy set, fuzzy transformation method, and influence diagram method to carry out probabilistic inferences on uncertain events. During the inference process, the occurrence probability of an event and the occurrence probability of other dependent events can be fully evaluated. Due to the advantages of intuition and easy understanding, the application scope of the fuzzy influence diagram is gradually expanding. Based on the 3D Hall structure model of system engineering, the risk identification method of overseas nuclear power projects was optimized by considering the idea of risk chain and risk map [30]. Xu et al. [31] used the fuzzy influence diagram to evaluate the construction safety risk of power transmission and transformation engineering construction projects. Quan et al. [32,33] conducted a qualitative risk analysis on the general contracting of overseas EPC power engineering projects using the fuzzy influence diagram method. Later, Zhang et al. [34] introduced the fuzzy influence diagram method into the risk assessment process of information security and introduced fuzzy theory to describe risk factors that are difficult to quantify and to obtain good results. Lin et al. [35] applied the fuzzy influence diagram method to the safety management of underground pipelines and focused on the analysis of the risks caused by external force damage. To the best knowledge of the authors, this is the first time using the risk chain model and fuzzy influence diagram method to evaluate dike safety. Theoretically, the fuzzy influence diagram method is based on fuzzy mathematics. The introduction of state fuzzy sets, frequency fuzzy sets, and membership degrees can fully consider the uncertainty of the risk factors and at the same time reduce the subjectivity of experts' judgment. It is suitable to perform the risk assessment for the dikes with uncertainty and complexity.

This present study aimed to (1) summarize the risk resources that affect dike safety based on the general experience of the dikes; (2) analyze the relationships between the risk resources and risk events; (3) construct a fuzzy influence diagram based on (1) and (2); and (4) take a fictitious dike as an example to illustrate the proposed method and apply the description term of IPCC to the result.

2. The Risk Chain Model and the Fuzzy Influence Diagram

2.1. The Risk Chain Model

A fuzzy influence diagram is a graph composed of multiple risk chains, and each risk chain can be regarded as an independent logical relationship. A risk chain is established by adding nodes such as risk sources, risk factors, and risk events in layers. Within an assessment system, different risk chains should ultimately point to the same risk consequence. The risk chain structure is shown in Figure 1.

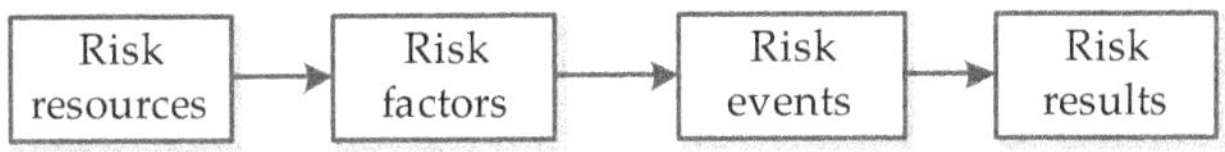

Figure 1. The risk chain structure.

2.2. Fuzzy Set and Membership Function

2.2.1. The Definition of Fuzzy Set and Membership

The introduction of fuzzy sets in this paper is limited to the case where the universe of discourse is a finite set, denoted by $\boldsymbol{A} = \{a_1, a_2, \cdots, a_n\}$. Mapping on the domain $\boldsymbol{A}$, $\mu_{\boldsymbol{X}}(\cdot) : \boldsymbol{A} \rightarrow [0, 1]$, then $a \rightarrow \mu_{\boldsymbol{X}}(a)$, the fuzzy set $\boldsymbol{X}$ was determined. $\mu_{\boldsymbol{X}}(\cdot)$ is the membership function of $\boldsymbol{X}$, and $\mu_{\boldsymbol{X}}(a)$ is the membership degree of a to the fuzzy set $\boldsymbol{X}$, denoted as:

$$\boldsymbol{X} = \{(a, \mu_{\boldsymbol{X}}(a)) | a \in \boldsymbol{A}\} \tag{1}$$

The algorithm of the fuzzy influence graph is based on fuzzy set operation and uses a directed graph to describe the state, frequency, and relationship between nodes, and then it carries out a risk assessment for the system.

2.2.2. The Operations on Fuzzy Sets

We briefly describe the three operation notations of the fuzzy set used in this paper. The symbol "$\times$" is a Cartesian product. If $\boldsymbol{X}$ and $\boldsymbol{Y}$ are fuzzy sets of two universes $\boldsymbol{A}$ and $\boldsymbol{B}$, respectively, then the fuzzy relation $\boldsymbol{R}$ from $\boldsymbol{X}$ to $\boldsymbol{Y}$ is:

$$\mu_{\boldsymbol{R}} = \mu_{\boldsymbol{X}\times\boldsymbol{Y}}(a_i, b_j) = \min[\mu_{\boldsymbol{X}}(a_i), \mu_{\boldsymbol{Y}}(b_j)] \tag{2}$$

A fuzzy relation is a special kind of fuzzy set, and the result is a fuzzy matrix.

The symbol "$\circ$" is called fuzzy synthesis, that is, dot product. If $\boldsymbol{R}$ is a fuzzy relation from $\boldsymbol{X}$ to $\boldsymbol{Y}$, similarly, if $\boldsymbol{S}$ is a fuzzy relation from $\boldsymbol{Y}$ to $\boldsymbol{Z}$, then the fuzzy composition of $\boldsymbol{R}$ and $\boldsymbol{S}$ is:

$$\mu_{\boldsymbol{R}\circ\boldsymbol{S}}(a_i, b_j) = \max\{\min[\mu_{\boldsymbol{R}}(a_i, b_j), \mu_{\boldsymbol{S}}(a_i, b_j)]\} \tag{3}$$

The result of fuzzy synthesis is still a fuzzy matrix, representing the fuzzy relationship from $\boldsymbol{X}$ to $\boldsymbol{Z}$.

The symbol "$\cup$" is called the "union" of fuzzy sets and has the following operations:

$$\mu_{\boldsymbol{R}\cup\boldsymbol{S}}(a_i, b_j) = \max[\mu_{\boldsymbol{X}}(a_i, b_j), \mu_{\boldsymbol{Y}}(a_i, b_j)] \tag{4}$$

It should be pointed out here that "Cartesian product" and "dot product" should operate on fuzzy sets of different universes, and "union" should operate on fuzzy sets in the same universe.

2.3. The Process of the Fuzzy Influence Diagram

The fuzzy influence diagram introduces the state fuzzy set and the frequency fuzzy set in the numerical layer to describe the data structure of the node and uses the fuzzy relationship to describe the relationship between the node variables in the function layer. The calculation process of the fuzzy influence diagram is explained below [36].

2.3.1. The Calculation of Independent Node Frequency Matrix

If $\boldsymbol{X}$ has no immediate predecessor, it is called an independent node, and it is assumed that the possible state vector of the independent node $\boldsymbol{X}$ is:

$$\boldsymbol{P_X} = \{P_{X_1}, P_{X_2}, \cdots, P_{X_n}\}^T \quad (5)$$

where $P_{X_1}, P_{X_2}, \cdots, P_{X_n}$ is the state fuzzy set defined by the risk assessor according to the actual situation. The frequency vector of an independent node $\boldsymbol{X}$ is:

$$\boldsymbol{f_X} = \left\{f_{X_1}, f_{X_2}, \cdots, f_{X_n}\right\}^T \quad (6)$$

where $f_{X_1}, f_{X_2}, \cdots, f_{X_n}$ is the state fuzzy set defined by the risk assessor according to the actual situation. The frequency vector of an independent node $\boldsymbol{X}$ is:

$$F_X = (f_{X_1} \times P_{X1}) \cup (f_{X_2} \times P_{X_2}) \cup \cdots \cup (f_{X_n} \times P_{X_n}) \quad (7)$$

2.3.2. The Calculation of the Frequency Matrix of Dependent Nodes

If $\boldsymbol{X}$ is made up of m random nodes $Y_1, Y_2, \cdots, Y_m$ as its immediate predecessors, then $\boldsymbol{X}$ is called a dependent node.

$$\boldsymbol{F_{XP}} = F_{Y_1} \cup F_{Y_2} \cup \cdots \cup F_{Y_m} \quad (8)$$

Define $\boldsymbol{R}_{XY_1}$ as the fuzzy relationship from node $\boldsymbol{Y}_1$ to dependent node $\boldsymbol{X}$:

$$\boldsymbol{R}_{XY_1} = \left(P_{Y_{11}} \times P_{X_i}\right) \cup \left(P_{Y_{12}} \times P_{X_i}\right) \cup \cdots \cup \left(P_{Y_{1n}} \times P_{X_i}\right) \quad (9)$$

where $P_{Y_{11}}, P_{Y_{12}}, \cdots, P_{Y_{1n}} \in P_{Y_1}$, $P_{X_i} \in \{P_{X_1}, P_{X_2}, \cdots, P_{X_n}\} = P_X$. The fuzzy relationship from node $\boldsymbol{Y}_m$ to node $\boldsymbol{X}$ is:

$$\boldsymbol{R}_{XY_m} = \left(P_{Y_{m1}} \times P_{X_i}\right) \cup \left(P_{Y_{m2}} \times P_{X_i}\right) \cup \cdots \cup \left(P_{Y_{mn}} \times P_{X_i}\right) \quad (10)$$

Then, the joint $\boldsymbol{R}_{XP}$ of the fuzzy relationship of all immediate predecessors of the dependent node $\boldsymbol{X}$ is:

$$\boldsymbol{R}_{XP} = \boldsymbol{R}_{XY_1} \cup \boldsymbol{R}_{XY_2} \cup \cdots \cup \boldsymbol{R}_{XY_m} \quad (11)$$

The frequency matrix of the dependent node $\boldsymbol{X}$ can be obtained by the dot product relationship:

$$\boldsymbol{F}_X = \boldsymbol{F}_{XP} \circ \boldsymbol{R}_{XP} \quad (12)$$

2.3.3. Result Analysis

The membership degree of the random result is selected from the frequency matrix $\boldsymbol{F}_X$ of the risk consequence node by Equation (13), and the probability value of the result is calculated by Equation (14):

$$\max\left(f_{X_i} \cdot \sum_{\Omega X} \mu_{X_i}\right) \quad (13)$$

$$P(X_i) = \frac{\mu_{X_i}}{\sum\limits_{\Omega X} \mu_{X_i}} \quad (14)$$

According to the probability description proposed by the Intergovernmental Panel on Climate Change (IPCC), this paper used seven levels for the result to give a corresponding language description of the possibility of the risk result [27,37], as shown in Table 1.

Table 1. The probabilistic result of IPCC.

Probability Range	Descriptive Term
<1%	Extremely unlikely
1~10%	Very unlikely
10~33%	Unlikely
33~66%	Medium likelihood
66~90%	Likely
90~99%	Very likely
>99%	Virtually certain

3. The Evaluation Process of the Fuzzy Influence Diagram

3.1. The Construction of the Frequency Fuzzy Set

For the convenience of calculation, combined with the experts' experience on the dike, the frequency vector of the risk node of the dike was uniformly simplified to $f = \{High, Medium, Low\}^T$, as shown in Table 2. The frequency was processed as follows, and 0 to 1 were decomposed into (0.0, 0.1, 0.2, 0.3, 0.4, 0.5, 0.6, 0.7, 0.8, 0.9, 1.0), indicating the occurrence frequency of risk nodes. From the definition of fuzzy sets and membership degrees, the three fuzzy sets and their membership degrees of the frequency vector f of the dike risk node can be expressed as follows:

$$High = \left\{\frac{0.7}{0.5}, \frac{0.8}{0.7}, \frac{0.9}{0.9}, \frac{1.0}{1.0}\right\} Medium = \left\{\frac{0.3}{0.2}, \frac{0.4}{0.8}, \frac{0.5}{1.0}, \frac{0.6}{0.8}, \frac{0.7}{0.2}\right\} Low = \left\{\frac{0.0}{1.0}, \frac{0.1}{0.9}, \frac{0.2}{0.7}, \frac{0.3}{0.5}\right\}$$

where the number above the symbol "—" is the frequency of node occurrence, and the number below represents the membership degree of the frequency. The number 0.7/0.5 means that the risk node is in the High Set, and the membership degree is 0.5 when the frequency of occurrence is 0.7. The number 0.7/0.2 means that the risk node is in the Medium Set, and the membership degree is 0.2 when the frequency is 0.7.

Table 2. The state and frequency of risk nodes.

Nodes	The Potential Risk State	The Frequency
Flooding	big middle small	low medium high
Illegal operation	big middle small	low medium high
Low skills	big middle small	low medium high
Reduced cohesion	big middle small	low high medium
Increase of water content	big middle small	medium low high

Table 2. *Cont.*

Nodes	The Potential Risk State	The Frequency
Void ratio change	big middle small	low medium high
Friction angle change	big middle small	low high medium
······	······	······

3.2. The Construction of the State Fuzzy Set

The state vector of the risk node in the fuzzy influence diagram is uniformly simplified as $\boldsymbol{P} = \{Increase\ max(IMAX),\ Increase\ medium(IMID),\ Increase\ \min(IMIN)\}^{\mathrm{T}}$. The language description of the fuzzy set in the state vector has the same meaning as the description of the relationship between the nodes in Table 3. The state fuzzify sets and their membership functions are expressed as follows:

$$IMAX = \left\{\frac{8\%}{0.5}, \frac{10\%}{0.8}, \frac{12\%}{1.0}\right\}\ IMID = \left\{\frac{4\%}{0.5}, \frac{6\%}{1.0}, \frac{8\%}{0.5}\right\}\ IMIN = \left\{\frac{0\%}{1.0}, \frac{2\%}{0.5}, \frac{4\%}{0.2}\right\}$$

It is more appropriate to describe the states of some parts of nodes in the fuzzy influence diagram with certainty. Hence, it is necessary to artificially fuzzify them. In this paper, the states of all risk source nodes were artificially fuzzified. The membership functions are as follows:

$$Big = \left\{\frac{big}{1.0}, \frac{middle}{0.0}, \frac{small}{0.0}\right\}\ Middle = \left\{\frac{big}{0.0}, \frac{middle}{1.0}, \frac{small}{0.0}\right\}\ Small = \left\{\frac{big}{0.0}, \frac{middle}{0.0}, \frac{small}{1.0}\right\}$$

Table 3. The fuzzy relationship of different risk nodes.

Relationship between Different Nodes	Relationship Description	
	The Degree of Change in Disadvantage for the Risk Nodes	**The Results Corresponding to the Degree of Change of the Risk Nodes**
Flooding → higher water level	big, middle, and small, respectively	Increase max, increase medium, increase min, respectively
Illegal operation → reduced drainage capacity	big, middle, and small, respectively	Increase max, increase medium, increase min, respectively
Low skills → reduced drainage capacity	big, middle, and small, respectively	Increase max, increase medium, increase min, respectively
Reduced cohesion → dike body damage	big, middle, and small, respectively	Increase max, increase medium, increase min, respectively
Water content → dike body damage	big, middle, and small, respectively	Increase max, increase medium, increase min, respectively
Reduced cohesion → dike foundation damage	big, middle, and small, respectively	Increase max, increase medium, increase min, respectively
water content → dike foundation damage	big, middle, and small, respectively	Increase max, increase medium, increase min, respectively
Void ratio → dike body damage	big, middle, and small, respectively	Increase max, increase medium, increase min, respectively
Friction angel → dike foundation damage	big, middle, and small, respectively	Increase max, increase medium, increase min, respectively
Void ratio → dike foundation damage	big, middle, and small, respectively	Increase max, increase medium, increase min, respectively
Friction angle → dike foundation damage	big, middle, and small, respectively	Increase max, increase medium, increase min, respectively
Higher water level → overtopping	Increase max, medium, min, respectively	Increase max, increase medium, increase min, respectively
Reduced drainage → overtopping	Increase max, medium, min, respectively	Increase max, increase medium, increase min, respectively

Table 3. *Cont.*

Relationship between Different Nodes	Relationship Description	
	The Degree of Change in Disadvantage for the Risk Nodes	**The Results Corresponding to the Degree of Change of the Risk Nodes**
Damage of dike body → seepage	Increase max, medium, min, respectively	Increase max, increase medium, increase min, respectively
Damage of dike foundation → seepage	Increase max, medium, min, respectively	Increase max, increase medium, increase min, respectively
Damage of dike body → scouring	Increase max, medium, min, respectively	Increase max, increase medium, increase min, respectively
Damage of dike foundation → scouring	Increase max, medium, min, respectively	Increase max, increase medium, increase min, respectively
Overtopping → dike failure	Increase max, medium, min, respectively	Increase max, increase medium, increase min, respectively
Seepage → dike failure	Increase max, medium, min, respectively	Increase max, increase medium, increase min, respectively
Scouring → dike failure	Increase max, medium, min, respectively	Increase max, increase medium, increase min, respectively

Take the "Flooding → higher water level" and "Higher water level → overtopping" as examples to illustrate Table 3. When the flooding is big, middle, and small, it can cause an increase in water level to maximum, medium, and minimum, respectively. When the increase of higher water level is maximum, medium, and minimum, it can cause a degree of overtopping to maximum, medium, and minimum, respectively.

3.3. Risk Assessment of One Dike Case

This paper sorted out the analytical factors and non-analytical factors that lead to dike failure and selected some factors that have an impact on the safety status of the dike as the risk source for analysis. The emergence of risk sources causes the occurrence of risk factors, which in turn leads to the occurrence of three common failure modes of dikes. The three failure modes comprehensively lead to dike failure (risk consequences). The constructed fuzzy influence diagram of dike failure is shown in Figure 2. According to the frequency fuzzy set, state fuzzy set, and fuzzy relationship constructed above, the calculation process of the risk chain "illegal operation-reduced drainage—overtopping—dike failure" was taken as an example, and the remaining evaluation process is similar to it.

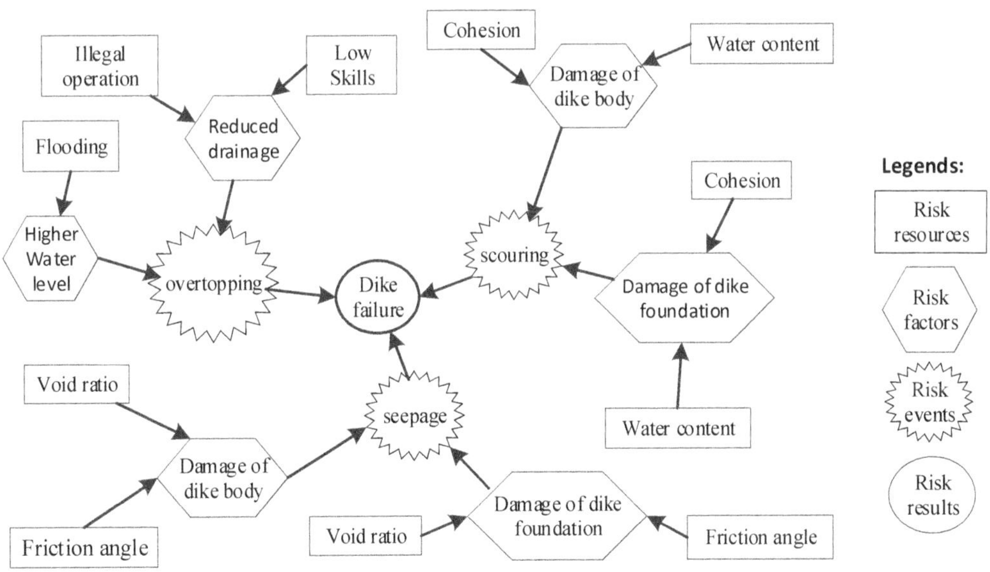

Figure 2. The fuzzy diagram.

Step 1, The frequency matrix $F_{illegal\_operation}$ of illegal operation risk source nodes is calculated by Equations (5)–(7):

$$F_{illegal\_operation} = \begin{bmatrix} & big & middle & small \\ 0.0 & 1.0 & 0.0 & 0.0 \\ 0.1 & 0.9 & 0.0 & 0.0 \\ 0.2 & 0.7 & 0.0 & 0.0 \\ 0.3 & 0.5 & 0.2 & 0.0 \\ 0.4 & 0.0 & 0.8 & 0.0 \\ 0.5 & 0.0 & 1.0 & 0.0 \\ 0.6 & 0.0 & 0.8 & 0.0 \\ 0.7 & 0.0 & 0.2 & 0.5 \\ 0.8 & 0.0 & 0.0 & 0.7 \\ 0.9 & 0.0 & 0.0 & 0.9 \\ 1.0 & 0.0 & 0.0 & 1.0 \end{bmatrix} \quad (15)$$

In the same way, the frequency matrix of low-risk source nodes of security skills can be obtained.

Step 2, The joint frequency matrix F_{union} of all immediate predecessor nodes of the node with reduced drainage capacity is calculated by Equation (8):

$$F_{union} = \begin{bmatrix} & big & middle & small \\ 0.0 & 1.0 & 0.0 & 0.0 \\ 0.1 & 0.9 & 0.0 & 0.0 \\ 0.2 & 0.7 & 0.0 & 0.0 \\ 0.3 & 0.5 & 0.2 & 0.0 \\ 0.4 & 0.0 & 0.8 & 0.0 \\ 0.5 & 0.0 & 1.0 & 0.0 \\ 0.6 & 0.0 & 0.8 & 0.0 \\ 0.7 & 0.0 & 0.2 & 0.5 \\ 0.8 & 0.0 & 0.0 & 0.7 \\ 0.9 & 0.0 & 0.0 & 0.9 \\ 1.0 & 0.0 & 0.0 & 1.0 \end{bmatrix} \quad (16)$$

Step 3, The fuzzy relationship matrix from the node of illegal operation risk source to the node of drainage capacity reduction risk factor is calculated by Equation (10):

$$R_{illegal\_operation \to reduced\_drainage} = \begin{bmatrix} & 0\% & 2\% & 4\% & 6\% & 8\% & 10\% & 12\% \\ big & 0.0 & 0.0 & 0.0 & 0.0 & 0.5 & 0.8 & 1.0 \\ middle & 0.0 & 0.0 & 0.5 & 1.0 & 0.5 & 0.0 & 0.0 \\ small & 1.0 & 0.5 & 0.2 & 0.0 & 0.0 & 0.0 & 0.0 \end{bmatrix} \quad (17)$$

In the same way, the fuzzy relation matrix of the lower drainage capacity node caused by the low skills node can be obtained.

Step 4, The joint matrix of the fuzzy relationship of all immediately preceding nodes of the node that leads to the reduction of drainage capacity is calculated by Equation (11):

$$\begin{aligned} R_{union} &= R_{illegal\_operation \to reduced\_drainage} \cup R_{low\_skills \to reduced\_drainage} \\ &= \begin{bmatrix} & 0\% & 2\% & 4\% & 6\% & 8\% & 10\% & 12\% \\ big & 0.0 & 0.0 & 0.0 & 0.0 & 0.5 & 0.8 & 1.0 \\ middle & 0.0 & 0.0 & 0.5 & 1.0 & 0.5 & 0.0 & 0.0 \\ small & 1.0 & 0.5 & 0.2 & 0.0 & 0.0 & 0.0 & 0.0 \end{bmatrix} \end{aligned} \quad (18)$$

Similarly, the joint matrix of other fuzzy relations can be obtained.

Step 5, The frequency matrix of the nodes with reduced drained capacity is calculated by Equation (12):

$$F_{reduced\_drainage} = F_{union} \circ R_{union} = \begin{bmatrix} & 0\% & 2\% & 4\% & 6\% & 8\% & 10\% & 12\% \\ 0.0 & 0.0 & 0.0 & 0.0 & 0.0 & 0.5 & 0.8 & 1.0 \\ 0.1 & 0.0 & 0.0 & 0.0 & 0.0 & 0.5 & 0.8 & 0.9 \\ 0.2 & 0.0 & 0.0 & 0.0 & 0.0 & 0.5 & 0.7 & 0.7 \\ 0.3 & 0.0 & 0.0 & 0.2 & 0.2 & 0.5 & 0.5 & 0.0 \\ 0.4 & 0.0 & 0.0 & 0.5 & 0.8 & 0.5 & 0.0 & 0.0 \\ 0.5 & 0.0 & 0.0 & 0.5 & 1.0 & 0.5 & 0.0 & 0.0 \\ 0.6 & 0.0 & 0.0 & 0.5 & 0.8 & 0.5 & 0.0 & 0.0 \\ 0.7 & 0.5 & 0.5 & 0.2 & 0.8 & 0.2 & 0.0 & 0.0 \\ 0.8 & 0.7 & 0.5 & 0.2 & 0.0 & 0.0 & 0.0 & 0.0 \\ 0.9 & 0.9 & 0.5 & 0.2 & 0.0 & 0.0 & 0.0 & 0.0 \\ 1.0 & 1.0 & 0.5 & 0.2 & 0.0 & 0.0 & 0.0 & 0.0 \end{bmatrix} \quad (19)$$

In the same way, the frequency matrix of the node of rising water level in front of the dikes can be obtained. Thus far, the frequency matrix, the frequency joint matrix, the fuzzy relationship matrix between the nodes, and the fuzzy relationship joint matrix of any node can be obtained by analogy with the calculation process above.

Corresponding to the fuzzy relationship between the nodes in Table 3, the fuzzy influence diagram is completely calculated by analogy to the above calculation process. Only the calculation results of the key nodes in the subsequent calculation process of the fuzzy influence diagram are given here.

The relationship matrix from the drainage capacity reduction node to the overtopping risk event node is as follows:

$$R_{reduced\_drainage \to overtopping} = \begin{bmatrix} & 0\% & 2\% & 4\% & 6\% & 8\% & 10\% & 12\% \\ 0\% & 1.0 & 0.5 & 0.0 & 0.0 & 0.0 & 0.0 & 0.0 \\ 2\% & 0.5 & 0.5 & 0.2 & 0.0 & 0.0 & 0.0 & 0.0 \\ 4\% & 0.2 & 0.2 & 0.5 & 0.5 & 0.5 & 0.0 & 0.0 \\ 6\% & 0.0 & 0.0 & 0.5 & 1.0 & 0.5 & 0.0 & 0.0 \\ 8\% & 0.0 & 0.0 & 0.5 & 0.5 & 0.5 & 0.5 & 0.5 \\ 10\% & 0.0 & 0.0 & 0.0 & 0.0 & 0.5 & 0.8 & 0.8 \\ 12\% & 0.0 & 0.0 & 0.0 & 0.0 & 0.5 & 0.8 & 1.0 \end{bmatrix} \quad (20)$$

The union of the fuzzy relationship matrices of all the immediate nodes of the overtopping risk event node is:

$$\begin{aligned} R_{union} &= R_{higher\_water \to overtopping} \cup R_{reduced\_drainage \to overtopping} \\ &= \begin{bmatrix} & 0\% & 2\% & 4\% & 6\% & 8\% & 10\% & 12\% \\ 0\% & 1.0 & 0.5 & 0.0 & 0.0 & 0.0 & 0.0 & 0.0 \\ 2\% & 0.5 & 0.5 & 0.2 & 0.0 & 0.0 & 0.0 & 0.0 \\ 4\% & 0.2 & 0.2 & 0.5 & 0.5 & 0.5 & 0.0 & 0.0 \\ 6\% & 0.0 & 0.0 & 0.5 & 1.0 & 0.5 & 0.0 & 0.0 \\ 8\% & 0.0 & 0.0 & 0.5 & 0.5 & 0.5 & 0.5 & 0.5 \\ 10\% & 0.0 & 0.0 & 0.0 & 0.0 & 0.5 & 0.8 & 0.8 \\ 12\% & 0.0 & 0.0 & 0.0 & 0.0 & 0.5 & 0.8 & 1.0 \end{bmatrix} \end{aligned} \quad (21)$$

The frequency matrix of overtopping risk event nodes can be obtained as:

$$F_{overtopping} = F_{union} \circ R_{union} = \begin{bmatrix} & 0\% & 2\% & 4\% & 6\% & 8\% & 10\% & 12\% \\ 0.0 & 0.0 & 0.0 & 0.5 & 0.5 & 0.5 & 0.8 & 1.0 \\ 0.1 & 0.0 & 0.0 & 0.5 & 0.5 & 0.5 & 0.8 & 0.9 \\ 0.2 & 0.0 & 0.0 & 0.5 & 0.5 & 0.5 & 0.7 & 0.7 \\ 0.3 & 0.2 & 0.2 & 0.5 & 0.5 & 0.5 & 0.5 & 0.5 \\ 0.4 & 0.2 & 0.2 & 0.5 & 0.8 & 0.5 & 0.5 & 0.5 \\ 0.5 & 0.2 & 0.2 & 0.5 & 1.0 & 0.5 & 0.5 & 0.5 \\ 0.6 & 0.2 & 0.2 & 0.5 & 0.8 & 0.5 & 0.5 & 0.5 \\ 0.7 & 0.5 & 0.5 & 0.5 & 0.8 & 0.5 & 0.2 & 0.2 \\ 0.8 & 0.7 & 0.5 & 0.2 & 0.2 & 0.2 & 0.0 & 0.0 \\ 0.9 & 0.9 & 0.5 & 0.2 & 0.2 & 0.2 & 0.0 & 0.0 \\ 1.0 & 1.0 & 0.5 & 0.2 & 0.2 & 0.5 & 0.0 & 0.0 \end{bmatrix} \tag{22}$$

The joint frequency matrix of the three risk event nodes of overtopping, scouring, and seepage is:

$$F_{union} = F_{overtopping} \cup F_{seepage} \cup F_{scouring} = \begin{bmatrix} & 0\% & 2\% & 4\% & 6\% & 8\% & 10\% & 12\% \\ 0.0 & 1.0 & 0.5 & 0.5 & 0.5 & 0.5 & 0.8 & 1.0 \\ 0.1 & 0.9 & 0.9 & 0.5 & 0.5 & 0.5 & 0.8 & 0.9 \\ 0.2 & 0.7 & 0.7 & 0.5 & 0.5 & 0.5 & 0.8 & 0.9 \\ 0.3 & 0.5 & 0.5 & 0.5 & 0.5 & 0.5 & 0.5 & 0.5 \\ 0.4 & 0.2 & 0.2 & 0.5 & 0.8 & 0.5 & 0.5 & 0.5 \\ 0.5 & 0.2 & 0.2 & 0.5 & 1.0 & 0.5 & 0.5 & 0.5 \\ 0.6 & 0.2 & 0.2 & 0.5 & 0.8 & 0.5 & 0.5 & 0.5 \\ 0.7 & 0.5 & 0.5 & 0.5 & 0.8 & 0.5 & 0.5 & 0.5 \\ 0.8 & 0.7 & 0.7 & 0.5 & 0.5 & 0.5 & 0.7 & 0.7 \\ 0.9 & 0.9 & 0.9 & 0.5 & 0.5 & 0.5 & 0.8 & 0.9 \\ 1.0 & 1.0 & 1.0 & 0.5 & 0.5 & 0.5 & 0.8 & 1.0 \end{bmatrix} \tag{23}$$

The fuzzy relationship matrix of the overtopping failure mode to the dike failure node is:

$$R_{overtopping \to dike\_failure} = \begin{bmatrix} & 0\% & 2\% & 4\% & 6\% & 8\% & 10\% & 12\% \\ 0\% & 1.0 & 0.5 & 0.0 & 0.0 & 0.0 & 0.0 & 0.0 \\ 2\% & 0.5 & 0.5 & 0.2 & 0.0 & 0.0 & 0.0 & 0.0 \\ 4\% & 0.2 & 0.2 & 0.5 & 0.5 & 0.5 & 0.0 & 0.0 \\ 6\% & 0.0 & 0.0 & 0.5 & 1.0 & 0.5 & 0.0 & 0.0 \\ 8\% & 0.0 & 0.0 & 0.5 & 0.5 & 0.5 & 0.5 & 0.5 \\ 10\% & 0.0 & 0.0 & 0.0 & 0.0 & 0.5 & 0.8 & 0.8 \\ 12\% & 0.0 & 0.0 & 0.0 & 0.0 & 0.5 & 0.8 & 1.0 \end{bmatrix} \tag{24}$$

$$R_{union} = \begin{bmatrix} & 0\% & 2\% & 4\% & 6\% & 8\% & 10\% & 12\% \\ 0\% & 1.0 & 0.5 & 0.0 & 0.0 & 0.0 & 0.0 & 0.0 \\ 2\% & 0.5 & 0.5 & 0.2 & 0.0 & 0.0 & 0.0 & 0.0 \\ 4\% & 0.2 & 0.2 & 0.5 & 0.5 & 0.5 & 0.0 & 0.0 \\ 6\% & 0.0 & 0.0 & 0.5 & 1.0 & 0.5 & 0.0 & 0.0 \\ 8\% & 0.0 & 0.0 & 0.5 & 0.5 & 0.5 & 0.5 & 0.5 \\ 10\% & 0.0 & 0.0 & 0.0 & 0.0 & 0.5 & 0.8 & 0.8 \\ 12\% & 0.0 & 0.0 & 0.0 & 0.0 & 0.5 & 0.8 & 1.0 \end{bmatrix} \tag{25}$$

The frequency matrix of risk consequence of dike failure nodes is:

$$F_{dike\_failure} = F_{union} \circ R_{union} = \begin{bmatrix} & 0\% & 2\% & 4\% & 6\% & 8\% & 10\% & 12\% \\ 0.0 & 1.0 & 0.5 & 0.5 & 0.5 & 0.5 & 0.8 & 1.0 \\ 0.1 & 0.9 & 0.5 & 0.5 & 0.5 & 0.5 & 0.8 & 0.9 \\ 0.2 & 0.7 & 0.5 & 0.5 & 0.5 & 0.5 & 0.8 & 0.9 \\ 0.3 & 0.5 & 0.5 & 0.5 & 0.5 & 0.5 & 0.5 & 0.5 \\ 0.4 & 0.2 & 0.2 & 0.5 & 0.8 & 0.5 & 0.5 & 0.5 \\ 0.5 & 0.2 & 0.2 & 0.5 & 1.0 & 0.5 & 0.5 & 0.5 \\ 0.6 & 0.2 & 0.2 & 0.5 & 0.8 & 0.5 & 0.5 & 0.5 \\ 0.7 & 0.5 & 0.5 & 0.5 & 0.8 & 0.5 & 0.5 & 0.5 \\ 0.8 & 0.7 & 0.5 & 0.5 & 0.5 & 0.5 & 0.7 & 0.7 \\ 0.9 & 0.9 & 0.5 & 0.5 & 0.5 & 0.5 & 0.8 & 0.9 \\ 1.0 & 1.0 & 0.5 & 0.5 & 0.5 & 0.5 & 0.8 & 1.0 \end{bmatrix} \tag{26}$$

By Equation (13), the row with the largest product value of the product of the frequency is selected, and the sum of the membership degrees of the row is calculated. In this example, it is the row corresponding to the frequency equal to 1.

$$\left\{ \frac{0\%}{1.0}, \frac{2\%}{0.5}, \frac{4\%}{0.5}, \frac{6\%}{0.5}, \frac{8\%}{0.5}, \frac{10\%}{0.8}, \frac{12\%}{1.0} \right\} \tag{27}$$

The change probability of the dike failure node calculated by Equation (14) is:

$$\begin{aligned} P(0\%, 2\%, 4\%, 6\%, 8\%, 10\%, 12\%) &= (1.0, 0.5, 0.5, 0.5, 0.5, 0.8, 1.0)/4.8 \\ &= (0.2083, 0.1042, 0.1042, 0.1042, 0.1042, 0.1667, 0.2083) \end{aligned} \tag{28}$$

The expected value of the increased risk can be calculated:

$$\begin{aligned} E_{failure} &= \frac{0\% \times 1 + (2\% + 4\% + 6\% + 8\%) \times 0.5 + 10\% \times 0.8 + 12\% \times 1}{4.8} \\ &= 6.25\% \end{aligned} \tag{29}$$

Referring to the descriptive term of the IPCC probability results (Table 1), it was concluded that the chance of this example dike was "very likely". Figures 3 and 4 show the probability distribution and cumulative distribution of the increased risk, respectively. The top three of the increased risk were 0%, 12%, and 10%, and the cumulative probability from 10% to 12% was 37.5%. The occurrence probability was relatively high, which was different from the result "very likely". Actually, the expected value of 6.25% of structure failure cannot be accepted for flood defense. The two conclusions from cumulative cure and the expected value had the same meanings, indicating that a risk exists. Hence, the probabilistic result of IPCC was not suitable for the dikes perfectly. Developing a similarly descriptive term for dikes is an urgent task for hydraulic experts.

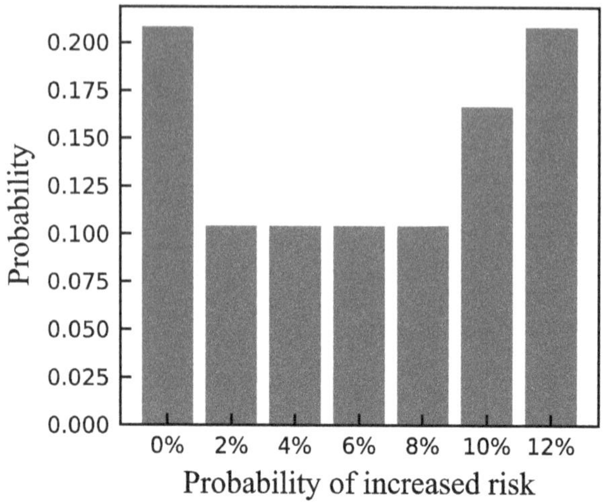

Figure 3. The distribution of the risk results.

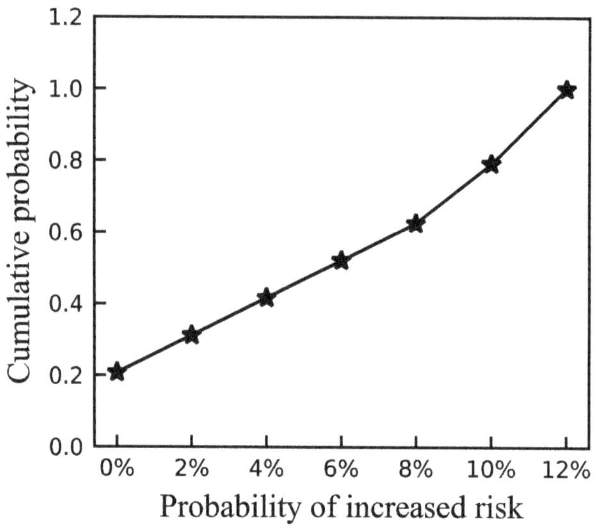

Figure 4. The cumulative probability of risk results.

4. Summary and Conclusions

Based on the influence diagram, this paper proposed a risk assessment method for dikes with a risk chain model. We summarized the common risk resources that affect the safety of dikes based on general experiences. The relationships between the different risk nodes were also deduced. The whole process of risk assessment was displayed by taking an example of a hypothesis dike. The probability of dike damage was 6.25%, illustrating that the chance of dike damage is "very likely" according to the descriptive term proposed by the Intergovernmental Panel on Climate Change (IPCC). The advantages of the proposed method are summarized as follows.

(1) All the analytical risk factors and non-analytical risk factors can theoretically be integrated into the influence diagram. The risk resources in Table 2 are only the common ones. The other risk resources can be further added as a supplement when this method is used to evaluate different dikes.

(2) This method is based on fuzzy mathematics and uses the membership degree function to describe the uncertainty of the analytical risk factors and non-analytical factors, which is more prone to reality.

(3) The different failure modes can be considered during the process of risk assessment at the same time.

(4) The risk assessment is obtained as a probabilistic result, which enables an intuitive perception of dike safety. We can provide reasonable suggestions for the engineering management and regular maintenance of dikes.

Expected for the risk assessment of dikes, the risk chain model and fuzzy influence diagram method can be applied to other hydraulic structures and geotechnical engineering. To obtain more accurate assessment results in engineering applications, more comprehensive risk sources should be selected, and field surveys or/and tests should be conducted to obtain real engineering data for calculation.

Author Contributions: Conceptualization, Q.Z.; formal analysis, X.W. and R.T.; funding acquisition, Q.Z.; methodology, X.W. and X.X.; writing original-draft, X.W.; writing review and editing, X.G. All authors have read and agreed to the published version of the manuscript.

Funding: This research was funded by the National Natural Science Foundation of China (No. 12002118); the Fundamental Research Funds for the Central Universities (No. 2019B65814); Postgraduate Research & Practice Innovation Program of Jiangsu Province (No. SJKY19_0422); the National Key Research and Development Program of China (No. 2017YFC1502603).

Institutional Review Board Statement: This study did not require ethical approval.

Informed Consent Statement: Informed consent was obtained from all subjects involved in the study.

Data Availability Statement: No new data were created or analyzed in this study.

Acknowledgments: The author Xiaobing Wang also gratefully acknowledges the support from China Scholarship Council (No. 202006710049).

Conflicts of Interest: The authors declare no conflict of interests.

References

1. Ministry of Water Resources, People's Republic of China. *2019 Statistic Bulletin on China Water Activities*; Water and Power Press: Beijing, China, 2019.
2. Mouri, G.; Minoshima, D.; Golosov, V.; Chalov, S.; Seto, S.; Yoshimura, K.; Nakamura, S.; Oki, T. Probability assessment of flood and sediment disasters in Japan using the Total Runoff-Integrating Pathways model. *Int. J. Disaster Risk Reduct.* **2013**, *3*, 31–43. [CrossRef]
3. Marijnissen, R.; Kok, M.; Kroeze, C.; Loon-Steensma, J. Re-evaluating safety risks of multifunctional dikes with a probabilistic risk framework. *Nat. Hazards Earth Syst. Sci.* **2019**, *19*, 737–756. [CrossRef]
4. Balistrocchi, M.; Moretti, G.; Ranzi, R.; Orlandini, S. Failure probability analysis of levees affected by mammal bioerosion. *Water Resour. Res.* **2021**, *57*, e2021WR030559. [CrossRef]
5. Agbaje, S.; Zhang, X.; Ward, D.; Dhimitri, L.; Patelli, E. Spatial variability characteristics of the effective friction angle of Crag deposits and its effects on slope stability. *Comput. Geotech.* **2022**, *141*, 104532. [CrossRef]
6. Mao, C.X. *Embankment Manual*; Water and Power Press: Beijing, China, 2009.
7. Xie, J.B.; Sun, D.Y. Model for analysis on reliability of levee overtopping and its application. *Water Resour. Hydropower Eng.* **2011**, *42*, 40–45.
8. Sharp, J.A.; Mcanally, W.H. Numerical modeling of surge overtopping of a levee. *Appl. Math. Model.* **2012**, *36*, 1359–1370. [CrossRef]
9. Yuan, S.Y.; Tang, H.W.; Li, L.; Pan, Y.; Amini, F. Combined wave and surge overtopping erosion failure model of HPTRM levees: Accounting for grass-mat strength. *Ocean Eng.* **2015**, *109*, 256–269. [CrossRef]
10. Saladdin, M.; O'Sullivan, J.; Abolfathi, S.; Dong, S.; Pearson, J. Distribution of individual wave overtopping volumes on a sloping structure with a permeable foreshore. *Coast. Eng. Proc.* **2020**, *36*, 54. [CrossRef]
11. Dong, S.; Abolfathi, S.; Salauddin, M.; Pearson, J.M. Spatial distribution of wave-by-wave overtopping at vertical seawalls. *Coast. Eng. Proc.* **2020**, *36*, 17. [CrossRef]
12. Dong, S.; Abolfathi, S.; Salauddin, M.; Pearson, J.M. Spatial distribution of wave-by-wave overtopping behind vertical seawall with recurve retrofitting. *Ocean Eng.* **2021**, *238*, 109674. [CrossRef]
13. Dong, S.; Salauddin, M.; Abolfathi, S.; Pearson, J. Wave impact loads on vertical seawalls: Effects of the geometrical properties of recurve retrofitting. *Water* **2021**, *12*, 2849. [CrossRef]

14. Salauddin, M.; O'Sullivan, J.J.; Abolfathi, S.; Peng, Z.; Dong, S.; Pearson, J.M. New insights in the probability distributions of wave-by-wave overtopping volumes at vertical breakwaters. *Sci. Rep.* **2022**, *12*, 16228. [CrossRef]
15. Guo, Y.; Zhu, D.; Zhang, F.; Lei, G.H.; Qiu, H.Y. Stability analysis of three-dimensional slopes under water drawdown conditions. *Can. Geotech. J.* **2014**, *51*, 1355–1364. [CrossRef]
16. Yeganeh-Bakhtiary, A.; Houshangi, H.; Hajivalie, F.; Abolfathi, S. A numerical study on hydrodynamics of standing waves in front of caisson breakwaters with WCSPH model. *Coast. Eng. J.* **2017**, *59*, 1750005. [CrossRef]
17. Guo, C.Y.; Li, D.Q.; Cao, Z.J.; Gao, G.H.; Tang, X.S. Efficient reliability sensitivity analysis for slope stability in spatial variable soils. *Rock Soil Mech.* **2018**, *39*, 2203–2210. [CrossRef]
18. Wang, X.B.; Xia, X.Z.; Zhang, Q. Reliability Analysis on Anti-sliding Stability of Levee Slope Based on Orthogonal Test and Neural Network. *J. Yangtze River Sci. Res. Inst.* **2019**, *36*, 89–93. [CrossRef]
19. Yeganeh-Bakhtiary, A.; Houshangi, H.; Abolfathi, S. Lagrangian two-phase flow modeling of scour in front of vertical breakwater. *Coast. Eng. J.* **2020**, *62*, 252–266. [CrossRef]
20. Wang, Z.F.; Zhang, Z.Q.; Yang, G.S. Risk calculation model for flood levee structure. *Shui Li Xue Bao* **1998**, *7*, 64–67.
21. Stark, T.D.; Jafari, N.H.; Leopold, A.L.; Brandon, T.L. Soil compressibility in transient unsaturated seepage analyses. *Can. Geotech. J.* **2014**, *51*, 858–868. [CrossRef]
22. Ni, X.D.; Zhu, C.M.; Wang, Y. Analyzing the behavior of seepage deformation by three-dimensional coupled continuum-discontinuum methods. *China Civ. Eng. J.* **2015**, *48*, 159–165.
23. Perlea, M.; Ketchum, E. Impact of non-analytical factors in geotechnical risk assessment of levees. In *Geo-Risk, Risk Assessment and Management*; American Society of Civil Engineers: Atlanta, GA, USA, 2011; pp. 1073–1081. [CrossRef]
24. Song, Z.P.; Guo, D.S.; Xu, T.; Hua, W.X. Risk assessment model in TBM construction based on nonlinear fuzzy analytic hierarchy process. *Rock Soil Mech.* **2021**, *42*, 1424–1433. [CrossRef]
25. Wang, Y.J.; Zhang, C.H.; Jin, F.; Zhang, W.H.; Wu, C.Y.; Ren, D.C. Research and application of comprehensive safety model and risk judgement system of embankment engineering. *J. Nat. Disasters* **2012**, *21*, 101–108.
26. Gu, C.S.; Wang, Z.L.; Liu, C.D. Experts' weight model assessing embankment safety. *Rock Soil Mech.* **2006**, *27*, 2099–2104.
27. Wang, T.; Liao, B.C.; Ma, X.; Fang, D.P. Using Bayesian network to develop a probability assessment approach for construction safety risk. *China Civ. Eng. J.* **2010**, *43*, 384–391.
28. Li, W.S.; Qin, X.Y.; Li, F.; Zhou, B. Analysis on probability of Stock Ammunition Accidents Based on FID. *China Saf. Sci. J.* **2009**, *19*, 62–69.
29. Howard, R.A.; Matheson, J.E. Influence diagrams. *Decis. Anal.* **2005**, *2*, 127–143. [CrossRef]
30. Yu, M. EPC Risk Identification and Analysis of Overseas Nuclear Power Project and Based on Risk Chain and Risk Map. *Nucl. Sci. Eng.* **2019**, *39*, 155–163.
31. Xu, S.Y. *Study on Construction Security Risk Assessment for Power Transmission Projection Based on Fuzzy Influence Diagram*; Southeast University: Nanjing, China, 2016.
32. Quan, J.; Huang, J.M.; Zhang, S.B.; Zeng, X.Y. Method of Risk Identification and Analysis Based on Risk Chain and Maps: A Case Study of an Overseas EPC Electric Power Project. *Energy Constr.* **2014**, *1*, 101–108.
33. Quan, J.; Huang, J.M.; Zhang, S.B.; Zeng, X.Y. Method of Risk Qualitative Assessment Based on Risk Chain and Fuzzy Influence Diagrams: A Case Study of an Overseas EPC Electric Power project. *South Energy Constr.* **2016**, *3*, 63–69.
34. Zhang, K.; Ge, L.; Wang, C.X.; Dai, F. Information Security Risk Assessment Based on Fuzzy Influence Diagram. *J. Zhengzhou Univ. (Eng. Sci.)* **2008**, *29*, 35–38.
35. Lin, S.; Lin, W.Y. Fuzzy Influence Diagrams Method and Risk Evaluation. *J. Tianjin Univ. Technol.* **2005**, *21*, 73–77.
36. Cheng, T.X.; Wang, P.; Zhang, W.B. Investigation on Fuzzy Influence Diagrams Evaluation Algorithm. *J. Syst. Eng.* **2004**, *19*, 177–182.
37. Patt, A.G.; Schrag, D.P. Using specific language to describe risk and probability. *Clim. Change* **2003**, *61*, 17–30. [CrossRef]

water

MDPI

Article

Study on the Influence of Water Level on Earth Dam Reinforced by Cut-Off Wall: A Case Study in Wujing Reservoir

Da Liu [1,2], Taiqing Lin [1], Jianglin Gao [1,*], Binghan Xue [3,*], Jianhua Yang [4], Congxin Chen [2], Weipeng Zhang [4] and Wenbin Sun [4]

1 Jiangxi Hydraulic Safety Engineering Technology Research Center, Jiangxi Academy of Water Science and Engineering, Nanchang 330029, China
2 State Key Laboratory of Geomechanics and Geotechnical Engineering, Institute of Rock and Soil Mechanics, Chinese Academy of Sciences, Wuhan 330029, China
3 School of Water Conservancy Engineering, Zhengzhou University, Zhengzhou 450001, China
4 School of Infrastructure Engineering, Nanchang University, Nanchang 330031, China
* Correspondence: gaojianglin126@126.com (J.G.); xuebinghan@zzu.edu.cn (B.X.)

Abstract: The construction of a cut-off wall is a common reinforcement method for earth rock dams. At present, compared with the in-depth study on homogeneous earth dams, more and more attention is being paid to the stability and deformation of earth dams strengthened by a concrete cut-off wall. In this study, aiming at the Wujing project of the earth dam strengthened by cut-off wall, the influence of the water level rise and fall on the stability of the dam slope, the deformation of the dam body, and the crack width on dam crest were analyzed by numerical calculation and in situ measurement. The analysis results show that when the reservoir encounters a sudden drawdown, the safety factor of the dam slope decreases sharply. The faster the sudden drawdown, the faster the safety factor decreases. When the reservoir water level rises, the dam's horizontal displacement shifts to the upstream direction, and the change of horizontal displacement of the downstream slope is significantly larger than that at the measuring point of the upstream slope. The water level of the reservoir rises, and the surface of the dam body rises, and the fluctuation of settlement deformation shows that the upstream side is larger than the downstream side, especially during the period of abrupt change in the reservoir water level. The longitudinal cracks on the dam crest show a tendency of shrinkage when the reservoir water level rises, and opening decreases with the decrease of deformation gradient increment and increases with the increase of gradient increment.

Keywords: earth rock dam; cut-off wall; dam safety; stability safety factor; water level fluctuation; deformation measurement; cracks on dam crest

Citation: Liu, D.; Lin, T.; Gao, J.; Xue, B.; Yang, J.; Chen, C.; Zhang, W.; Sun, W. Study on the Influence of Water Level on Earth Dam Reinforced by Cut-Off Wall: A Case Study in Wujing Reservoir. *Water* **2023**, *15*, 140. https://doi.org/10.3390/w15010140

Academic Editor: Giuseppe Oliveto

Received: 26 September 2022
Revised: 12 December 2022
Accepted: 25 December 2022
Published: 30 December 2022

1. Introduction

Most of the reservoir earth rock dam projects in China were built in the 1960s–1970s. Due to the limitations of technical and historical conditions at that time, there are many problems, such as poor project quality and many hidden diseases [1–3]. Therefore, dam seepage has become the most common hidden danger of earth rock dams [4–6]. At present, there are few studies on whether the construction of concrete cut-off walls in the reinforcement project of earth rock dams will affect the stability and the deformation of the dam.

The deformation of earth dams reinforced by concrete cut-off walls has certain particularity. First, most of the reinforced earth dams are old dams that have been in operation for decades. The consolidation of the dam soil has been basically completed, and the consolidation deformation and self-weight load deformation can be ignored [7,8]. However, the additional cut-off wall will change the seepage field and stress field in the dam, resulting in the reconstruction of the stress deformation of the dam [9]. Secondly, the factors that cause deformation are multi-year cyclic loads, such as reservoir water level load, flooding, and drainage processes [10–12]. The operation of the reservoir often experiences the working condition of water level fluctuation [13–15]. Water level fluctuation will cause the transient change of soil pore water pressure [16,17]. When the reservoir water level drops too fast, the soil drainage time is insufficient, and the pore water pressure in the dam body cannot dissipate quickly. Thus, the dam slope will form a downward trend under the action of seepage force, which causes slope instability, dam deformation, dam crest cracking, and other failure modes [18,19]. Therefore, it is of great significance for the design and application of seepage walls and the operation and management of reservoir dam projects in practical projects, especially for many small- and medium-sized reservoirs under construction or built, to systematically analyze the influence of seepage wall conditions and water level fluctuation on dam slope stability, dam deformation, and dam crest cracks.

At present, the research on slope stability analysis has been relatively mature [20–23]. The limit equilibrium method is widely used because of its clear mechanical concept and computational stability [24,25], which divides the sliding soil into several soil strips, and each soil strip and the whole meet the conditions of force and moment balance. That is, according to the shear strength criterion, the stability of the slope is solved by analyzing the balance under the limit state. The limit equilibrium method includes two aspects of calculation: calculating the safety factor of anti-sliding stability on a hypothetical sliding surface and finding a critical sliding surface with the lowest safety factor among the possible sliding surfaces [26]. According to the shear strength criterion, the stability of the slope is solved by analyzing the equilibrium under the limit state [27,28].

In this study, the Wujing project of the earth dam strengthened by the cut-off wall is taken as the research object, in which the stability of the dam slope, deformation of the dam body, and crack width on the dam crest are analyzed by means of limit equilibrium method and in-situ deformation measurement. It has been found that when the reservoir encounters a sudden drawdown, the safety factor will also decrease sharply. The horizontal displacement and settlements of the dam are affected by the change in the upstream reservoir water level. With the rise in the reservoir water level, the longitudinal cracks on the dam crest show a tendency of shrinkage, while the cracks show a tendency of opening when the reservoir water level drops.

2. Project Overview

Wujing reservoir is located in Wujing village, Zhaoxian Town, Wanli District, Nanchang City, Jiangxi Province, China, 3 km away from the urban area of Wanli, which is a small reservoir mainly for water supply and with comprehensive benefits of flood control. On the basis of the original homogeneous earth dam, high-pressure jet grouting was used to build a concrete cut-off wall in the dam in November 2011 to May 2012. Cracks were first found on the dam crest pavement in 2015, and continued to extend from 2015 to 2018. The cracks are about 132 m long and close to the dam axis (concrete cut-off wall). The average width of the cracks is 3 mm, the widest part is 40 mm, and the maximum dislocation is about 4 mm, as shown in Figure 1.

Figure 1. Dam crest of Wujing reservoir.

The dam is a homogeneous earth dam. After the construction of the concrete high-pressure jet grouting cut-off wall, the dam crest elevation is 192.55 m, the maximum dam height is 45.4 m, and the dam crest is 181 m long and 8 m wide. The normal pool level is 190 m, the design flood level is 190 m, the check flood level is 190.16 m, and the total storage capacity is 4.48 million m^3. From 183 m elevation to dam crest elevation, precast block slope protection is adopted for the upstream slope of the dam, with a slope ratio of 1:2; From the elevation of 170 to 183 m, the upstream slope of the dam adopts dry masonry slope protection, with a slope ratio of 1:2.5. Below the elevation of 170 m, the upstream slope of the dam is a block stone prism with a slope ratio of 1:1.25. From the elevation of 170 m to the elevation of the dam crest, the downstream slope of the dam is protected by concrete grid turf. Below the elevation of 171 m, the downstream slope of the dam is sloped for drainage, as shown in Figure 2.

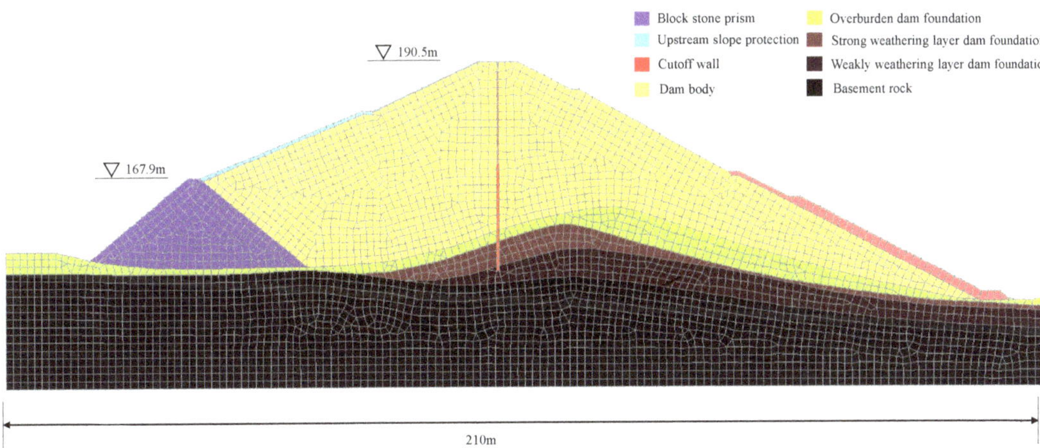

Figure 2. Numerical model of Wujing reservoir dam.

3. Study on the Influence of Sudden Drawdown on Dam Slope Stability

3.1. Seepage and Slope Stability Analysis Method

The earth rock dam and foundation are assumed to be isotropic porous media, the seepage in the dam body conforms to Darcy's law, and the seepage control equation is:

$$\frac{\partial}{\partial x}(k_x\frac{\partial H}{\partial x})+\frac{\partial}{\partial y}(k_y\frac{\partial H}{\partial y})+Q=\frac{\partial \Theta}{\partial t} \tag{1}$$

where H is the total head, k_x and k_y are the horizontal and vertical permeability coefficients, cm/s or m/d, Q is the flow of water on the boundary, cm^3/s or m^3/d, and Θ is the volumetric moisture content.

Among them, the unsaturated percolation model was adopted from the Fredlund and Xing model, and its expression is:

$$\frac{\theta}{\theta_s}=\frac{1-\frac{\ln(1+u_s/u_r)}{\ln(1+10^6/u_r)}}{\left\{\ln\left[e+(u_s/a)^b u_s/u_r\right]\right\}^c} \tag{2}$$

where θ is the volume water content, u_{se} is the saturated volume water content, u_s is the suction, kPa, c is the soil property parameter of the residual water content function, u_r is the matrix suction corresponding to the water content, kPa, and a and b are the fitting parameters.

The limit equilibrium method was used to calculate the stability of dam slope. The solution idea is based on the Mohr Coulomb shear strength theory. The slope within the potential sliding surface is divided into several soil strip according to a certain proportion, and the static balance equation is established according to the limit equilibrium conditions between the soil strips. Considering the overall moment balance of sliding soil, the safety factor F_S of slope is calculated, and the stability of slope is evaluated according to the equation.

The moment equilibrium relationship is given by:

$$F_S=\frac{\sum(c\Delta LR+RN\tan\phi)}{\sum WL_W-\sum NL_N} \tag{3}$$

Or expressed by horizontal force balance relation:

$$F_S=\frac{\sum(c\Delta LR\cos\alpha+RN\tan\varphi\cos\alpha)}{\sum N\sin\alpha} \tag{4}$$

where:

$$N=\frac{W+\lambda f(x)\sum(\frac{c\Delta LR\cos\alpha}{F_S})-\frac{c\Delta L\sin\alpha}{F}}{(\cos\alpha+\frac{\sin\alpha}{F_S})-\lambda f(x)(\frac{\cos\alpha\tan\varphi}{F_S}-\sin\alpha)} \tag{5}$$

where c is the cohesion, kPa, ϕ is internal friction angle of soil mass, °. ΔL are the respective soil lengths on the sliding surface, m. L_W is the moments of each soil strip against the center of the slip surface, Nm. L_N are the moments corresponding to their normal for the midpoint of each soil strip at the sliding surface, Nm. α is the angle between the tangent lines of each soil and the horizontal plane, R is the moment length to the center of the circle, m. N is the normal phase force of the soil strip, kN. W is the weight of soil strip, kN. λ is coefficient of variation of interaction force between soil strips, and the f(x) in Equation (5) is the function of interaction force between soil strips.

3.2. Numerical Model and Parameters

The axis of cut-off wall of Wujing reservoir dam is located on the dam axis with a thickness of 0.25 m and a wall top elevation of 190.5 m. Stone prism is used to fix the foot in the upstream. According to relevant design data, site investigation report and topographic mapping data of dam slope of Wujing Reservoir, a typical section of design is selected to establish a 2D dam model by GeoStudio software to analyze stability of dam slope with sudden drawdown, as shown in Figure 2. The boundary and initial conditions of the upstream face of the dam are the design water level and steady-state seepage field, and the dam foundation is the fully constrained boundary condition.

The Wujing reservoir dam is filled with silty sand and the foundation is granite rock mass, in which the weathered rock mass of the dam foundation mainly concentrates on the downstream side of the dam. According to the engineering geological report, the calculation material parameters of the dam and rock mass foundation are as follows:

When simulating the saturated–unsaturated seepage flow field, the saturated/unsaturated model is used for the dam body, and the saturated model is used for the dam foundation. The concrete cut-off wall is constructed by high-pressure jet grouting, and the measured anti-seepage coefficient of the cut-off wall is 5×10^{-7} cm/s.

3.3. Stability Analysis of Dam Slope under Sudden Drawdown

When the reservoir water level drops in a short time, it is easy to cause dam slope collapse, which is an important reason for the instability of earth rock dams. The process of sudden drawdown is analyzed and simulated under the condition of sudden drawdown of Wujing reservoir with the water level under normal storage conditions of 190 m, and the safety factor of dam slope stability before and after adding the cut-off wall is calculated. The simulation process of sudden drawdown is assumed as: water level drops from 190 m (The normal water level) to 160 m, lasting 10, 6, and 3 days, with descending speeds of 3, 5, and 10 m/day, respectively. The initial condition is the steady-state analysis result under long-term immersion. Figures 3 and 4 have shown the typical most dangerous sliding surface calculated by GeoStudio with the parameter in Table 1 and its safety coefficient before and after sudden drawdown reaches the lowest level respectively with descending speed of 10 m/day by the mean of the limit equilibrium method, and the arrow indicates the effect of upstream reservoir water level.

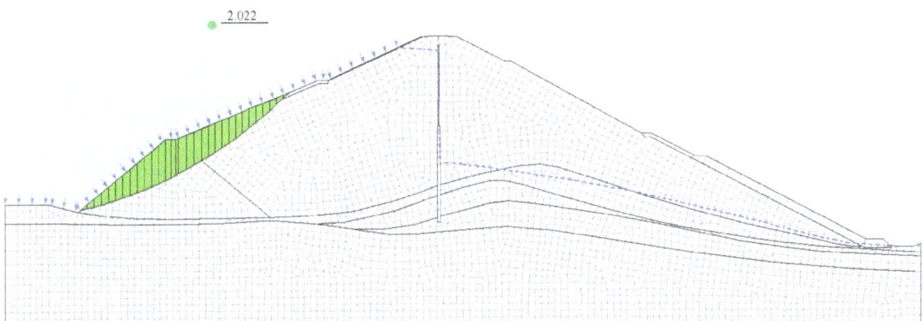

Figure 3. The most dangerous sliding surface of upstream dam slope before sudden drop of water level.

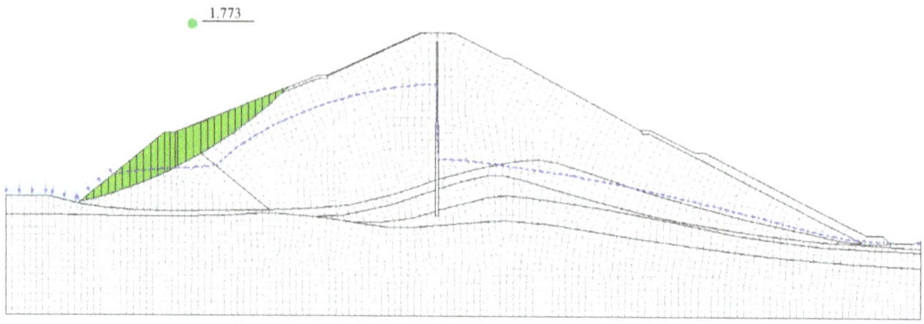

Figure 4. The most dangerous sliding surface of upstream dam slope reaches a stable state after sudden drop of water level.

Table 1. Model calculation parameter.

Material	Permeability Coefficient/cm/s	Density/kg/m³	Cohesion/kPa	Friction Angle
Silty sand of dam body	5.0×10^{-5}	1700	20	20
Dam foundation overburden	5.0×10^{-4}	1800	15	28
Strongly weathered rock mass of dam foundation	1.0×10^{-4}	2400	15	30
Weakly weathered rock mass of dam foundation	1.0×10^{-5}	2600	20	33
Basement rock	1.0×10^{-6}	2700	30	35

The safety factor of upstream dam slope decreases with the sudden drawdown. When the water level drops to the lowest level, the safety factor reaches the lowest level, then shows a slow upward trend and finally remains stable. The reason is that the water head difference changes sharply due to the decrease in water level, which makes the stability of the slope decrease. At this time, the suction of the inner matrix of the rocks and soils on the dam slope does not rise in time. With the water level no longer dropping, the suction of the matrix inside the dam slope gradually recovers as the pore water pressure dissipates and the shear strength of the soil body gradually rises, thus the safety factor of stability increases. The stability safety factor duration curve of upstream and downstream dam slopes is shown in Figure 5.

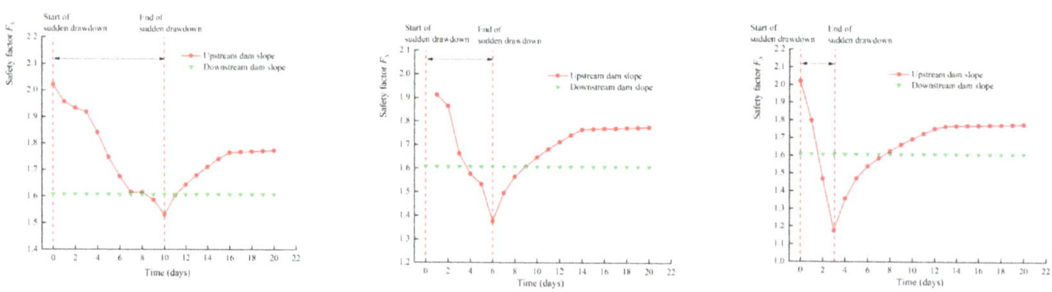

(1) 3 m/d (2) 5 m/d (3) 10 m/d

Figure 5. Duration curve of dam slope stability safety factor under different sudden drawdown speeds.

According to the calculation parameters in the Wujing engineering geological report, the stability safety factor of the dam slope is about 2.0 without water level change, which is generally safe and stable. Only at the extreme sudden drawdown can the dam slope be unsafe. The faster the speed of sudden drawdown speed is, the faster the safety factor of upstream dam slope stability will decrease, which will be disadvantageous to the stability of the dam slope. With the end of the sudden drop, the pore water pressure of the dam body soil gradually dissipates, which makes the coefficient of stability safety rise again and finally tends to a stable value.

4. Measuring and Analysis of Dam Deformation under Water Level Rise and Fall

4.1. Arrangement of Measuring Points

Deformation measurements are carried out on the existing observation piers for surface displacement observation of the dam. Surface displacement and settlement of three transverse sections of the dam are observed, and each transverse section has 2 observation points, a total of 6 observation points, as shown in Figure 6.

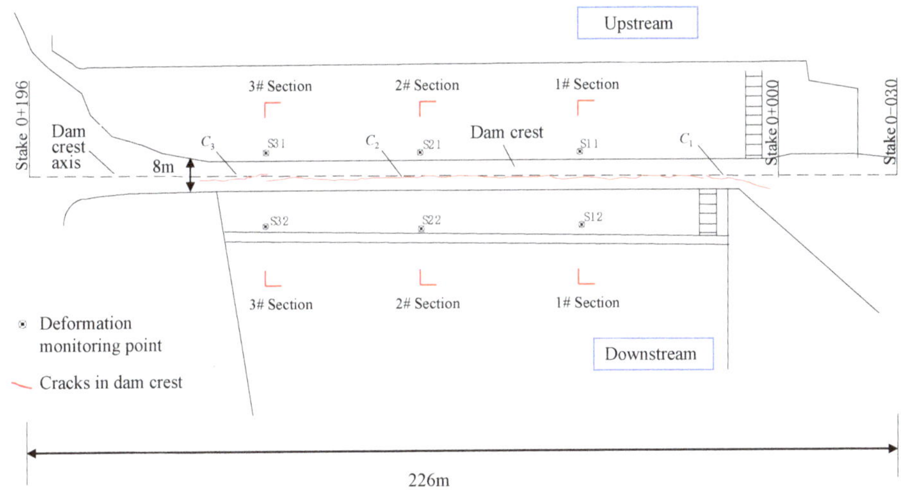

Figure 6. Layout diagram of deformation measurement points of Wujing dam.

4.2. Horizontal Displacement

The horizontal displacement of the dam is measured by observing piers. The results of horizontal displacement measurement for three measuring sections are given in Figures 7–9, in which data on reservoir water level and rainfall are also given. The horizontal displacement obtained by the monitoring is listed in Table 2.

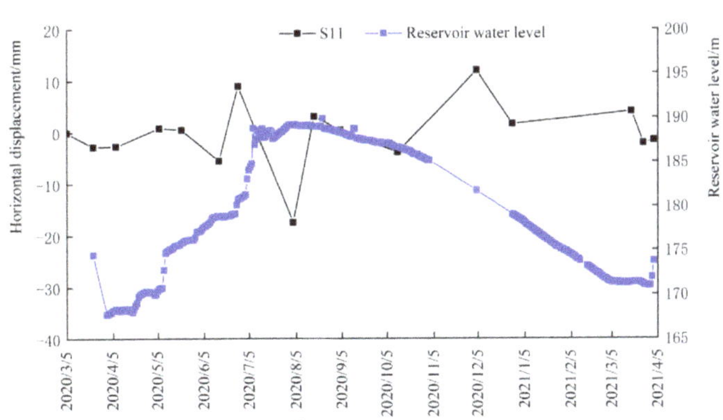

Figure 7. 1 # section horizontal displacement duration curve (upstream is −, downstream is +).

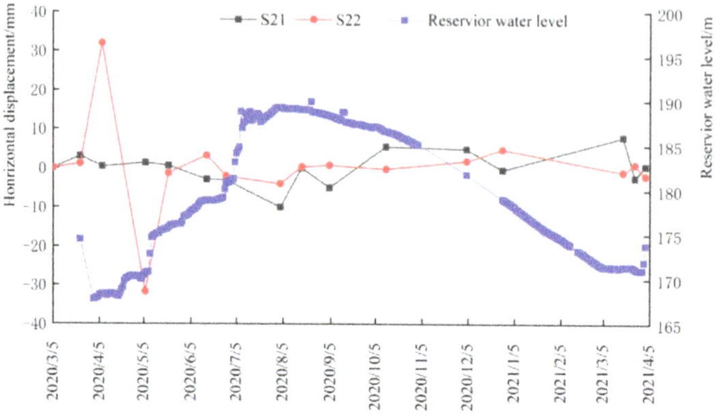

Figure 8. 2 # section horizontal displacement duration curve (upstream is −, downstream is +).

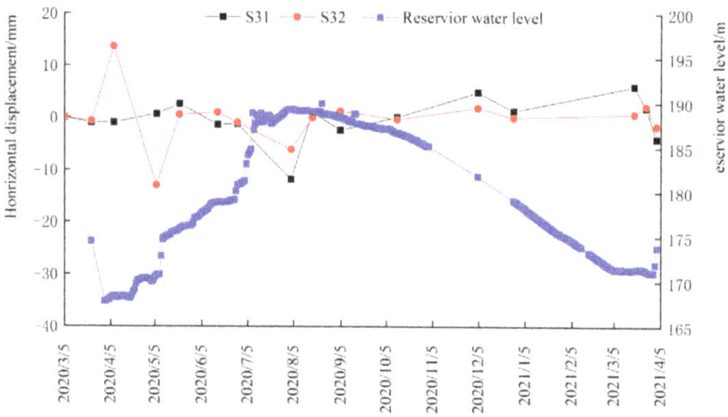

Figure 9. 3 # section horizontal displacement duration curve (upstream is −, downstream is +).

Table 2. Horizontal displacement.

Monitoring Time	1# Section		2# Section		3# Section	
	S11	S12	S21	S22	S31	S32
5 March 2020	0.00		0.00	0.00	0.00	0.00
23 March 2020	−2.75		3.07	1.17	−0.99	−0.67
7 April 2020	−2.66		0.38	32.05	−0.97	13.70
6 May 2020	0.84		1.31	−31.72	0.69	−12.90
21 May 2020	0.53		0.62	−1.33	2.58	0.55
15 June 2020	−5.46		−2.93	3.22	−1.31	1.03
28 June 2020	8.95		−3.04	−2.01	−1.23	−0.96
3 August 2020	−17.46		−9.96	−4.01	−11.80	−6.05
17 August 2020	3.13	/	−0.02	0.32	0.99	0.06
4 September 2020	0.41		−5.00	0.71	−2.37	1.26
12 October 2020	−3.88		5.52	−0.21	0.10	−0.28
4 December 2020	11.93		4.90	1.83	4.90	1.91
28 December 2020	1.57		−0.46	4.81	1.36	0.07
18 March 2021	4.04		7.98	−1.01	6.04	0.76
26 March 2021	−2.17		−2.50	0.98	1.94	2.25
2 April 2021	−1.54		0.54	−1.95	−3.96	−1.54

The horizontal displacement of the dam body at Section 1 # (S11–S21) is only the data on the upstream side, and the downstream measurement data are missing due to the failure of observation pier S21. The maximum displacement amplitude of water level at the upstream measuring point occurred from the end of June to the beginning of August 2020. From the deformation of measuring points and the variation law of the reservoir water level, as the water level rises from March 2020 to August 2020, except for the 8.95 value monitored on June 28, the horizontal displacement of the dam was almost smaller than 0, which generally tended to change to the upstream. As the water level drops from August 2020 to April 2021, the horizontal displacement of the dam was almost all greater than 0, and the horizontal displacement of the dam generally tended to change to the downstream.

From the deformation laws of the three sections, it can be seen that the deformation laws of the measuring points of the 2 # and 3 # sections are similar. As the water level rises from March 2020 to August 2020, if we do not consider the point with huge extreme deformation, it may be caused by measurement error, i.e., 32.05 and 13.70 mm measured on April 7 and 31.72 mm measured on May 6 in S22 and S32, the horizontal displacement of the dam was almost smaller than 0, which generally tended to change to the upstream. As the water level dropped from August 2020 to April 2021, the horizontal displacement of the dam was almost all greater than 0, and the horizontal displacement of the dam generally tends to change to the downstream. On the whole, the horizontal displacement of the dam body was affected by the change of the water level of the upstream reservoir, the water level of the upstream reservoir rises, and the dam body shifted to the upstream direction (the measured value is negative). On the contrary, the water level of the upstream reservoir decreased, and the dam body generally tended to shift to the downstream (the measured value was positive), and the horizontal displacement amplitude of the downstream dam slope measuring points of the two sections was significantly greater than that of the upstream slope measuring points.

4.3. Settlements

The results of settlements measurement for three measuring sections are given in the Figures 10–12. The horizontal displacement obtained by monitoring are listed in Table 3.

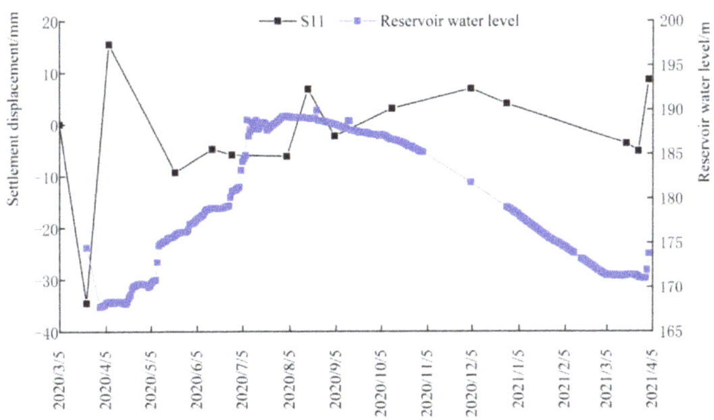

Figure 10. 1 # section settlement displacement duration curve (downward is +, upward is −).

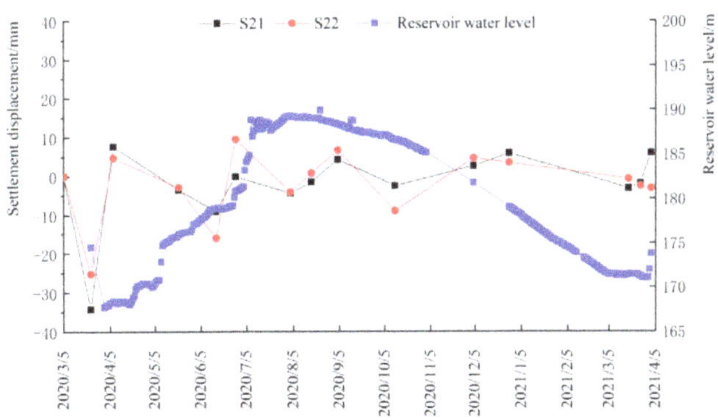

Figure 11. 2 # section settlement displacement duration curve (downward is +, upward is −).

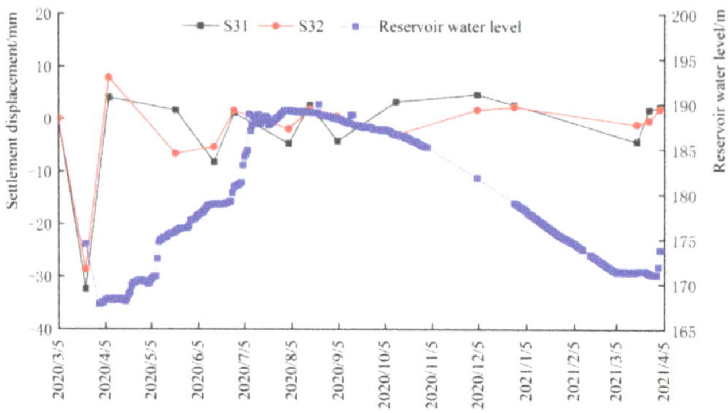

Figure 12. 3 # section settlement displacement duration curve (downward is +, upward is −).

Table 3. Settlement.

Monitoring Time	1# Section		2# Section		3# Section	
	S11	S12	S21	S22	S31	S32
5 March 2020	0.00	0.00	0.00	0.00	0.00	0.00
23 March 2020	−34.50	−38.30	−34.20	−25.19	−32.40	−28.60
7 April 2020	15.50	4.40	7.70	4.79	4.10	7.90
21 May 2020	−9.30	−1.70	−3.50	−2.90	1.70	−6.60
15 June 2020	−4.80	−9.80	−9.00	−16.00	−8.20	−5.30
28 June 2020	−5.90	6.80	−0.10	9.50	1.10	1.60
3 August 2020	−6.20	−11.20	−4.40	−4.10	−4.70	−1.90
17 August 2020	6.80	5.40	−1.50	0.80	2.60	1.80
4 September 2020	−2.30	8.70	4.30	6.70	−4.20	0.50
12 October 2020	3.10	−9.90	−2.50	−8.90	3.20	−3.00
4 December 2020	6.90	6.30	2.60	4.70	4.60	1.70
28 December 2020	4.00	−0.10	5.90	3.50	2.60	2.30
18 March 2021	−3.70	3.50	−3.10	−0.70	−4.30	−1.00
26 March 2021	−5.20	−1.60	−1.90	−2.50	1.70	−0.30
2 April 2021	8.60	−1.00	6.00	−3.10	2.20	1.90

From the settlement laws of the three sections, it can be seen that as the water level rises from March 2020 to August 2020, except for the settlement monitored on April 7, the horizontal displacement of the dam was almost smaller than 0, which generally tended to change to the upward. As the water level drops from August 2020 to April 2021, the horizontal displacement of the dam was almost all greater than 0, and the horizontal displacement of the dam generally tended to change to the downward. During the continuous change of reservoir water level, the two monitoring points in the same section basically show the same change trend. The change in the dam surface settlement was related to the fluctuation of the upstream reservoir water level. When the upstream reservoir water level rises, the dam surface settlement value was -, indicating that the surface had a trend of rising. On the contrary, when the water level of the upstream reservoir dropped, the surface settlement value of the dam body was +, showing a downward subsidence trend.

5. Crack Propagation Law Based on Deformation Gradient Method

5.1. Deformation Gradient Method

The deformation gradient method was to predict dam cracks based on dam settlement observation data. As shown in Figure 13, if there were two observation points a and b at the same elevation of the dam body, and the horizontal distance between the two points was Δy, and if the cumulative settlement measured by T_i on a certain calculation date was

Z_a and Z_b, respectively, then the deformation gradient of a and b on the date T_i was defined as γ, asshown in Equation (5).

$$\gamma \approx \tan\gamma = \frac{\Delta Z}{\Delta y} \times 100 = \frac{Z_a - Z_b}{|y_a - y_b|} \times 100 \tag{6}$$

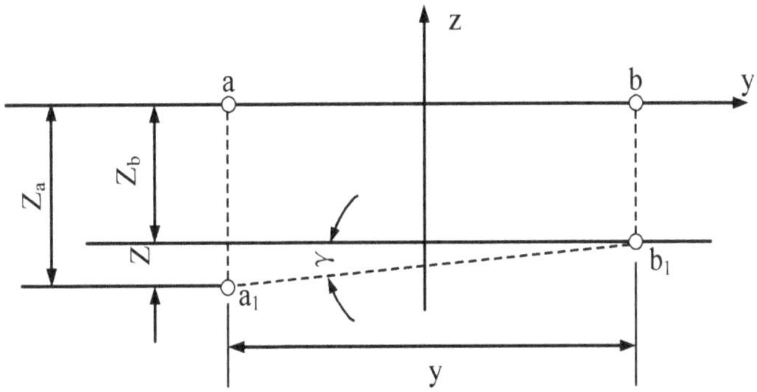

Figure 13. Schematic diagram of deformation gradient method.

When the monitoring point of the same cross-section was not at an elevation, the deformation gradient could be modified. As shown in Figure 14, assuming 2 observation points a and b at different elevations of the dam, the horizontal distance between the two points was Δy, the initial difference in height between the two points was ΔZ_1, and the difference in height on a certain calculation date was ΔZ_2, then the modified deformation gradient of points a and b was defined as:

$$\Delta\gamma \approx |\tan\gamma_1 - \tan\gamma_2| \times 100 \tag{7}$$

$$\tan\gamma_1 = \frac{\Delta Z_1}{\Delta y} \tag{8}$$

$$\tan\gamma_2 = \frac{\Delta Z_2}{\Delta y} \tag{9}$$

$$\Delta Z_2 = Z_a - Z_b + \Delta Z_1 \tag{10}$$

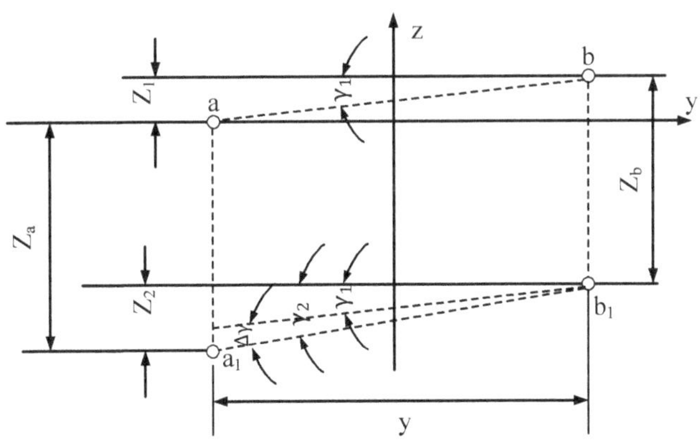

Figure 14. Schematic diagram of modified deformation gradient method.

It is worth noting that when the initial elevation difference $\Delta Z_1 = 0$, i.e., a and b are at the same elevation, the modified deformational gradient agrees with the calculation result of Equation (5).

5.2. Analysis of Crack Width Propagation Law

Longitudinal cracks of dam crest are distributed along dam axis in plane in the range of dam 0+005~dam 0+148 stake, as shown in Figure 6. The crack width was measured with the vernier caliper. The measured crack width changes with time as shown in Figure 15.

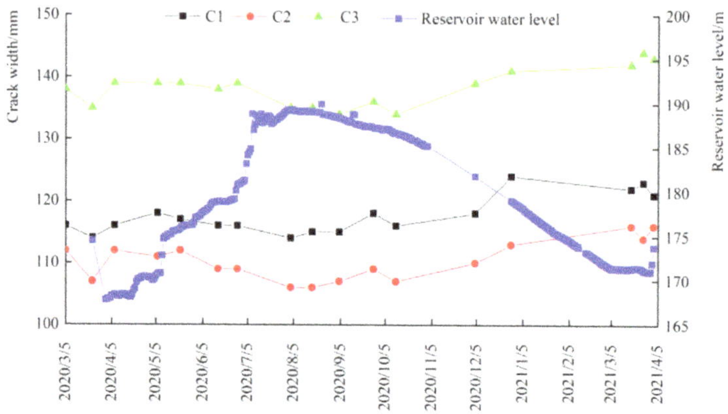

Figure 15. Time history curve of crack width and reservoir water level.

Measuring data from 5 March 2020 to 2 April 2020 show that the overall change of crack width is relatively stable. During the rising process of reservoir water level, the development and change of crack width showed a certain contraction trend, and then the development and change of crack width showed a slowing trend of opening during the falling process of reservoir water level.

According to the monitoring results of Wujing reservoir, the longitudinal crack of the dam crest was analyzed by the deformation gradient method. Table 4 is the result of calculating the deformation gradient according to the monitoring data from March 2020 to March 2021.

Table 4. Gradient calculation results.

Monitoring Time	Settlement Z/mm						Horizontal Distance of Observation Points on the Same Section/m	Deformation Gradient/%		
	1# Section		2# Section		3# Section					
	S31	S32	S21	S22	S11	S12		γ_1	γ_2	γ_3
5 March 2020	0.0	0.0	0.0	0.0	0.0	0.0	9.00	0	0	0
23 March 2020	−32.4	−28.6	−34.2	−25.19	−34.5	−38.3		0.0422	0.1001	0.0422
7 April 2020	4.1	7.9	7.7	4.79	15.5	4.4		0.0422	0.0323	0.1233
21 May 2020	1.7	−6.6	−3.5	−2.9	−9.3	−1.7		0.0922	0.0067	0.0844
15 June 2020	−8.2	−5.3	−9	−16	−4.8	−9.8		0.0322	0.0778	0.0556
28 June 2020	1.1	1.6	−0.1	9.5	−5.9	6.8		0.0056	0.1067	0.1411
3 August 2020	−4.7	−1.9	−4.4	−4.1	−6.2	−11.2		0.0311	0.0033	0.0556
17 August 2020	2.6	1.8	−1.5	0.8	6.8	5.4		0.0089	0.0256	0.0156
4 September 2020	−4.2	0.5	4.3	6.7	−2.3	8.7		0.0522	0.0267	0.1222
12 October 2020	3.2	−3	−2.5	−8.9	3.1	−9.9		0.0689	0.0711	0.1444
4 December 2020	4.6	1.7	2.6	4.7	6.9	6.3		0.0322	0.0233	0.0067
28 December 2020	2.6	2.3	5.9	3.5	4.0	−0.1		0.0033	0.0267	0.0456
18 March 2021	−4.3	−1.0	−3.1	−0.7	−3.7	3.5		0.0367	0.0267	0.0800

Notes: The initial elevation difference is 4.3198 m for 1# section, 4.4942 m for 2# section, and 4.1936 m for 3# section.

Figure 16 shows the variation curve of dam deformation gradient increment and crack width. The monitoring results showed that the variation rule of 3 cracks with deformation gradient is basically the same. Except for the monitoring in June, other monitoring calculation results showed that the change of increment of deformation gradient was basically consistent with that of crack width, i.e., the crack width decreased with the decrease in gradient and increased with the increase in gradient increment.

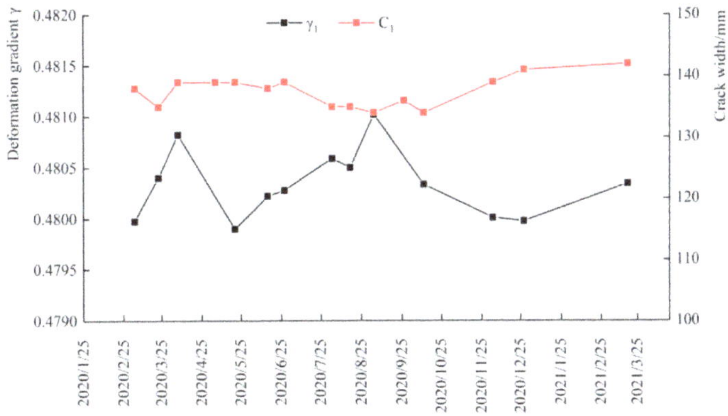

(**1**) 1# section

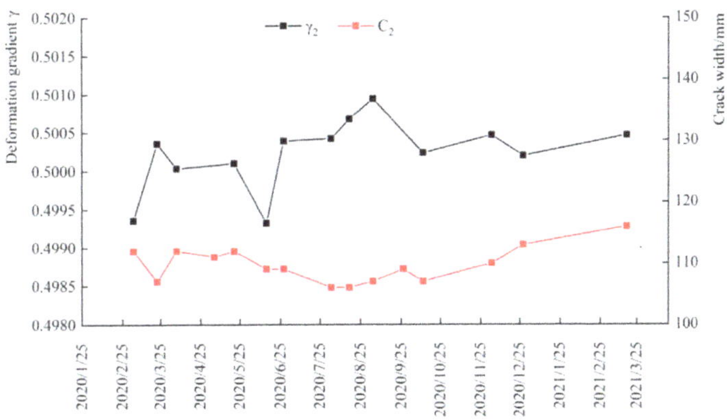

(**2**) 2# section

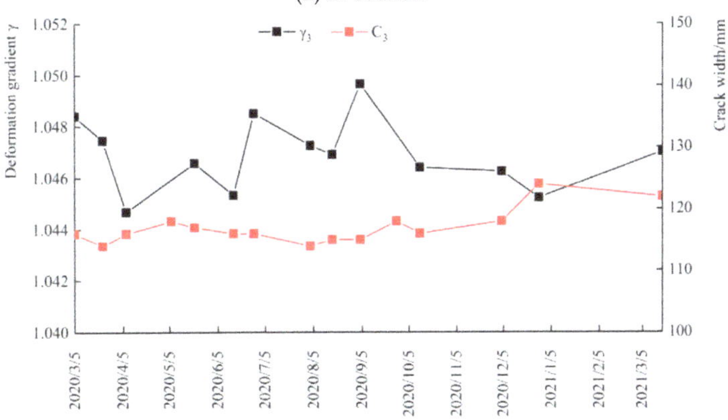

(**3**) 3# section

Figure 16. Time history curve of crack width and deformation gradient.

The relationship between the expansion of dam crest cracks and the uneven settlement of the dam body on both sides of the cracks is shown in Figure 16. The longitudinal cracks on the dam crest of Wujing Reservoir close with the rise of reservoir water and open with the fall of reservoir water, indicating that the soil mass on the upstream slope of the reservoir water fluctuation dam undergoes cyclic expansion and contraction deformation, thus affecting the opening and closing deformation of existing cracks.

6. Discussion and Conclusions

The deformation of the dam is closely related to the change in the reservoir water level, whether for horizontal deformation or settlement deformation. However, due to the different states of different dams, the impact of the dam on the change in the water level is different, as shown in Table 5.

Table 5. Study on the effect of water level fluctuation on dam deformation.

Dam	Dam Type	Research Method	Whether the Water Level Fluctuation Affects the Dam Deformation	How Water Level Fluctuation Affects Dam Deformation
Chengbihe [7]	Earth dam	In situ measurements	Yes	Settlement on the dam's upstream side
Yamula [29]	Earth dam	In situ measurements	Yes	Vertical deformation. The rising water level increases the subsidence velocity
An ultra-high arch dam [30]	Concrete arch dam	In situ measurements	Yes, but the influence of water level on dam deformation is hysteretic.	With the rise of the water level, the cluster center area also rises, indicating that the trend of the expansion and upward movement of the maximum deformation area of the dam body
Jiangya [31]	Gravity dam	In situ measurements	Yes	The reservoir impoundment is the predominant cause of the uplift of dam foundation.
An earth embankments of dams [32]	Earth dam.	Numerical simulation	Yes	The horizontal deformation of the dam toe is higher for a higher rising rate
Princeville [33]	Earth levee	Numerical simulation	Yes	The Factor of safety is affected by rate of rise/drawdown of the water level

In this study, by means of numerical analysis and in situ measuring, the influence of water level rise and fall on the Wujing earth dam reinforced by the cut-off wall has been analyzed, and the following conclusions can be drawn:

1. Based on the established numerical calculation model of Wujing reservoir, and the geological report parameters, the safety factor of dam slope stability was obtained by the limit equilibrium method, which was about 2.0. When the reservoir encountered a sudden drawdown, the safety factor also decreased sharply. The faster the sudden drawdown was, the faster the safety factor decreased, and the more unfavorable it was to dam slope stability. With the end of the sudden drawdown, the pore water pressure of the dam body soil gradually dissipated, so that the safety factor of stability increased again and finally tended to be stable;
2. The horizontal displacement of the dam was affected by the change in the upstream reservoir water level. It was found that the upstream reservoir water level rose and the horizontal displacement of dam shifted to the upstream direction. The upstream reservoir water level dropped, the horizontal displacement of the dam shifted to downstream, and the change of horizontal displacement of downstream slope was significantly larger than that at measuring point of upstream slope;
3. The settlement deformation of the dam body was related to the fluctuation of the reservoir water level, in which the water level of upstream reservoir rose, and as surface of the dam body rose, conversely, it tended to sink. The fluctuation of the settlement deformation shows that the upstream side was larger than the downstream side, especially during the period of abrupt change in reservoir water level;
4. With the rise in the reservoir water level, the longitudinal cracks on the dam crest showed a tendency of shrinkage, while the cracks showed a tendency of opening when the reservoir water level dropped. The change in the deformation gradient increment was basically consistent with the change in the crack opening, that is, the crack opening decreased with the decrease in the deformation gradient increment and increased with the increase in the gradient increment.

Author Contributions: Conceptualization, D.L., J.G., J.Y. and C.C.; Methodology, J.G.; Software, J.Y., C.C., W.Z. and W.S.; Validation, J.G.; Investigation, T.L., B.X. and J.Y.; Resources, C.C.; Data curation, D.L. and T.L.; Writing—original draft, D.L.; Writing—review & editing, D.L. and B.X.; Visualization, B.X., W.Z. and W.S.; Funding acquisition, D.L. All authors have read and agreed to the published version of the manuscript.

Funding: This work is supported by the Jiangxi Provincial Natural Science Foundation (20212BAB214044, 20212BAB214045), and the Jiangxi Provincial Department of water resources Foundation (202224ZDKT08).

Data Availability Statement: The data presented in this study are available on request from the corresponding author. The data are not publicly available due to privacy restrictions.

Acknowledgments: The authors wish to express their thanks to all supporters.

Conflicts of Interest: The authors declare no conflict of interest.

References

1. Wang, T.; Chen, J.; Li, P.; Yin, Y.; Shen, C. Natural tracing for concentrated leakage detection in a rockfill dam. *Eng. Geol.* **2019**, *249*, 1–12. [CrossRef]
2. Chen, S.; Gu, C.; Lin, C.; Wang, Y.; Hariri-Ardebili, M.A. Prediction, monitoring, and interpretation of dam leakage flow via adaptative kernel extreme learning machine. *Measurement* **2020**, *166*, 108161. [CrossRef]
3. Mozafari, M.; Milanović, P.; Jamei, J. Water leakage problems at the Tangab Dam Reservoir (SW Iran), case study of the complexities of dams on karst. *Bull. Eng. Geol. Environ.* **2021**, *80*, 7989–8007. [CrossRef]
4. Pan, Y.; Liu, Y.; Chen, E.J. Probabilistic investigation on defective jet-grouted cut-off wall with random geometric imperfections. *Géotechnique* **2019**, *69*, 420–433. [CrossRef]
5. Wang, K.; Li, Z.; Zheng, H.; Xu, X.; He, H. A theoretical model for estimating the water-tightness of jet-grouted cut-off walls with geometric imperfections. *Comput. Geotech.* **2021**, *138*, 104316. [CrossRef]
6. Shepherd, D.A.; Kotan, E.; Dehn, F. Plastic concrete for cut-off walls: A review. *Constr. Build. Mater.* **2020**, *255*, 119248. [CrossRef]
7. Luo, J.; Zhang, Q.; Li, L.; Xiang, W. Monitoring and characterizing the deformation of an earth dam in Guangxi Province, China. *Eng. Geol.* **2019**, *248*, 50–60. [CrossRef]
8. Yavaşoğlu, H.H.; Kalkan, Y.; Tiryakioğlu, I.; Yigit, C.; Özbey, V.; Alkan, M.N.; Bilgi, S.; Alkan, R.M. Monitoring the deformation and strain analysis on the Ataturk Dam, Turkey. *Geomat. Nat. Hazards Risk* **2018**, *9*, 94–107. [CrossRef]
9. Bestuzheva, A.S.; Anakhaev, K.K. Seepage Through a Homogeneous Earth-Fill Dam with a Cut-Off Wall on a Permeable Foundation. *Power Technol. Eng.* **2020**, *54*, 147–153. [CrossRef]
10. Barzaghi, R.; Cazzaniga, N.E.; De Gaetani, C.I.; Pinto, L.; Tornatore, V. Estimating and Comparing Dam Deformation Using Classical and GNSS Techniques. *Sensors* **2018**, *18*, 756. [CrossRef]
11. Lai, J.; Liu, H.; Qiu, J.; Chen, J. Settlement Analysis of Saturated Tailings Dam Treated by CFG Pile Composite Foundation. *Adv. Mater. Sci. Eng.* **2016**, *2016*, 7383762. [CrossRef]
12. Cheng, X.; Li, Q.; Zhou, Z.; Luo, Z.; Liu, M.; Liu, L. Research on a Seepage Monitoring Model of a High Core Rockfill Dam Based on Machine Learning. *Sensors* **2018**, *18*, 2749. [CrossRef] [PubMed]
13. Ri, Y.N.; Han, U.C.; Jang, U.J.; Jong, D.Y.; Kim, C.U. Study on Stability Reduction Characteristics of Earth and Rockfill Dams under Rapid Drawdown Using Fully Coupled Seepage-Stress Analysis. *Adv. Civ. Eng.* **2022**, *2022*, 7954991. [CrossRef]
14. Azadi, A.; Irani, A.E.; Azarafza, M.; Bonab, M.H.; Sarand, F.B.; Derakhshani, R. Coupled Numerical and Analytical Stability Analysis Charts for an Earth-Fill Dam under Rapid Drawdown Conditions. *Appl. Sci.* **2022**, *12*, 4550. [CrossRef]
15. Sun, G.; Lin, S.; Jiang, W.; Yang, Y. A Simplified Solution for Calculating the Phreatic Line and Slope Stability during a Sudden Drawdown of the Reservoir Water Level. *Geofluids* **2018**, *2018*, 1859285. [CrossRef]
16. Zhang, Y.; Zhang, Z.; Xue, S.; Wang, R. Stability analysis of a typical landslide mass in the Three Gorges Reservoir under varying reservoir water levels. *Environ. Earth Sci.* **2020**, *79*, 42. [CrossRef]
17. Meng, Q.; Qian, K.; Zhong, L.; Gu, J.; Li, Y.; Fan, K.; Yan, L. Numerical Analysis of Slope Stability under Reservoir Water Level Fluctuations Using a FEM-LEM-Combined Method. *Geofluids* **2020**, *2020*, 6683311. [CrossRef]
18. Zhang, H.; Jing, Y.; Chen, J.; Gao, Z.; Xu, Y. Characteristics and causes of crest cracking on a high core-wall rockfill dam: A case study. *Eng. Geol.* **2022**, *297*, 106488. [CrossRef]
19. Wang, Y.; Li, J.; Wu, Z.; Chen, J.; Yin, C.; Bian, K. Dynamic Risk Evaluation and Early Warning of Crest Cracking for High Earth-Rockfill Dams through Bayesian Parameter Updating. *Appl. Sci.* **2020**, *10*, 7627. [CrossRef]
20. Kaur, A.; Sharma, R. Slope stability analysis techniques: A review. *Int. J. Eng. Appl. Sci. Technol.* **2016**, *1*, 52–57.
21. Firincioglu, B.S.; Ercanoglu, M. Insights and perspectives into the limit equilibrium method from 2D and 3D analyses. *Eng. Geol.* **2021**, *281*, 105968. [CrossRef]
22. Arvin, M.R.; Zakeri, A.; Shoorijeh, M.B. Using Finite Element Strength Reduction Method for Stability Analysis of Geocell-Reinforced Slopes. *Geotech. Geol. Eng.* **2019**, *37*, 1453–1467. [CrossRef]

23. Zhang, K.; Cao, P.; Meng, J.; Li, K.; Fan, W. Modeling the Progressive Failure of Jointed Rock Slope Using Fracture Mechanics and the Strength Reduction Method. *Rock Mech. Rock Eng.* **2015**, *48*, 771–785. [CrossRef]
24. Dong-Ping, D.; Liang, L.; Jian-Feng, W.; Lian-Heng, Z. Limit equilibrium method for rock slope stability analysis by using the Generalized Hoek–Brown criterion. *Int. J. Rock Mech. Min. Sci.* **2016**, *89*, 176–184. [CrossRef]
25. Liu, S.; Su, Z.; Li, M.; Shao, L. Slope stability analysis using elastic finite element stress fields. *Eng. Geol.* **2020**, *273*, 105673. [CrossRef]
26. Siacara, A.; Beck, A.; Futai, M. Reliability analysis of rapid drawdown of an earth dam using direct coupling. *Comput. Geotech.* **2020**, *118*, 103336. [CrossRef]
27. Duncan, J.; Wright, S. The accuracy of equilibrium methods of slope stability analysis. *Eng. Geol.* **1980**, *16*, 5–17. [CrossRef]
28. Kong, X.; Cai, G.; Cheng, Y.; Zhao, C. Numerical Implementation of Three-Dimensional Nonlinear Strength Model of Soil and Its Application in Slope Stability Analysis. *Sustainability* **2022**, *14*, 5127. [CrossRef]
29. Bayrak, T. Modelling the relationship between water level and vertical displacements on the Yamula Dam, Turkey. *Nat. Hazards Earth Syst. Sci.* **2007**, *7*, 289–297. [CrossRef]
30. Liu, Y.; Zheng, D.; Georgakis, C.; Kabel, T.; Cao, E.; Wu, X.; Ma, J. Deformation Analysis of an Ultra-High Arch Dam under Different Water Level Conditions Based on Optimized Dynamic Panel Clustering. *Appl. Sci.* **2022**, *12*, 481. [CrossRef]
31. Wu, F.; Qi, S.; Lan, H. Mechanism of uplift deformation of the dam foundation of Jiangya Water Power Station, Hunan Province, P.R. China. *Hydrogeol. J.* **2005**, *13*, 451–466. [CrossRef]
32. Khalilzad, M.; Gabr, M.A.; Hynes, M.E. Deformation-Based Limit State Analysis of Embankment Dams Including Geometry and Water Level Effects. *Int. J. Géoméch.* **2015**, *15*, 04014086. [CrossRef]
33. Jadid, R.; Montoya, B.; Bennett, V.; Gabr, M. Effect of repeated rise and fall of water level on seepage-induced deformation and related stability analysis of Princeville levee. *Eng. Geol.* **2020**, *266*, 105458. [CrossRef]

water

MDPI

Article

Focal Mechanisms and Stress Field Characteristics of Microearthquakes in Baihetan Reservoir in the Downstream Area of Jinsha River

Wei Guo [1,2] and Cuiping Zhao [1,*]

1 Institute of Earthquake Forecasting, China Earthquake Administration, Beijing 100036, China
2 Seismological Bureau of Inner Mongolia Autonomous Region, Hohhot 010010, China
* Correspondence: zhaocp@ief.ac.cn

Abstract: The Baihetan Reservoir was impounded on 6 April 2021, after which the water level rose significantly. Notably, after one week of impoundment, microseismic activities were prominent around the reservoir area, which was highly associated with the water level change. From 6 April 2021 to 31 December 2021, over 7000 microearthquakes were recorded by the seismic stations in the vicinity of the reservoir, including 12 $M_L > 3$ events. The maximum was the 21 December 2021 M_L3.9 earthquake in Qiluogou town, Sichuan. The post-impoundment seismic events were clustered in Hulukou town in the Qiaojia Basin, with an overall "Y-shaped" pattern. In this study, taking advantage of the high-frequency waveform matching approach, the pre- and post-impoundment focal mechanism solutions totaling 207 $M_L > 2$ earthquakes are successfully obtained. The impoundment-induced stress change is analyzed, and the iterative joint inversion method is used to invert the stress field. Major results and conclusions include the following: (1) After impoundment, the number of normal fault earthquakes remarkably increased in the reservoir area; (2) Impoundment has led to a vertical compressive stress field and horizontal tensile stress field in the area where microearthquakes occurred. It is necessary to pay close attention to possible moderate-to-strong earthquakes in the future.

Keywords: Baihetan Reservoir; reservoir-induced earthquakes; focal mechanism; stress field

Citation: Guo, W.; Zhao, C. Focal Mechanisms and Stress Field Characteristics of Microearthquakes in Baihetan Reservoir in the Downstream Area of Jinsha River. *Water* **2023**, *15*, 709. https://doi.org/10.3390/w15040709

Academic Editors: Rui Pang, Binghan Xue, Xiang Yu, Yantao Zhu and Xiaoying Lin

Received: 5 January 2023
Revised: 7 February 2023
Accepted: 8 February 2023
Published: 10 February 2023

1. Introduction

The downstream area of the Jinsha River is located on the eastern margin of the Sichuan-Yunnan block, and is characterized by complex geological structures and frequent seismic activity. The regional topography is high in the west and low in the east, producing a huge drop and, hence, an abundant hydraulic energy resource. From south to north, 4 cascade hydropower stations (Wudongde, Baihetan, Xiluodu, Xiangjiaba) were successively constructed (Figure 1). In particular, the Baihetan station is situated at the junction zone of several large geological structures with different strikes. It possesses the largest reservoir storage capacity and the highest cascade benefit among the four stations, and is the second-largest hydropower station in the world (ranked only behind the Three Gorges station). This station has a height of 289 m, a storage capacity of up to 206.27×10^8 m^3, and an installed gross capacity of up to 1600 million kilowatts (kw). The Xiangjiaba, Xiluodu, Wudongde, and Baihetan Reservoirs began their initial impoundment in October 2012, May 2013, January 2020, and April 2021. Notably, a process of seismicity enhancement dominated by microearthquakes appeared after the impoundment of the Xiluodu Reservoir [1,2]. In this study, we obtained high-quality observational seismic data 5 years before and 9 months after the impoundment of the Baihetan reservoir through the seismic arrays continuously deployed in this region since 2016. A concentration of small earthquakes occurred in Hulukou town immediately after the impoundment. Baihetan Reservoir was impounded on 6 April 2021, after which the water level rose significantly.

Notably, after impoundment, microseismic activities were prominent around the reservoir area, highly associated with water level change. From 6 April 2021 to 31 December 2021, a total of over 7000 earthquakes were recorded by the seismic stations in the vicinity of the reservoir, including 12 events with a magnitude greater than 3. The earthquakes were dominantly clustered in Hulukou town and distributed along the main Jinsha River and the Heishuihe tributary [3].

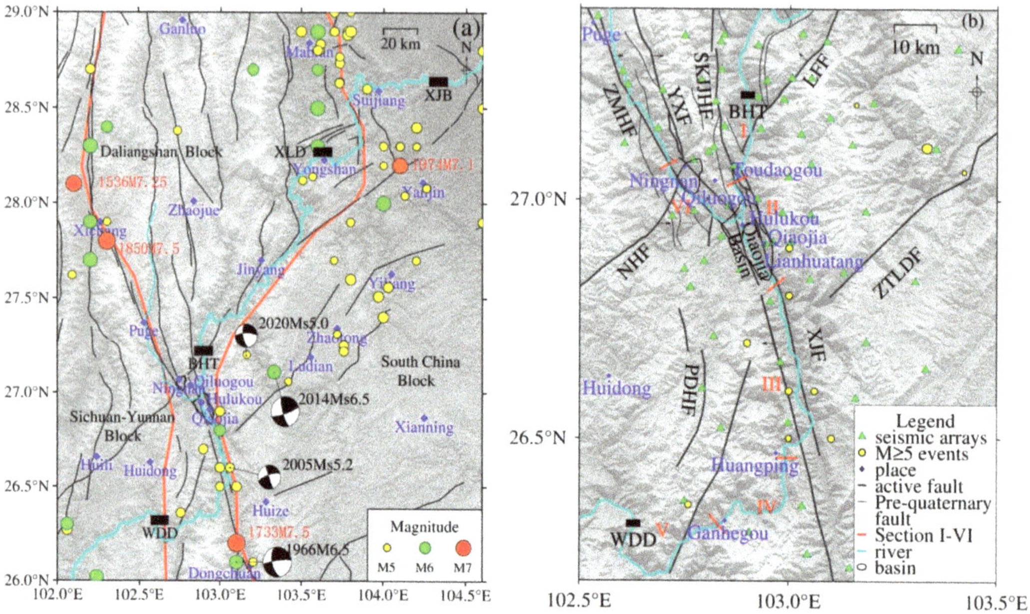

Figure 1. Tectonic background, historical earthquakes and distribution of seismic stations. (**a**) Tectonic background and historical earthquakes in the downstream area of Jinsha River; (**b**) Tectonic background, historical earthquakes, and distribution of seismic stations in the Baihetan Reservoir. Black rectangles represent hydropower stations. XJB: Xiangjiaba hydropower stations; BHT: Baihetan hydropower stations; XLD: Xiluodu hydropower stations; WDD: Wudongde hydropower stations. Black lines represent faults, obtained from activefault-datacenter.cn. LFF: Lianfeng fault; SKJJHF: Sikaijiaojihe fault; YXF: Yuexi fault; ZMHF: Zemuhe fault; NHF: Ninghui fault; PDHF: Puduhe fault; XJF: Xiaojiang fault.

Investigating the seismic focal mechanism and tectonic stress field before and after reservoir impoundment is essential for a more thorough understanding of the various characteristics of seismic type and stress state, providing significant evidence for further probing the mechanism of reservoir-induced earthquakes and analyzing the risk of larger earthquakes in the future. Over the past decades, numerous studies regarding the focal mechanism of earthquakes in typical large reservoirs have been conducted. For example, through studying the 19 March 1962 M6.1 earthquake in the Xinfengjiang Reservoir, Wang et al., (2002) [4] proposed that strike-slip was the main mechanism for the seismogenic faults of the mainshock and aftershocks within 18 months. The mechanism altered to dip-slip after 58 years, as indicated by He et al., (2018) [5] after calculating the focal mechanism of $Ms > 1.5$ earthquakes in the same region. They also suggested that the main type is still a strike-slip mechanism outside the reservoir area. The long-lasting diffusion effect increased the pore pressure, resulting in deeper earthquakes and hence the change of seismic type. Yao et al., (2017) [6] analyzed the variation characteristics of seismicity and related focal mechanisms throughout different impoundment periods of the Three Gorges Reservoir. They suggested that seismicity was closely associated with water level change.

During the impoundment period with 135 m, 156 m, and 175 m water levels, strike-slip and normal mechanisms were the most dominant. In addition, the P- and T-axes were discretely distributed, inconsistent with the regional tectonic stress field. By contrast, multiple larger earthquakes occurred in the later stage of the 175 m impoundment, and the P- and T-axis distribution was accordant with the stress field.

It was suggested that the small regional earthquakes were triggered by the reservoir, while the large earthquakes were controlled by regional faults and tectonic stress fields. Similar investigations in the Longtan Reservoir in Guangxi [7,8] suggested a variety of earthquake types in the early stage of impoundment. After 51 months, earthquakes were primarily seen in the shallow layers and were mostly related to reverse faults. The difference of seismicity and seismic properties between shallow and deep layers can be attributed to the difference in their intrinsic features, such as the tectonic stress field, rock mechanics properties, and permeability. The study area of this paper, which is the downstream area of the Jinsha River, has also been investigated by some researchers. Xiluodu Reservoir was impounded on 4 May 2013. As of 31 October 2013, Diao et al., (2014) [2] obtained the focal mechanisms of over 700 earthquakes before and after the impoundment of the Xiluodu Reservoir (2007–2013). They reported that the focal mechanisms of small earthquakes at the initial stage of impounding were characterized by complicated spatial fault plane orientation, various fracturing types, and unstable stress states. Duan and Zhao (2019) [9] further calculated the focal mechanism solutions of earthquakes (2016–2018) in this area, obtained the maximum principal compressive stress σ_1 axis and the maximum tensile stress σ_3 axis, and compared them with the results calculated by Diao et al., (2014) before and at the initial stage of impounding. The results showed that the spatial distribution of the stress axis is closer to that before impoundment. Therefore, Duan and Zhao (2019) concluded that as of 2018, the stress state had gradually recovered to the pre-impoundment state. The pre-impoundment focal mechanism in Baihetan and Wudongde Reservoirs is dominated by the strike-slip type.

The seismicity and focal mechanism solution before and after the water impoundment of Baihetan Reservoir provided an important example for investigating reservoir-induced earthquakes in Alpine canyon reservoir regions. Based on the precise earthquake location and seismicity analysis, together with the assistance of waveform data collected by the continuously completed seismic array since 2013, the focal mechanism of small earthquakes in the Baihetan Reservoir area was successfully obtained using the waveform matching method. Furthermore, the pre- and post-impoundment seismicity features and the corresponding variation characteristics of the epicenter and stress field were analyzed to provide data for determining the future large earthquake risk and other relevant studies [10].

2. Tectonic Background and Data

The Baihetan Reservoir in the lower reaches of the Jinsha River is located in the transitional zone extending from the strongly uplifted Qinghai Tibet Plateau and the eastern edge of the Western Sichuan Plateau to the Yunnan Guizhou Plateau and Sichuan Basin. The dam site is located east of the Kangdian axis within the Yangtze Huaihe Platform [11]. The left bank of the dam site is Ningnan County, Sichuan Province, and the right bank is Qiaojia County, Yunnan Province. The regional geological structure is complex. Deep and large fractures have developed, and some intersect the Jinsha River. The regional large fault east of the Zemuhe-Xiaojiang fault zone, driven by the northwest-southeast principal compressive stress field, is characterized by the northwest to southeast thrusting nappe, while the strike-slip movement occurs under the traction of the left lateral strike-slip movement of the Xiaojiang fault zone. The blocks enclosed by the regional fault zone, driven by the Sichuan Yunnan rhombic block wedged in the south-southeast direction, show different degrees of rotation, resulting in the Ningnan Basin and Qiaojia Basin. The Zemuhe fault strikes north-northwest, extending from the north of Xichang to the Ningnan and Qiaojia, and the fault dips northeast with a high dip angle. The Lianfeng-Qiaojia fault developed along the axis of the Lianfeng anticline and extends northeast along

Qiaojia-Dazhai Lianfeng-Yanjin, and the fault dips northwest with a high dip angle. The Sikaijiaojihe fault strikes nearly north-south, and it is in a right-step oblique sequence along Sikai-Toudaogou-Jinshajiang and Xiaojiang fault zones. The Yuexi fault intersects with the Zemuhe fault in the northwest area of Ningnan town, both of which are sinistral strike faults.

The Zemuhe fault zone is characterized by obvious segmentation. Around Ningnan, it is rifted into a basin. Each segment is composed of several small feather-like structures, and the southeast part, the Ningnan-datong fault, shows frequent seismic activity. The overall strike and dip of the Ningnan-datong fault are N40°W and NE, with a dip angle of more than 60° and a length of about 20 km. The reservoir water is closely connected with the Ningnan-datong fault. The Xiaojiang fault extends from Qiaojia to the south along the Jinsha River and Xiaojiang River valley, trending north-northwest and dipping to the west in the section of Qiaojia, and turning to nearly north-south in the southeast of Dongchuan. The Qiaojia Basin along the Xiaojiang fault extends along the Jinsha River valley, 13 km long from north to south and 3–5 km wide. Hulukou Town, located at the junction of the Zemuhe fault, Sikaijiaojihe fault, and the north section of the Xiaojiang fault zone, is also where the Jinsha River and the northwest branch of the Heishui River converge. Due to the relatively open terrain in the valley, it is completely submerged after impounding. The length of the Zemuhe fault zone inundated from Hulukou along the Heishuihe branch is about 20 km, and the length of the Xiaojiang fault zone inundated by reservoir water from the Qiaojia Basin to the tail of the Xiaojiang branch is about 50 km. The Xiaojiang fault zone and Jinsha River experience historically frequent seismic activity. Since 1770 BC, there have been 9 earthquakes above *Ms*5.0 in the reservoir area. The Menggu *Ms*6.0 earthquake of 15 May 1935 in the south of Qiaojia was only 10 km from the reservoir area. The 3 August 2014 Ludian *Ms*6.5 earthquake and 18 May 2020 Qiaojia *Ms*5.0 earthquake that occurred in recent years were also near the reservoir area (Figure 1).

In this study, using the observational data (2013.8.1–2021.12.31) recorded by the dense seismic array in the downstream area of the Jinsha River, the pre- and post-impoundment seismicity and focal mechanism solution features of earthquakes are analyzed in the vicinity of the Baihetan Reservoir area (26.2°~27.4° N, 102.5°~103.5° E). The seismic array consists of a total of 169 seismic stations, including 74 stations deployed by the research group of the present paper in the downstream area of the Jinsha River, 62 stations that belong to the Qiaojia seismic array deployed by the Institute of Geophysics, and 33 stations from the Sichuan, Yunnan, and Guizhou Regional Seismic Network. The seismometers are CMG-3ESPC, CMG-3T, CMG-40T, Trillium 120P, BBVS-60, GL-PCS60, etc. These instruments have a frequency band range of 50 Hz~60 s or 50 Hz~120 s, with a sampling rate of 100 Hz. The 63 stations from the Jinsha River dense array are distributed throughout the Baihetan Reservoir area (Figure 1), with a station interval smaller than 10 km and the ability to monitor $M_L < 1.0$ microearthquakes. Such extensive and widespread station distribution has laid a solid foundation for studying the seismicity and related focal mechanisms before and after the Baihetan Reservoir impoundment.

After comprehensively analyzing the regional geological structure, lithology, hydrogeology, and seismicity, the main river channel (i.e., from the Baihetan Dam to the Wudongde Dam) was divided into five sections according to different risk levels of reservoir-induced earthquakes. Based on the observational results in this study, we added the sixth section (the Heishuihe tributary section), in which the increase in water level in the Jinsha River triggered several microearthquakes. Finally, six reservoir sections were selected for monitoring and investigation (Figure 1b).

3. Method and Parameter Setting

The seismic activity in the Baihetan Reservoir is dominated mostly by microearthquakes, and the magnitudes are generally less than M_L3.0. In this study, the high-frequency waveform matching method [12,13] is employed to obtain the focal mechanism solutions of these microearthquakes. This method realizes the maximum matching degree between the seismic phases and amplitude of both observational and theoretical waveforms. Furthermore, the method constrains the matching degree using the P-wave first motion and the S/P amplitude ratio, thereby constructing the objective function that contains four types of constraints and searching for the optimal solution using the grid-searching approach. The Green function utilizes the discrete wavenumber method for calculation [14,15], and the objective function is as follows:

$$\begin{aligned} &\text{maximize}[J(x, y, z, dip, rake, ts)] \\ &= \sum_{n=1}^{N} \sum_{j=1}^{3} \{\alpha_1 \max(\widetilde{d_j^n} \otimes \widetilde{v_j^n}) - \alpha_2 ||\widetilde{d_j^n} - \widetilde{v_j^n}|| + \alpha_3 f[pol(\widetilde{d_j^n}), pol(\widetilde{v_j^n})] + \alpha_4 h[rat(\frac{S(d_j^n)}{P(d_j^n)}), rat(\frac{S(v_j^n)}{P(v_j^n)})]\} \end{aligned} \quad (1)$$

where $\widetilde{d_j^n}$ represents the normalized data, and $\widetilde{v_j^n}$ is the normalized theoretical waveform. $\alpha_1 \sim \alpha_4$ are the weighting factors of each term. The optimal value is selected considering that each term cannot dominate the objective function. Specifically, the first term calculates the maximum cross-correlation coefficient between the normalized data and normalized theoretical waveform. The negative sign in the second term is used to minimize the amplitude difference. These two terms are not independent, and their combination can better constrain the waveform similarity. The third term calculates and determines if the P-wave first motion polarity of the observational data is consistent with that of the theoretical waveform. The fourth term measures the consistency of the S/P amplitude ratio between the observational waveform and the theoretical waveform. In addition, the weighting factors of the waveform, the P-wave first motion, and the amplitude ratio are selected as 4, 2, and 0.5, respectively. The matching frequency band is set as 2–4 Hz. During the inversion of focal mechanism solutions, a layered velocity model constructed by the artificial depth-measuring profile [16] is employed to compute the Green function.

4. Results and Analyses

4.1. Seismicity in the Reservoir Area

The Baihetan Reservoir started its impoundment on 6 April 2021, and the water level gradually increased from 658 m (Figure 2a). As of 13 April, the water level had reached 690 m (Figure 2b), and the number of microearthquakes dramatically increased from below 100 (Since 2016) to over 400 times monthly (Figure 2c). As of 30 September 2021, the water level had reached the maximum (816 m), and both the monthly and daily earthquake frequencies reached their peaks (Figure 2d). After that, with the gradual drop in the water level, earthquake frequency also showed strong fluctuation. As of 31 December 2021, 7401 above M_L0 earthquakes had been recorded by the authors of the present study, including 5400 M_L0.0–0.9 events, 1858 M_L1.0–1.9 events, 141 M_L2.0–2.9 events, and 12 M_L3.0–3.9 events. The largest one, the 21 December 2021 M_L3.9 earthquake, occurred in Ningnan, Sichuan.

The post-impoundment seismic events in the reservoir area were clustered in Hulukou town, 37 km away from the dam, with an overall "Y-shaped" pattern. The three seismic branches of the "Y" pattern lie along the Sikaijiaojihe and Jinsha River, the Zemuhe fault and Heishuihe tributary, and the northern section of Xiaojiang fault and the Jinsha River, with a length of about 30 km, respectively. After impoundment, the seismic events first occurred in the area of Hulukou town and then developed northward to section I (Figure 3b). The seismicity stopped about 7 km from the dam along the Sikaijiaojihe fault and the Jinsha River towards the dam, forming three small belts spreading north-northwest from the riverside. The southern section of the Sikaijiaojihe fault is divided into two branches, the western and eastern parts, both of which dip steeply to the west. The branch fault is distributed in the same direction. The focal depth of these belts is relatively shallow, concentrated at 0–8 km (Figure 3d). After one week of impoundment, earthquakes along the north section of the Xiaojiang fault and in the Qiaojia Basin occurred in section II (Figure 3b). The northern section of the Xiaojiang fault is hydraulically connected with the reservoir water. Several concealed water-filled fault zones parallel the Xiaojiang fault developed in the Qiaojia Basin. The focal depth of earthquakes in section II is deeper than that of reservoir section I, concentrated at 0–10 km. It is characterized by the continuous deepening of the focal depth southward (Figure 3d). The seismic activity along the north section of the Xiaojiang fault and the Jinsha River towards the south stopped near the Ms6.0 earthquake in Qiaojia South Menggu on 15 May 1930. After one month of impoundment, earthquakes began to occur in section VI (Figure 3b). The earthquakes were distributed in belts along the Zemuhe fault or Heishuihe tributary (Figure 3d). The focal depth was concentrated at 0–9 km. The farther away from the Heishuihe tributary, the shallower the focal depth (about 3 km). The focal depth of earthquakes near the river and the fault was relatively deeper (about 8 km). The seismic activity along the NW Zemuhe fault and the branch direction of the Heishui River stopped near the south of Ningnan.

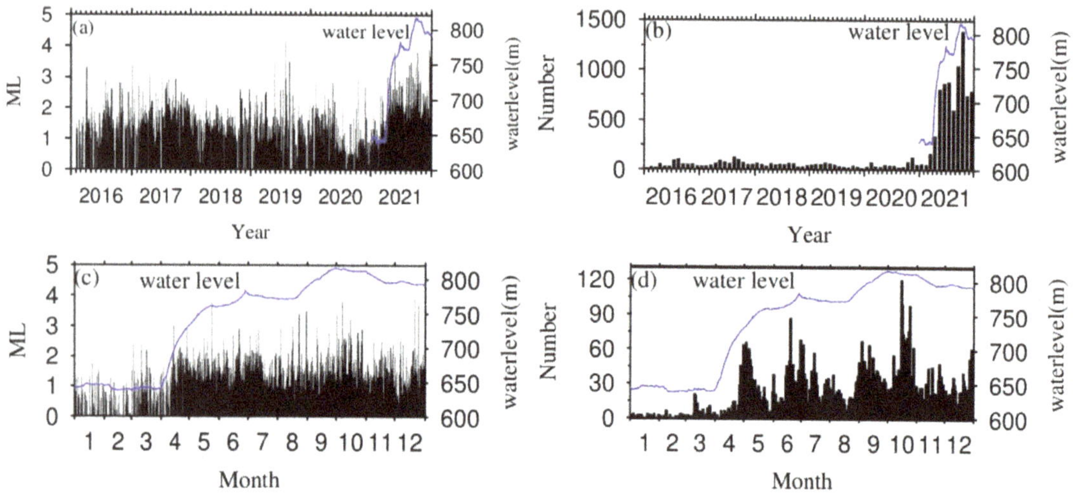

Figure 2. Magnitude, monthly frequency of seismicity, and water level sequence diagram in the Baihetan Reservoir. (**a**) Magnitude-time plot; (**b**) Monthly frequency-time plot; (**c**) Magnitude and water level (blue line) since January 2021; (**d**) Daily frequency and water level (blue line) of seismicity since January 2021.

4.2. Focal Mechanism Solution

The 207 $M_L \geq 2$ focal mechanism solutions before and after the impoundment of Baihetan Reservoir were obtained using the high-frequency waveform matching method. Figure 4 shows the waveform matching results of the 22 June 2021 M_L 2.2 earthquake in Ningnan, Sichuan, and the 24 December 2021 M_L 3.7 earthquake in Qiaojia, Yunnan. In this study, the classification method proposed by Zoback (1992) [17] was utilized (Table 1) to compile the focal mechanism solutions in the Baihetan Reservoir, according to the plunge of the P, B, and T-axes.

To evaluate the stability of the present study's results, the mean value and standard deviation of the first 100 best solutions of the two events were calculated (Figure 4) (Table 2). The results show the errors of strike, dip, and rake were all around 10°, consistent with the stability testing results of Li et al., (2011) [12].

Figure 5 displays the pre- and post-impoundment earthquake locations and the corresponding mechanism solutions of reservoir sections I, II, and VI. In general, normal fault-induced earthquakes are the most dominant after reservoir impoundment, which is similar to the feature that occurred in the downstream Xiluodu Reservoir during its water level increase stages. In the following, the focal mechanism features are analyzed section by section.

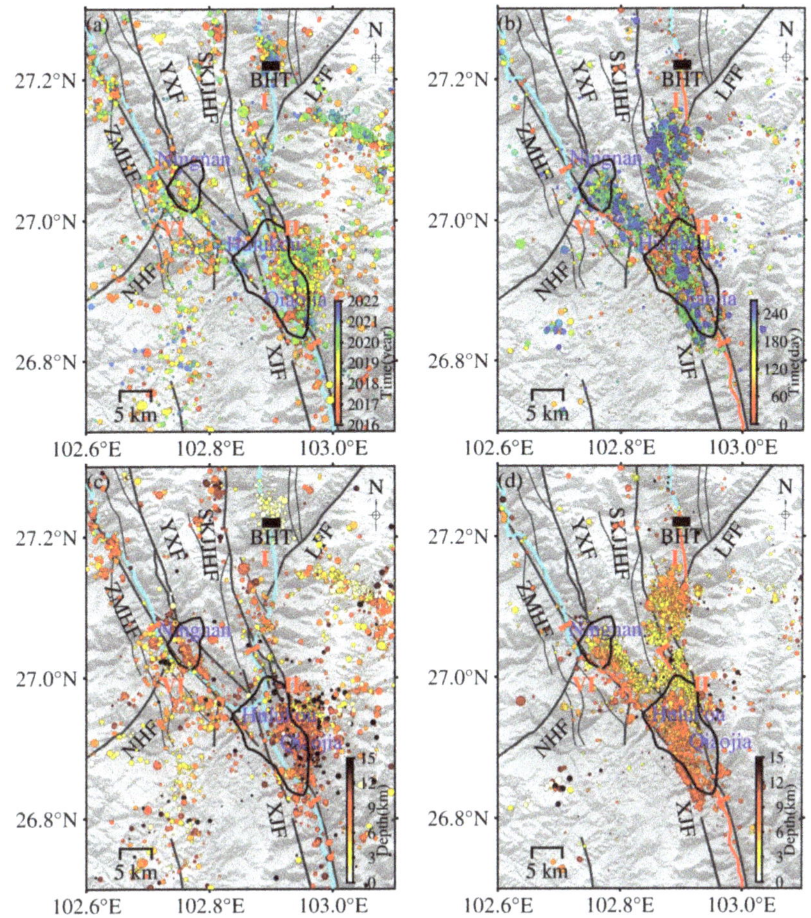

Figure 3. Spatial and temporal distribution of earthquakes in the reservoir area before and after impoundment. (**a**) Distribution of earthquakes that occurred before impoundment; (**b**) Distribution of earthquakes that occurred after impoundment; (**c**) Distribution of earthquakes with different depths before impoundment; (**d**) Distribution of earthquakes with different depths after impoundment. The red curves in (**b**,**d**) represent the reservoir sections where the post-impoundment water rise exceeded 30 m; black trap areas are the Qiaojia Basin and Ningnan Basin, respectively. Red short lines and Roman characters represent the I, II, and VI sections, respectively. LFF: Lianfeng fault; SKJJHF: Sikaijiaojihe fault; YXF: Yuexi fault; ZMHF: Zemuhe fault; NHF: Ninghui fault; XJF: Xiaojiang fault.

Reservoir section I (from Hulukou to the front of the dam): Before impoundment, there were few seismic activities from the Hulukou to the front of the dam, and although there were various types of focal mechanisms, their nodal plane strikes are consistent with the strike of the Sikaijiaojihe fault (Figure 5a). After one week of impoundment, the small earthquake began to increase, forming three parallel small earthquake belts extending from the Jinsha River in a northwest direction, the focal depths of the earthquakes were 3–5 km, and the largest earthquake (M_L3.9) occurred on 21 December 2021 in Qiluogou Town, Ningnan County, Sichuan Province. Among the 35 $M_L \geq 2$ mechanism solutions obtained (Figure 5b), the proportions of normal fault type, thrust fault type, strike-slip fault type, and other types of earthquake events account for 54.28%, 8.57%, 25.71%, and 11.44%, respectively. This suggests that in addition to the obvious dominance of normal fault-induced earthquakes, a few small earthquakes of strike-slip and thrust types existed. Among them, the $M_L3.9$ earthquake was the strike-slip type, and the strikes of the two nodal planes were northwest and northeast, respectively. This reservoir section is mainly composed of carbonate rocks. No large-scale karst pipeline system, karst caves, underground rivers, funnel hot springs, etc., were found in the karst hydrogeological surveys. Therefore, it is speculated that the earthquakes in this region were mainly caused by the reservoir water gravity on the shallow cracks or small-scale secondary structures or the rapid fluid penetration as the water level rose.

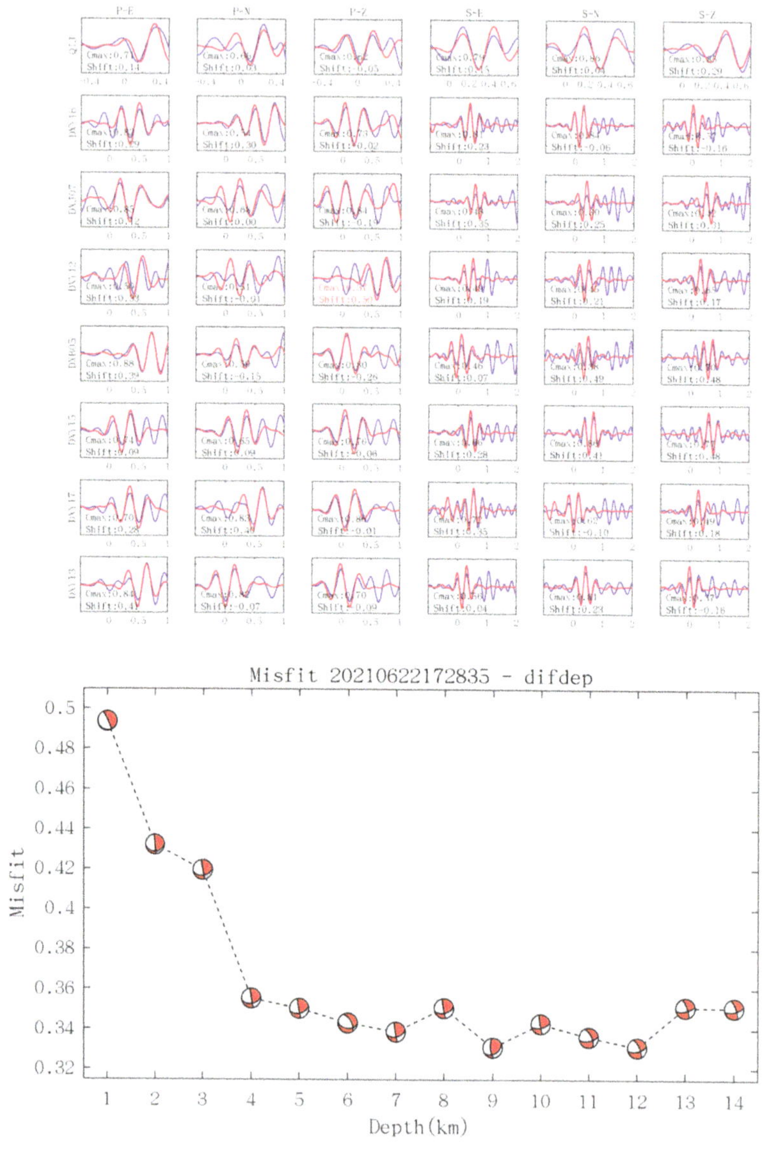

(a)

Figure 4. *Cont.*

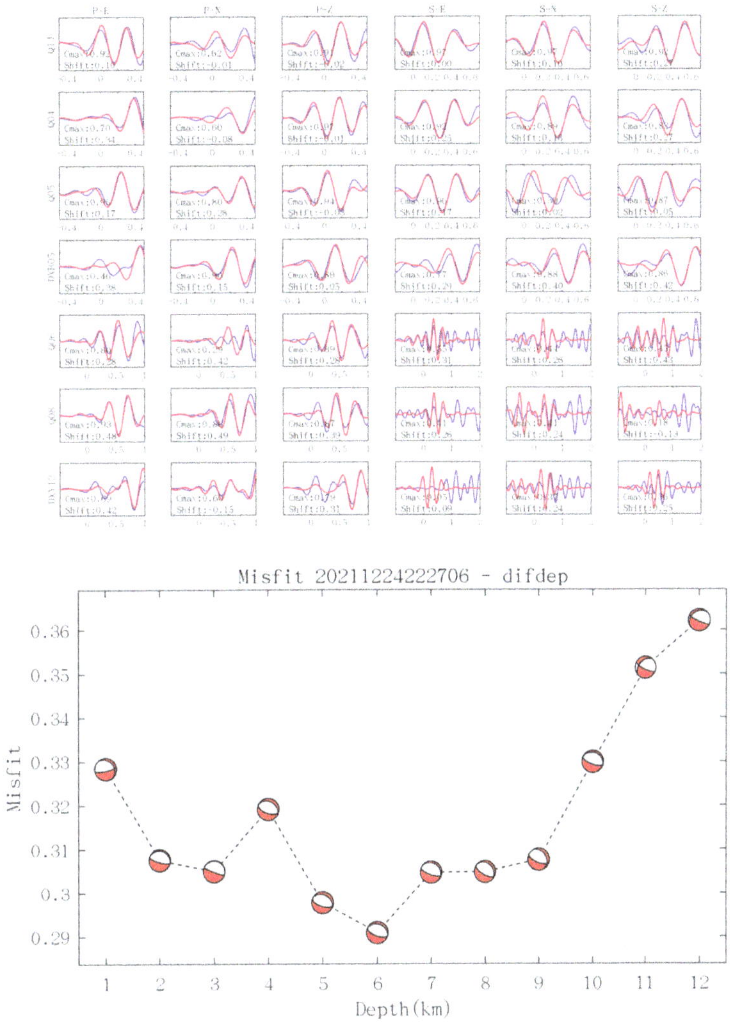

(**b**)

Figure 4. Focal mechanism solutions and waveform fitting. (**a**) The 22 June 2021 M_L 2.2 earthquake; (**b**) The 24 December 2021 M_L 3.7 earthquake. The uppermost plot in Figure 4 is the waveform matching figure. Each row is the matching condition of the observed waveform (blue) and the theoretical waveform (red) recorded by a station. From left to right, the 6 waveforms represent the EW component of the P-wave, the NS component of the P-wave, the Z component of the P-wave, the EW component of the S-wave, the NS component of the S-wave, and the Z component of the S-wave. The lower left corner shows the maximum correlation coefficient (Cmax) and the time shift (Shift). The lowermost plot shows the variation of the waveform fitting mismatch value with the change of focal depth.

Reservoir Section II (from Hulukou southward to Xiaojiang Fault): Reservoir Section II is the broadest zone in the Baihetan Reservoir area. The Xiaojiang fault is controlled by stress from the southeast and south-southeast directions of the Sichuan-Yunnan rhombic block and the northwest and northwest-west stress from the South China block [18]. The fault is a sinistral and normal fault with a north-northwest strike. Prior to the impoundment, there were mainly seismic activities in the northern section of the Xiaojiang fault, and the earthquakes were distributed along the fault in the north-northwest direction. The mechanism solutions of 48 earthquakes (Figure 5c) were obtained, of which the normal fault type, thrust fault type, strike-slip type, and other types of seismic events account for 46.92%, 6.25%, 12.5%, and 34.33%, respectively. The largest proportion of normal fault type is related to the fact that Qiaojia is a pull-apart basin, consistent with the nature of the northern section of the Xiaojiang fault zone. After one week of impoundment, the earthquakes were distributed along the Jinsha River, Qiaojia Basin, and the northern section of the Xiaojiang fault in a south-southeast direction. The corresponding focal depths were concentrated at 0–10 km, deeper than those north of Hulukou, and showed a southward deepening pattern. Jiang et al., (2014) [19] summarized the characteristics of reservoir earthquakes in the Chinese Mainland. They found that at the initial stage of reservoir impoundment, with the development of the impoundment process, the focal depth had a gradually deepening trend, mainly related to the fluid infiltration and the gradual deepening of the depth of the changes in pore pressure. After several years, the focal depth tended to be stable. If the focal depth showed an obvious downward trend, attention should be paid to the possibility of a major earthquake. By investigating the depth change of the Koyna Reservoir before several major earthquakes, Rostogi et al., (1999) [20] found that about one month before major earthquakes, the focal depths often showed a statistically credible downward trend. After impoundment, the proportions of normal fault type, thrust fault type, strike-slip fault type, and other types of earthquakes in section II were 36.23%, 15.94%, 26.09%, and 21.74%, respectively (Figure 5d). It is prominent that although the proportion of thrust and strike-slip earthquakes increased, the proportion of normal fault earthquakes was still the largest. Particularly, this reservoir section is characterized by the most complicated geological structure and hydrogeological conditions in the entire reservoir area. The ground surface on both sides of the river valley is composed of the Quaternary fluvial sedimentary layer, and the underlying formations consist of carbonate rocks. In the Qiaojia Basin, there are several concealed water-filled faults parallel to the strike of the Xiaojiang fault. On both sides of the water-filled faults are karst and basalt water-filling bodies. Hydraulic connections may exist between the deep faults and the karst water-filling bodies. The small earthquakes distributed in belts along the Jinsha River may have been generated by the stress and strain readjustment under the physical or chemical effects (e.g., self-weight stress, reservoir water loading, wedging, and pore water pressure) of the reservoir bank slope or rock interfaces. Besides, this section is also the area with the highest seismic intensity after impoundment. In total, three $M_L \geq 3.7$ earthquakes occurred in Qiaojia, Yunnan (2021.05.24 M_L 3.8, 2021.10.07 M_L 3.8, 2021.12.24 M_L 3.7), and the corresponding mechanism solutions were (strike 170°, dip 70°, rake −30°), (strike 345°, dip 30°, rake −70°), (strike 290°, dip 25°, rake −90°). It is suggested by the authors that the genesis of seismicity after impoundment in section II is associated with the joint effect of a regional tectonic stress field and reservoir impoundment.

Reservoir Section VI (from Hulukou to Nningnan along the Zemuhe fault and Heishuihe tributary): When the water level of the Heishuihe tributary rose by 30 m, seismic activity began to appear in the northwest-most of Ningnan Basin. Then, the seismic activity gradually propagated toward Hulukou in a southeast direction, forming a belt-shaped seismic distribution consistent with the north-northwest-trending fault zone basin on the east side of the Heishui River. The focal depths of the earthquakes are concentrated at 0–9 km (Figure 3d). The focal depths became shallower (about 3 km) as the earthquakes moved away from the Heishui River eastward, while those near the Heishuihe tributary and the Zemuhe fault were relatively deep (about 8 km) and the same those occurred there before loading. The seismic activity in this section of the reservoir ceased near Ningnan. Among these seismic events, the largest earthquake was the 2021.08.30 M_L3.5 earthquake in Ningnan, Sichuan. The event had a focal mechanism solution of (strike 315°, dip 65°, rake −20°) and was a strike-slip earthquake, consistent with the northwest sinistral strike-slip feature of the Zemuhe fault. Before impoundment, the proportions of normal fault type, thrust fault type, strike-slip fault type, and other types of $M_L \geq 2$ earthquake events were 26.32%, 5.26%, 47.37%, and 21.05%, respectively (Figure 5e). The proportions after impoundment became 50.00%, 0.00%, 14.28%, and 35.72%, respectively (Figure 5f). It is obvious that the number of normal fault earthquakes in section VI increased significantly after impoundment. According to the geological background around this section, it is suggested by the authors that with the increase of the reservoir water level, water loading may lead to the increase of the maximum principal compressive stress in the vertical direction, producing small-scale normal fault activities. Meanwhile, the gravity of water and the water-induced softening of the structural surface may also promote the generation of micro-cracks in the rock mass, further causing the collapse of the free surface at the riverside. The seismic events here belong to reservoir-induced earthquakes associated with the shallow micro-cracks.

Table 1. Types of focal mechanism solutions.

Type	Plunge of P-Axis σ_P/°	Plunge of B-Axis σ_B/°	Plunge of T-Axis σ_T/°
Normal fault	≥52		≤35
Normal-strike-slip	$40 \leq \sigma_P < 52$		≤20
Strike-slip	<40	≥45	≤20
Thrust-strike-slip	≤20		$40 \leq \sigma_T < 52$
Thrust fault	≤35		≥52
Others		$20 < \sigma_P, \sigma_B, \sigma_T < 45$ or $40 \leq \sigma_P, \sigma_T \leq 50$	

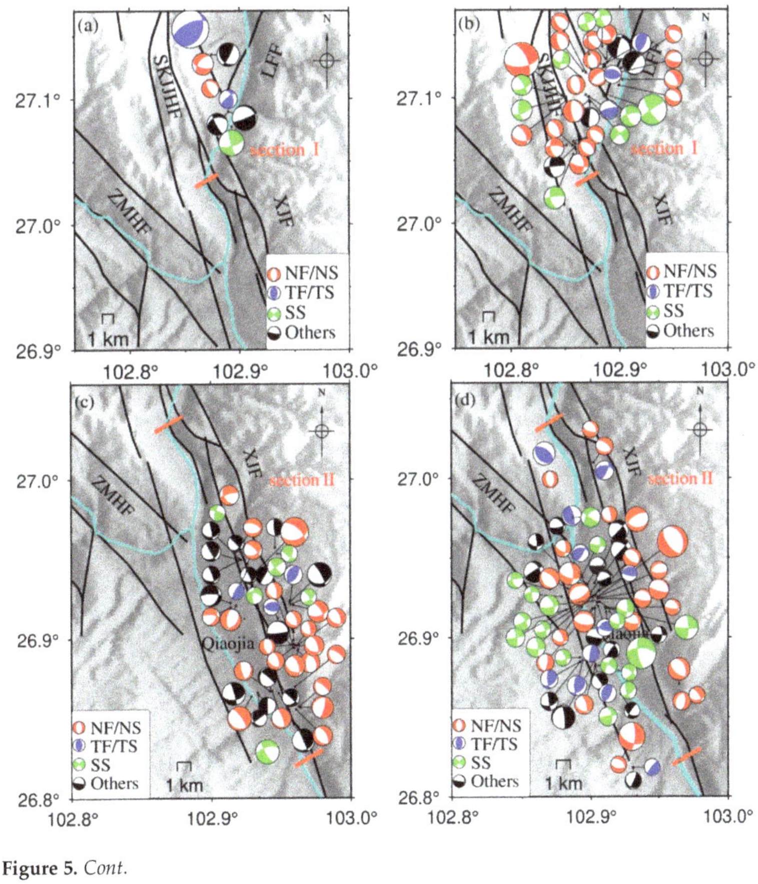

Figure 5. *Cont.*

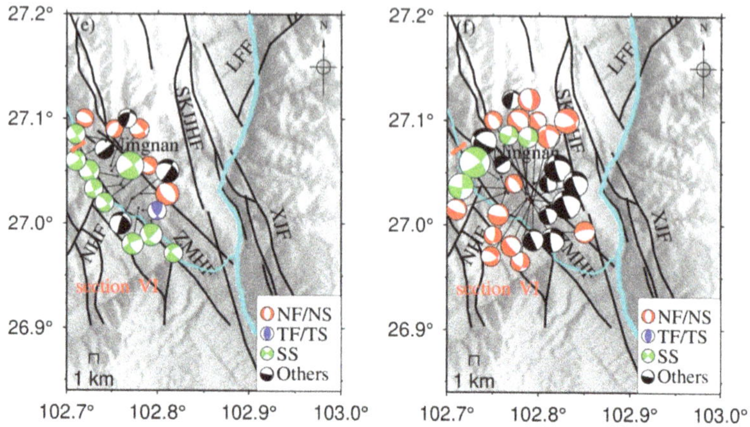

Figure 5. Focal mechanism solutions of $M_L \geq 2$ earthquakes around the Baihetan Reservoir. (**a**,**b**) Section I; (**c**,**d**) Section II; (**e**,**f**) Section VI. Left figure (**a**,**c**,**e**): Pre-impoundment (1 August 2013–5 April 2021); Right figure (**b**,**d**,**f**): Post-impoundment (6 April 2021–31 December 2021).LFF: Lianfeng fault; SKJJHF: Sikaijiaojihe fault; ZMHF: Zemuhe fault; NHF: Ninghui fault; XJF: Xiaojiang fault.

Table 2. Mean and standard deviation of the first 100 best solutions.

Seismic Event	Index	Strike/°	Dip/°	Rake/°
2021.06.22 M_L 2.2	Optimal solution	175	70	55
	Mean value	173	79	42
	Standard deviation	5	7	10
2021.12.24 M_L 3.7	Optimal solution	290	25	−90
	Mean value	308	28	−74
	Standard deviation	13	4	11

Figure 6 shows the statistics of the nodal plane strike of the earthquakes that occurred in the three reservoir sections, respectively. It is obvious that the dominant nodal plane strikes in section I are concentrated from northwest to north-northwest, consistent with the strike of the nearest large-scale fault zone (Sikaijiaojihe fault). The dominant nodal planes of the mechanism solution of earthquakes in section II are concentrated from northwest to north-northeast, which accords well with the strike of the Xiaojiang fault. For section VI, the dominant nodal plane strikes are also concentrated in the direction from northwest to north-northeast, consistent with the strike of the Zemuhe fault. The overall pattern of the focal mechanism solutions indicates that most of the microearthquakes occurred near the river channel after impoundment and were generally distributed along valley basins controlled by large fault zones or smaller-scale structures.

Figure 6. Rose diagrams of nodal strikes for focal mechanism in three regions of Baihetan Reservoir.

4.3. Stress Field

Furthermore, the iterative joint inversion method [21] was used to invert the stress field of the three sections. The pre- and post-impoundment P and T -axes of the earthquake and the principal stress axis added with 100 times of random noise are shown in Figure 7 and Table 3, respectively. Based on the focal mechanism solutions of small earthquakes in the northern section of the Xiaojiang fault, Yan (2015) [22] found that the azimuth of the maximum principal compressive stress σ_1 axis in this area was around 120°, and the dip was 23°. The results also showed that the azimuth of the maximum tensile stress σ_3 axis was 215°, and the dip was around 14°. The azimuth of the medium principal compressive stress σ_2 axis was around 335°, and the dip was around 63°. The above studies indicate that this area was mainly affected by near-horizontal compressive stress and tensile stress before the impoundment of the Baihetan Reservoir. According to Figure 7 and Table 3, the post-impoundment azimuth and dip angle of the maximum tensile stress axes of the three reservoir sections are almost in the same south-southwesterly direction, and nearly horizontal. The maximum principal compressive stress axes are all nearly vertical, revealing that the strong vertical principal compressive stress and the horizontal tensile stress field control both sides of the river. The R-value is expressed as $R = (\sigma_2\text{-}\sigma_1)/(\sigma_3\text{-}\sigma_1)$, indicating that the post-impoundment stress field of the three sections tends to be dominated by tensile stress. Among the three sections, section I had the highest water level rise in front of the dam. The earthquakes in section VI before impoundment were mainly distributed along the Zemuhe fault. The earthquakes after impoundment generally occurred in the Zemuhe fault zone and the Ningnan Basin on the east side of the Heishui River.

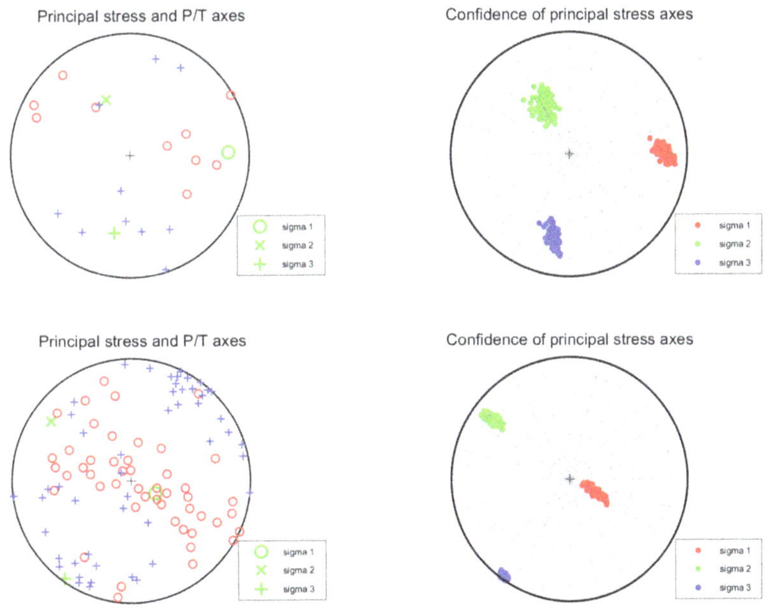

(a)

Figure 7. *Cont.*

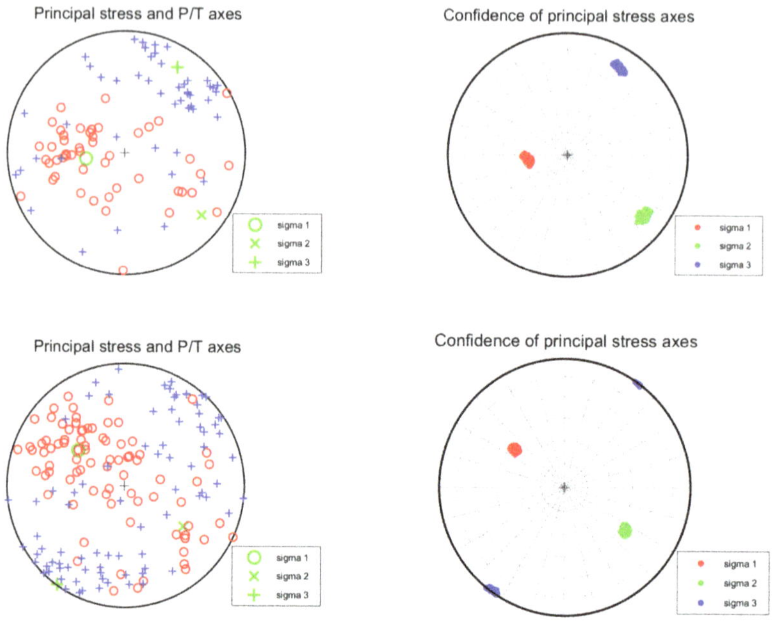

(**b**)

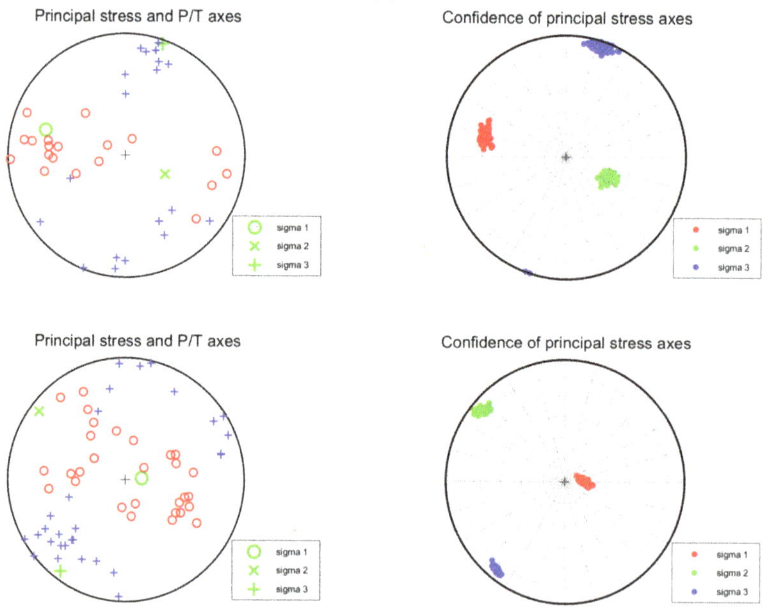

(**c**)

Figure 7. Spatial distribution map of P, T, and stress axes in three sections of the Baihetan Reservoir area after impoundment. (**a**) Section I; (**b**) Section II; (**c**) Section VI.

Table 3. Number of focal mechanisms and the inversion results of the stress field in each reservoir section before and after impoundment.

Section	Stress Field	σ_1(/°) Azimuth	σ_1(/°) Dip	σ_2(/°) Azimuth	σ_2(/°) Dip	σ_3(/°) Azimuth	σ_3(/°) Dip	R	N
I	Before impoundment	88.5	18.4	336.3	48.6	192.2	35.5	0.58	8
	After impoundment	116.9	71.5	305.9	18.3	215.0	2.7	0.26	35
II	Before impoundment	262.0	62.8	128.5	19.5	31.8	18.2	0.60	48
	After impoundment	306.4	49.8	124.2	40.2	215.1	1.1	0.55	69
VI	Before impoundment	287.0	29.9	114.6	59.9	18.9	3.3	0.66	19
	After impoundment	84.5	78.4	307.5	8.5	216.3	7.8	0.55	28

Overall, the Baihetan Reservoir is currently in the initial stage of impoundment. The considerable rise in seismicity frequency is related to the high water level and its drastic change following 6 April 2021. Additionally, a normal fault is the most dominant focal mechanism solution of the earthquakes, similar to what happened in the Three Gorges Reservoir and Xiluodu Dam within 1 year of impoundment.

5. Conclusions

As of 31 December 2021, the water level in Hulukou town had risen by nearly 158 m. In this study, the high-frequency waveform matching method is employed, and the focal mechanism solutions of 207 $M_L \geq 2.0$ earthquakes are obtained in the vicinity of the Baihetan Dam. By combining seismicity, focal mechanism solutions, regional geological settings, and the impoundment-induced water-level change, the major conclusions are obtained as follows:

(1) The distribution of pre-impoundment earthquakes in the reservoir area is mainly controlled by regional stress fields and faults. As the impoundment proceeds, the number of earthquakes increases significantly, especially small and microearthquakes. These earthquakes are clustered in Hulukou town and dominantly distributed along the north-northeastern Sikaijiaoji and Jinsha Rivers, the north-northwestern Zemuhe fault and Heishuihe tributary, and the south-southeastern Xiaojiang fault and Jinsha River, showing an overall "Y-shaped" pattern. In addition, the number of earthquakes in the Ludian aftershock zone enormously decreases;

(2) After the impoundment, microearthquakes are observed in three reservoir sections, where normal fault-induced earthquakes are the most predominant. The dominant distribution of the fracture planes obtained from the inversion of the mechanism solution is consistent with the direction of the local main structures or fault zones, indicating that they are controlled by the local tectonic environment;

(3) In the three reservoir sections with post-impoundment microearthquakes, the azimuth angle and dip angle of the maximum tensile stress axis are consistent; both are in the south-southwestern direction and nearly horizontal. whereas the maximum primary compressive stress axes are nearly vertical, suggesting the effect of post-impoundment vertical compressive stress and horizontal tensile stress fields in the areas with microearthquakes;

(4) Pre-existing fissures and structures in the reservoir area are the prerequisites for inducing earthquakes, and the water level change is an essential external factor that influences earthquake occurrence;

(5) The occurrence of earthquakes is closely related to the drastic increase in impoundment loading and the water-level-change-induced elastic stress on the side slope, joints, fissures, and other small-scale structures. Such features accord well with the features observed in the pre-impoundment period of the downstream area of the Xiluodu Reservoir. Due to the continuous influence of reservoir impoundment on

the surrounding geological environment, and considering complex regional tectonic structures and the occurrence of strong historical earthquakes, it is necessary to pay close attention to the possibility of moderate-to-strong earthquakes in the reservoir area in the future.

Author Contributions: Conceptualization, C.Z.; methodology, W.G.; software, W.G.; validation, W.G. and C.Z.; formal analysis, C.Z.; investigation, W.G.; resources, C.Z.; data curation, W.G.; writing—original draft preparation, WG.; writing—review and editing, C.Z.; visualization, W.G. and C.Z.; supervision, C.Z.; project administration, C.Z.; funding acquisition, C.Z. All authors have read and agreed to the published version of the manuscript.

Funding: This research was funded by the National Key Research and Development Program of China (2021YFC3000703) and the Special Fund of the Institute of Earthquake Forecasting, China Earthquake Administration (CEAIEF20220401).

Data Availability Statement: The data presented in this study are not publicly available due to confidentiality agreements.

Acknowledgments: The authors would like to thank Li Junlun from the University of Science and Technology of China provided a program for the inversion of earthquake focal mechanism solutions. The topography data used in this study was obtained from SRTMGL3 (https://lpdaac.usgs.gov/products/srtmgl3v003/ (accessed on 22 June 2021)), and figures were plotted using GMT [23]. Also, we would like to express our appreciation to the reviewers of this paper for their valuable comments.

Conflicts of Interest: The authors declare no conflict of interest.

References

1. Cheng, W. The reservoir induced earthquake problems in the high intensity areas. *Recent Dev. World Seismol.* **2013**, *4*, 10–18. (In Chinese)
2. Diao, G.; Wang, R.; Feng, X.; Wang, X.; Feng, Z.; Zhang, H.; Chen, W.; Li, Y.; Wang, L. Analysis of the focal mechanism after impoundment in the head area of Xiluodu Reservoir. *Seismol. Geol.* **2014**, *36*, 644–657. (In Chinese)
3. Guo, W.; Zhao, C.; Zuo, K.; Zhao, C. Characteristics of seismicity before and after impoundment of Baihetan dam in the lower reaches of Jinsha River. *Chin. J. Geophys.* **2022**, *65*, 4569–4671. (In Chinese)
4. Wang, M.; Yang, M.; Hu, Y.; Li, Z.; Chen, Y.; Jin, Y.; Feng, R. A preliminary study on the focal mechanism of the Xinfengjiang Reservoir earthquake and its causes. *Sci. China* **2002**, *19*, 85–97. (In Chinese)
5. He, L.; Sun, X.; Yang, H.; Qin, J.; Shen, Y.; Ye, X. Upper Crustal Structure and Earthquake Mechanism in the Xinfengjiang Water Reservoir, Guangdong. *China. J. Geophys. Res. Solid Earth* **2018**, *123*, 3799–3813. [CrossRef]
6. Yao, Y.; Wang, Q.; Liao, W.; Zhang, L.; Chen, J.; Li, J.; Yuan, L.; Zhao, Y. Influences of the Three Gorges Project on seismic activities in the reservoir area. *Sci. Bull.* **2017**, *62*, 1089–1098. [CrossRef] [PubMed]
7. Chen, H.; Zhao, C.; Xiu, J.; Chen, Z. Focal mechanism and stress field characteristics of reservoir induced earthquakes in Longtan reservoir area. *Seismol. Geol.* **2009**, *31*, 686–698. (In Chinese)
8. Yan, C.; Zhou, B.; Lu, L.; Sun, W.; Wen, X. Focal mechanism of small and medium earthquakes in Longtan reservoir area after impoundment. *Chin. J. Geophys.* **2015**, *58*, 4207–4222. (In Chinese)
9. Duan, M.; Zhao, C. Characteristics of focal mechanism in the downstream reservoir area of Jinsha River. *Seismol. Geol.* **2019**, *41*, 1155–1171. (In Chinese)
10. Chrysanidis, T.; Mousama, D.; Tzatzo, E.; Alamanis, N.; Zachos, D. Study of the Effect of a Seismic Zone to the Construction Cost of a Five-Story Reinforced Concrete Building. *Sustainability* **2022**, *14*, 10076. [CrossRef]
11. Ren, J. *Geotectonic Evolution of China*, 1st ed.; Science Press: Beijing, China, 1980; pp. 1–116. (In Chinese)
12. Li, J.; Zhang, H.; Kuleli, H.; Toksoz, M. Focal mechanism determination using high-frequency waveform matching and its application to small magnitude induced earthquakes. *Geophys. J. Int.* **2011**, *184*, 1261–1274. [CrossRef]
13. Li, J.; Kuleli, H.; Zhang, H.; Toksoz, M. Focal mechanism determination of induced microearthquakes in an oil field using full waveforms from shallow and deep seismic networks. *Geophysics* **2011**, *76*, 87–101. [CrossRef]
14. Bouchon, M. A simple method to calculate Green's functions for elastic layered media. *Bull. Seismol. Soc. Am.* **1981**, *71*, 959–971. [CrossRef]
15. Bouchon, M. A Review of the Discrete Wavenumber Method. *Pure Appl. Geophys.* **2003**, *160*, 445–465. [CrossRef]
16. Wang, C.; Mooney, W.D.; Wang, X.; Wu, J.; Lou, H.; Wang, F. Three-dimensional velocity structure of crust and upper mantle in Sichuan-Yunnan region. *Acta Seismol. Sin.* **2002**, *24*, 1–16. (In Chinese) [CrossRef]
17. Zoback, M.L. First- and second-order patterns of stress in the lithosphere: The world stress map project. *J. Geophys. Res.* **1992**, *97*, 703–728. [CrossRef]
18. Qian, X.; Qin, J.; Liu, L. Research on modern tectonic stress field in Yunnan area. *Seismol. Geol.* **2011**, 33, 91–106. (In Chinese)

19. Jiang, H.; Zhang, X.; Shan, X. *Research on the Statistical Characteristics and Prediction Methods of Reservoir Earthquakes in Mainland China*, 1st ed.; Seismological Press: Beijing, China, 2014; pp. 1–312. (In Chinese)
20. Rastogi, B.K.; Prantik, M. Foreshocks and nucleation of small- to moderate-sized Koyna earthquakes (India). *Bull. Seismol. Soc. Am.* **1999**, *89*, 829–836. [CrossRef]
21. Vavryčuk, V. Iterative joint inversion for stress and fault orientations from focal mechanisms. *Geophys. J. Int.* **2014**, *1*, 69–77. [CrossRef]
22. Yan, C. Focal Mechanism and Stress Field Inversion Method for Small Earthquakes and Its Application. PhD Thesis, Institute of Geophysics, China Earthquake Administration, Beijing, China, 2015. (In Chinese).
23. Wessel, P.; Smith, W.H.F.; Scharroo, R.; Luis, J.; Wobbe, F. Generic Mapping Tools: Improved Version Released. *Eos Trans. Am. Geophys. Union* **2013**, *94*, 409–410. [CrossRef]

water

Article

Dynamic Reliability Analysis of Layered Slope Considering Soil Spatial Variability Subjected to Mainshock–Aftershock Sequence

Huaiming Zhou [1], Gan Wang [2,3], Xiang Yu [3] and Rui Pang [2,4,*]

1 China Communications Investment Nanjing Co., Ltd., Nanjing 210018, China
2 School of Hydraulic Engineering, Faculty of Infrastructure Engineering, Dalian University of Technology, Dalian 116024, China
3 College of Water Conservancy Science and Engineering, Zhengzhou University, Zhengzhou 450001, China
4 State Key Laboratory of Coastal and Offshore Engineering, Dalian University of Technology, Dalian 116024, China
* Correspondence: pangrui@dlut.edu.cn

Abstract: The slope instability brought on by earthquakes frequently results in significant property damage and casualties. At present, the research on displacement response of a slope under earthquake has mainly emphasized the action of the mainshock, without accounting for the impact of an aftershock, and the spatial variability of material parameters is often neglected. The spatial variability of parameters is fully accounted for in this paper, and dynamic reliability of permanent displacement (D_P) of a slope produced by the mainshock–aftershock sequence (MAS) is studied. A slope reliability analysis method is proposed based on the Newmark displacement method and the generalized probability density evolution method (GPDEM) to quantify the effect of the spatial variability of materials parameters on dynamic reliability. Firstly, the parameter random field is generated based on the spectral representation method, and the randomly generated parameters are assigned to the finite element model (FEM). In addition, the random simulation method of MAS considering the correlation between aftershock and mainshock is adopted based on the Copula function to generate the MAS. Then, the D_P of slopes caused by the MAS considering the spatial variability is calculated based on the Newmark method. The impacts of the coefficient of variation (*COV*) and aftershock on the D_P of slope is analyzed by means of mean values. Finally, the effect of *COV* and aftershock on the reliability of D_P is explained from a probabilistic point of view based on the GPDEM. The results revealed that with the increase in the *COV*, the mean of the D_P of the slope shows a trend of increasing gradually. The D_P of slope is more sensitive to the coefficient of variation of friction angle (COV_F). The mean D_P of the slope induced by the MAS is larger compared to the single mainshock, and the PGA has a significant impact on the D_P.

Keywords: slope reliability analysis; mainshock–aftershock sequence; spatial variability; Newmark; permanent displacement

Citation: Zhou, H.; Wang, G.; Yu, X.; Pang, R. Dynamic Reliability Analysis of Layered Slope Considering Soil Spatial Variability Subjected to Mainshock–Aftershock Sequence. *Water* **2023**, *15*, 1540. https://doi.org/10.3390/w15081540

Academic Editor: Georg Umgiesser

Received: 16 March 2023
Revised: 8 April 2023
Accepted: 12 April 2023
Published: 14 April 2023

1. Introduction

The instability of slopes brought on by earthquakes is a significant geological risk. Strong earthquakes have a significant impact on large-scale geological disasters, such as landslides and debris flows brought on by slope instability, which frequently result in catastrophic losses and negative social repercussions [1]. It is reported that the Chi-Chi earthquake initiated in excess of 10,000 landslides and slope instability in 11,000 m^2 in Central Taiwan [2]. Around 20,000 people perished in the 2008 Wenchuan earthquake as a result of a large number of landslides and slope instability issues, which made up nearly half of all earthquake fatalities [3,4]. In the Yushu earthquake in 2010, the earthquake created more than 2000 landslides, resulting in a direct economic loss of about CNY 600,000,

8 deaths, and 14 injuries [5]. These aforementioned disaster consequences show that reasonable consideration of dynamic response and sliding displacement of slopes induced by strong earthquakes is very essential for predicting the potential damage possibility of ground motion and conducting rapid seismic risk assessment.

A large amount of historical seismic data has shown the occurrence of strong earthquakes is frequently complemented by multiple aftershocks [6,7]. In the two months after the Wenchuan earthquake in 2008, more than 20,000 aftershocks were triggered, including dozens of strong aftershocks of magnitude 5 or greater. In the three days after the 2013 Lushan earthquake, there were more than 3000 aftershocks and about 4 aftershocks with magnitude stronger than 5 [8]. The spatiotemporal distribution characteristics of aftershocks and mainshocks also play an essential part in influencing the dynamic response of building structures [9]. Hence, it is worthwhile to effectively examine the dynamic response of a slope induced by the MAS. At present, researchers have paid attention to the damage of building structures subjected to the MAS, and more attention has been paid to the dynamic response caused by the combined action of the mainshock and the aftershock with the maximum magnitude. Pang et al. [10] discussed the susceptibility of a CFRD with a height of over 200 m produced by the MAS based on the multiple analysis. Zhou et al. [11] investigated the association between the intensity parameters of ground motion and the structural destruction under the MAS, and established the damage prediction model of the MAS based on the optimal parameters. Based on the Copula theory, Shen et al. [12] established a sequential random model of ground motion that can better represent the spatial correlation of sequential earthquakes. However, there are few studies on the response characteristics of slope caused by mainshock–aftershock sequence at present. In addition, the repeated method is mostly used to construct artificial mainshock–aftershock sequences in the above research, which can neither truly reflect the characteristics of the real mainshock–aftershock sequence nor properly consider the association between the aftershock and mainshock intensity. Therefore, the effect of an aftershock on the structural dynamic response cannot be reasonably responded to. The researchers cannot really grasp the safety of the slope when it is further subjected to an aftershock after the mainshock's initial damage due to the lack of research on the response of the slope induced by the MAS. Therefore, the advanced stochastic simulation method of the MAS requires of further investigation.

In order to consider the seismic slip danger of slopes, many methods have been developed, such as statistical analysis, the permanent displacement method, the pseudo-static method, the safety factor method, and the stress–deformation method [13–16]. Compared with the safety factor method, the permanent displacement caused by seismic action can assess the damage condition and seismic performance of the slope more reasonably [17–19]. In addition, traditional slope stability analysis methods are in general primarily based on deterministic analysis, which considers a series of elements affecting the slope stability as definitive factors. However, a many disaster results and geotechnical tests have revealed the apparent stochastic nature of variables affecting slope stability, such as external load, performance of materials, and model geometry [20–22]. Calculating and analyzing the stability of slopes by using deterministic methods can create many errors. The adoption of reliability theory has provided the opportunity of quantitative consideration of uncertainties in recent years [23,24]. By establishing extreme state equations, metrics such as probability of failure and reliability are employed to describe the safety of systems, thus providing a more complete guide for engineering design. Some traditional probabilistic methods, such as the First Order Second Moment method [25], the Monte Carlo method [26], the response surface method [27], and their improved forms, have been used to analyze the reliability of results and proven to be effective [28]. However, these methods have the characteristics of having difficulty obtaining the random dynamic information of the structure, a huge calculation scale, and a need to be coupled with the structural response analysis and continuous sample training and iteration. Therefore, it is demanding and challenging to apply the seismic random dynamic response and probability analysis of slopes with strong nonlinearity, complex problems, and a large calculation scale. The GPDEM is a newly developed method for probabilistic analysis [29,30]. At present, the GPDEM has been implemented for dynamic reliability assessment of bridges, slopes, earth-rock dams, and other engineering structures, and its efficiency has been confirmed by comparing it with the MCS method [30–34]. However, the feasibility of GPDEM in the evaluation of slope D_P caused by a mainshock–aftershock sequence needs further verification.

Soil parameters have significant spatial variability due to differences in depositional conditions, loading history, and other geological processes [35–37]. Moreover, the soil parameters at different spatial locations have a certain relevance and are not completely independent, which makes the slope stability research more complex. In slope reliability analysis, two methods are generally adopted to imitate the variability of soil parameters. Assuming that the parameters are spatially homogeneous, the probability distribution model is applied to describe the inherent variability, which is called random variable model [38,39]. The random variable model assumes that the parameters in the study area are perfectly correlated and that differences in the physical and mechanical properties of the local and overall geotechnical properties at different points in space cannot be considered. This obviously does not conform to the actual situation of geotechnical engineering and cannot meet the needs of objective analysis and evaluation of the spatial variation characteristics of geotechnical parameters. The random field theory was first proposed and adopted by Cornel to describe the random characteristics of parameters in 1972 [40]. On this basis, the theory was gradually developed and refined by Vanmarcke [41]. The main idea is to treat the soil parameters at a certain location in the space as random variables subject to certain statistical laws and describe the spatial variability of soil parameters at different locations through variance reduction function, correlation distance, correlation function, etc. In contrast, the spatial variability of soil material parameters would be better depicted by the random field theory.

In this paper, a slope reliability analysis method based on the GPDEM and the Newmark displacement method is proposed to quantify the effect of spatial variability of soil parameters on dynamic reliability. The MAS and random field are generated by the random simulation method of MAS and the spectral representation method (SRM), and the D_P of the slope is obtained by a nonintrusive analysis. Firstly, the impact of COV on a slope's dynamic stability is investigated from the mean value of D_P, and then the influence of COV and PGA on a slope's dynamic reliability is explained from a probabilistic point of view by combining the GPDEM. The flowchart of the evaluation framework is depicted in Figure 1.

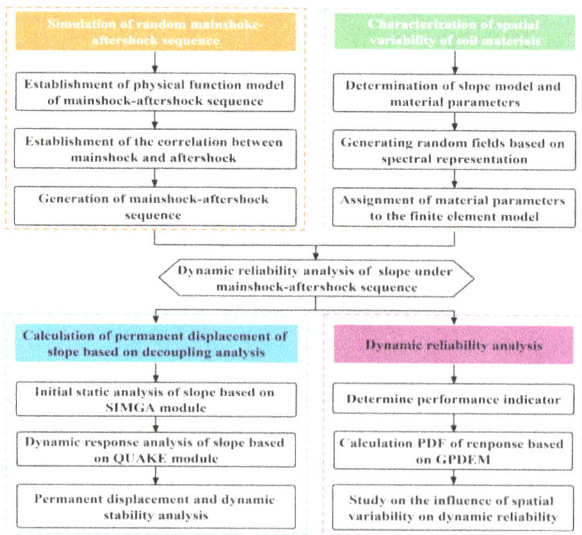

Figure 1. The flowchart of the evaluation framework.

2. Generalized Probability Density Evolution Method

This paper considers the spatial variability of layered slope material parameters. Therefore, the slope dynamic equation under the action of MAS may be represented as:

$$\overline{M}(\Theta)\ddot{X}(t) + C(\Theta)\ddot{X}(t) + K(\Theta)X(t) = -\overline{M}\ddot{X}_g(\Theta, t) \tag{1}$$

where, K, C, and $\overline{M}$ represent the stiffness matrices, damping, and effective mass of the structure, respectively, and their basic parameters may be random.

For convenience, the solution of Equation (1) can be formulated as:

$$X(t) = H(\Theta, t) \tag{2}$$

where $H = (H_1, H_2, \ldots, H_n)^T$; n is the number of degrees of freedom. In addition, it is worth noting that it can be regarded as a variable Θ and t.

Accordingly, variables such as acceleration, velocity, strain, and stress could also be expressed in the form similar to Equation (2). Therefore, in order to be more general, we can uniformly express the physical quantities of interest in the following form:

$$Z(t) = H_z(\Theta, t) \tag{3}$$

where $Hz = (H_{z,1}, H_{z,2}, \ldots, H_{z,m})^T$.

Under the framework of probability conservation, the generalized probability density evolution equation of the stochastic process may be represented as:

$$\frac{\partial p_{z\Theta}(z,\theta,t)}{\partial_t} + \sum_{l=1}^{m} \dot{Z}_l(\theta,t)\frac{\partial p_{z\Theta}(z,\theta,t)}{\partial z_l} = 0 \tag{4}$$

where $p_{z\Theta}(z,\theta,t)$ is the joint probability density function of the system (z,Θ), and m is the dimension of the equation, independent of the number of degrees of freedom of the system n.

$$\frac{\partial p_{z\Theta}(z,\theta,t)}{\partial t} + \dot{Z}(\theta,t)\frac{\partial p_{z\Theta}(z,\theta,t)}{\partial z_l} = 0 \tag{5}$$

$$p_{z\Theta}(z,\theta,t)|_{t=t_0} = \delta(z - z_0)p_\Theta \tag{6}$$

$$p_{z\Theta}(z,\theta,t)|_{z_j \to \pm\infty} = 0,\ j = 1,2,3,\ldots,m \tag{7}$$

If only a single physical quantity is considered, Equation (4) is simplified to the form of Equation (5). Combining the initial conditions represented by Equation (6) and the boundary conditions represented by Equation (7), the structural reliability is finally obtained through mathematical processing. For simple problems, the analytical solution can be obtained in this way. For complex problems, such as large complex nonlinear systems, the numerical solution can only be obtained by mathematical methods, which could be achieved by the procedure as follows:

(1) Point selection and probability assignment in probability space.

Discrete representative points are selected by some means in Ω_Θ of random variables (such as the number-theory method, the point-selecting method by cutting the ball, the quasi-rotational symmetry point method, and the GF-deviation method).

(2) Deterministic solutions for dynamic systems.

The physical Equations (1) and (3) are solved, and the velocity of the required physical quantity is found for each given $\Theta = \theta_q$.

(3) Solving probability density evolution equation.

After the representative points are selected and the probabilities are assigned in step (1), the Equation (4) is transformed into:

$$\frac{\partial p_{z\Theta}(z,\theta_q,t)}{\partial t}+\sum_{j=1}^{m}\dot{Z}_j(\theta_q,t)\frac{\partial p_{z\Theta}(z,\theta_q,t)}{\partial z_j}=0 \tag{8}$$

The corresponding initial conditions are transformed into:

$$p_{z\Theta}(z,\theta_q,t)\big|_{t=t_0}=\delta(z-z_0)P_q \tag{9}$$

The result of the partial differential equation can be obtained by substituting $\dot{Z}(\theta_q,t_m)$, as obtained in step (2), into Equations (8) and (9).

(4) Cumulative summation.

The result of $p_z(z,t)$ is acquired by the summation of all the above single results $p_{z\Theta}(z,\theta_q,t)$.

$$p_z(z,t)=\sum\nolimits_{q=1}^{n_{sel}} p_{Z\Theta}(z,\theta_q,t) \tag{10}$$

3. Simulation of Random Field and Random Main Aftershock Sequence

3.1. Spectral Representation Method

Many methods are currently employed to decompose random fields, for example, the midpoint method [42], spectral representation method [43], the spatial averaging method [44], the K-L decomposition method [45], and other methods. The spectral representation method has gradually become a widely used random field simulation method because of its good speed and accuracy in convergence to the objective function, and the generated sample function has ergodicity in all states.

By using the spectral representation method [43], the establishment of one-dimensional stationary random field could be expressed as:

$$\hat{f}(x_1)=\sqrt{2}\sum_{n=0}^{N-1}A_n\cos(\kappa_n x+\Phi_n) \tag{11}$$

where Φ_n is the independent phase angle uniformly distributed within the region of [0, 2π]; A_n is the amplitude; and κ_n is the frequency.

$$A_n=\sqrt{2S_{ff}(\kappa_n)\Delta\kappa} \tag{12}$$

$$\kappa_n=n\Delta\kappa=n\frac{\kappa_u}{N} \tag{13}$$

where κ_u denotes the number of truncated waves; $n = 0, 1, 2, \ldots, N-1$.

In Equation (12), $S_{ff}(\kappa_n)$ is the power spectrum function. The relationship between $S_{ff}(\kappa_n)$ and autocorrelation function can be acquired by Fourier transform, as shown below:

$$S_{ff}(\kappa)=\frac{1}{2\pi}\int_{-\infty}^{\infty}\rho_{ff}(\xi)e^{-i\kappa\xi}d\xi \tag{14}$$

Note that:

$$A_0=0 \text{ or } S_{ff}(\kappa_0)=0 \tag{15}$$

The two-dimensional stationary random field can be characterized as:

$$f(x_1,x_2)=\sqrt{2}\sum_{n_1=0}^{N_1-1}\sum_{n_2=0}^{N_2-1}\left[A_{n_1n_2}\cos\left(\kappa_{1n_1}x_1+\kappa_{2n_2}x_2+\Phi^{(1)}_{n_1n_2}\right)+\tilde{A}_{n_1n_2}\cos\left(\kappa_{1n_1}x_1-\kappa_{2n_2}x_2+\Phi^{(2)}_{n_1n_2}\right)\right] \tag{16}$$

where $\Phi^{(1)}_{n_1 n_2}$ and $\Phi^{(2)}_{n_1 n_2}$ are individual random phase angles uniformly distributed within region of [0, 2π]; $A_{n_1 n_2}$ and $\widetilde{A}_{n_1 n_2}$ are amplitude; κ_{1n_1} and κ_{1n_2} are frequency.

$$A_{n_1 n_2} = \sqrt{2S_{f_0 f_0}(\kappa_{1n_1}, \kappa_{2n_2})\Delta\kappa_1\Delta\kappa_2} \tag{17}$$

$$A_{n_1 n_2} = \sqrt{2S_{f_0 f_0}(\kappa_{1n_1}, -\kappa_{1n_2})\Delta\kappa_1\Delta\kappa_2} \tag{18}$$

$$K_{1n_1} = n_1\Delta\kappa_1 = \frac{\kappa_{1u}}{N_1} \tag{19}$$

$$K_{2n_2} = n_2\Delta\kappa_2 = \frac{\kappa_{2u}}{N_2} \tag{20}$$

where κ_{1u} and κ_{2u} denote the number of truncated waves and meet the following relationship:

$$\begin{cases} -\kappa_{1u} \le \kappa_1 \le \kappa_{1u} \\ -\kappa_{2u} \le \kappa_2 \le \kappa_{2u} \end{cases} \tag{21}$$

In Equation (17), $S_{f_0 f_0}(\kappa_1, \kappa_2)$ is the power spectrum function. The relationship between $S_{f_0 f_0}(\kappa_1, \kappa_2)$ and autocorrelation function can be obtained through Fourier transform, as shown below:

$$S_{f_0 f_0}(\kappa_1, \kappa_2) = \frac{1}{(2\pi)^2}\int_{-\infty}^{\infty} R_{f_0 f_0}(\xi_1, \xi_2)e^{-i(\kappa_1\xi_1 + \kappa_2\xi_2)}d\xi_1\xi_2 \tag{22}$$

3.2. Generation of Parametric Random Fields Based on Spectral Representation Method

In the random field simulation of slope strength parameters, since the value of strength parameters is usually positive, lognormal random field is employed to simulate the spatial difference and correlation of material parameters. The logarithmic stationary random field of slope strength parameters is established based on Equation (16).

$$\omega(x,z) = \exp\left(\xi_{\ln}\cdot\sum_{i=0}^{M-1}\sum_{j=0}^{N-1}\sigma_{ij}\left[V_{ij}(\theta)\cos(\omega_{1i}x + \omega_{2i}z) + W_{ij}(\theta)\sin(\omega_{1i}x + \omega_{2j}z) + \lambda_{\ln}\right]\right) \tag{23}$$

where the $V_{ij}(\theta)$ and $W_{ij}(\theta)$ are mutually independent and obey the standard normal distribution; ω_{1i} and ω_{2j} are frequency coordinate values. x and z are the horizontal and vertical coordinate values of space. $\xi_{\ln}$ and $\lambda_{\ln}$ are the logarithmic standard deviation and logarithmic mean of parameters. σ_{ij} is the standard deviation of $i * M + j + 1$.

$$\begin{cases} \lambda_{\ln} = \ln(\mu) - 0.5\cdot\ln(1 + \mathrm{cov}^2) \\ \xi_{\ln} = \sqrt{\ln(1 + \mathrm{cov}^2)} \end{cases} \tag{24}$$

$$\sigma_{ij} = \sqrt{4S_{\omega\omega}(\omega_{1i}, \omega_{2j})\cdot\Delta\omega_1\cdot\Delta\omega_2} \tag{25}$$

where $S_{\omega\omega}$ is the power spectral density function corresponding to the correlation function, which can be obtained by two-dimensional Fourier transform of the autocorrelation function.

$$S_{\omega\omega}(\omega_1, \omega_2) = \frac{1}{(2\pi)^2}\int_{-\infty}^{+\infty}\int_{-\infty}^{+\infty}\rho(x,z)e^{-i(\omega_1 x + \omega_2 z)}dxdz \tag{26}$$

where $\rho(x,z)$ is the autocorrelation function. The Gaussian autocorrelation function with good stability and continuity is used for calculation. $\Delta\omega_1$ and $\Delta\omega_2$ are the discrete intervals of the frequency coordinate axes ω_1 and ω_2, respectively.

3.3. Random Simulation of Mainshock–Aftershock Sequence (MAS)

Due to the limited number of measured records, it is necessary to generate the MAS ground motion through artificial simulation for seismic analysis of engineering structures. The existing method for constructing the MAS is to develop the magnitude relationship between the mainshock and aftershock and then separate and adjust the actual ground motion records (or artificial ground motion) to obtain the time histories of the MAS. However, in addition to the magnitude, the mainshock and aftershock are intimately associated in respect to source, propagation path, and local site impact, i.e., they are highly correlated in terms of spectrum characteristics, ground motion intensity, and duration. Obviously, a single magnitude parameter cannot accurately reflect the characteristics of the MAS. In addition, by adjusting the recorded ground motion or adopting the ground motion model of single shock, the changes of the amplitude, duration, and frequency spectrum of ground motion in the process of seismic wave propagation cannot be well reflected. Therefore, a random MAS simulation method accounting for the relevance between aftershock and mainshock based on Copula function is adopted to generate the MAS. This approach is characterized in greater depth in previous studies [46], and the primary steps of the approach can be simplified as follows:

(1) Establishment of a physical random function model of the MAS.
(2) The real MASs are collected from the PEER to determine the physical parameters in the physical random function model of the mainshock–aftershock sequence.
(3) Select a representative set of points of seismic parameters according to the GF difference. Then, establish the relevance between the aftershock and mainshock parameters based on the Copula theory.
(4) Generate of a series of random MASs by using the narrowband harmonic superposition method.

4. Nonintrusive Analysis of Slope Dynamic Reliability

The biggest advantage of noninvasive randomness analysis is that the process of deterministic analysis and randomness analysis are independent of each other. The FE method is adopted to perform deterministic analysis without modifying the finite element kernel; therefore, the integration of deterministic analysis and stochastic analysis is realized, which significantly improves the reliability of the stochastic analysis results. By combining dynamic reliability analysis with finite element batch processing, this paper proposes a nonintrusive analysis frame of slope reliability considering spatial variability subjected to the MAS and compiles the interface program between dynamic reliability analysis and GeoStudio finite element software.

(1) Establish the slope of the FE model, divide the model mesh, set the boundary conditions, define the load loading method, define the material properties, and assign the elements in the SIGMA/W module with the parameter averages. Then, establish the corresponding relationship between the elements, groups, and material properties. Additionally, establish the stability analysis model in SLOPE/W, and save the FEM as a file with the extension name of ".xml".
(2) The slope strength parameters are simulated by the spectral representation method. N groups of data of parameters will be generated, and the parameters in the "xml" file will be replaced in batches with the newly generated n groups of data through MATLAB programming to obtain n new ".xml" files.
(3) Use the UE text editor software to directly use GeoStudio to batch calculate the stability of the n new ".xml" files obtained in step (2). Then, output the calculation result files corresponding to each group of parameters.
(4) The calculation results corresponding to all parameter groups are extracted in batch, and the D_P is statistically analyzed.

5. Model Establishment and Material Parameters

5.1. Finite Element Model

In this study, the FEM adopted is a simplified layered soil slope based on geological data and field survey along the G317 Sichuan–Tibet Highway. The two-dimensional FEM was adopted to carry out the dynamic reliability analysis of layered soil slope. As demonstrated in Figure 2, the layered soil slope model is 40 m long and 24 m high. The FEM has been used in other studies to research the failure mode of a slope through numerical simulation and model tests [47]. In this study, two different types of layered soil slopes are used to research the dynamic reliability of the slope considering the spatial variability subjected to the MAS. The size of the grid is chosen to be 0.5 m, which ensures both computational efficiency and accuracy. Other information about the finite element model is introduced in detail in previous studies [6]. According to different soil layer distribution types, the two layered soil slopes are clayey soil–gravel soil–sandy soil–foundation soil and clayey soil–sandy soil–gravel soil–foundation soil. (The soil mass is arranged from top to bottom.)

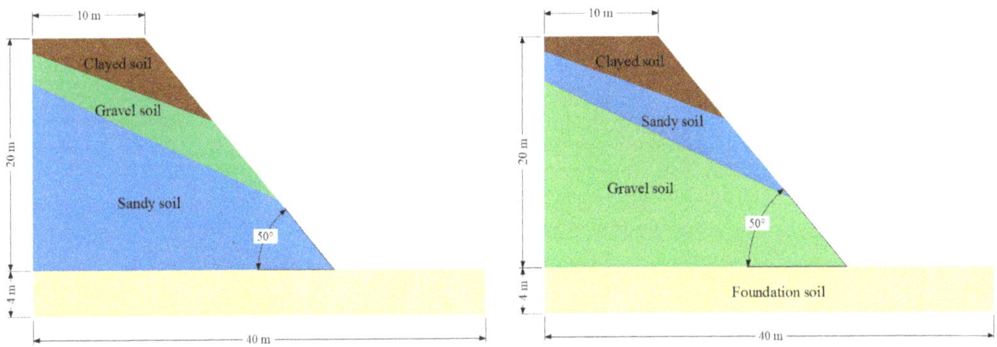

Figure 2. Size of finite element model: (**a**) Case 1; (**b**) Case 2.

During the initial static analysis and dynamic response analysis of the layered soil slope, the bottom of the FEM is constrained both horizontally and vertically. In addition, the right and left boundaries of the FEM are restrained horizontally during initial static analysis of the slope but not during the dynamic analysis.

5.2. Calculation Parameters

Various constitutive models are adopted to characterize the mechanical properties of different soil materials of the layered soil slope. The three layers of soil above the foundation soil are described by the equivalent linear model. The equivalent linear model is applied to the foundation soil because it is compacted. The correlation between damping ratio, shear modulus, and shear strain is presented in Figure 3. The material calculation parameters are the same as those employed by Huang [37], as shown in Table 1.

Table 1. Basic parameters of slope soil materials.

Materials	c (kPa)	φ (°)	E (MPa)	γ (kN/m^3)	v
Clayey soil	70.24	24.00	86	22.16	0.35
Sandy soil	13.65	32.50	60	17.23	0.32
Gravelly soil	18.23	38.50	73	19.55	0.3
Foundation soil	200	35.02	800	25.14	0.25

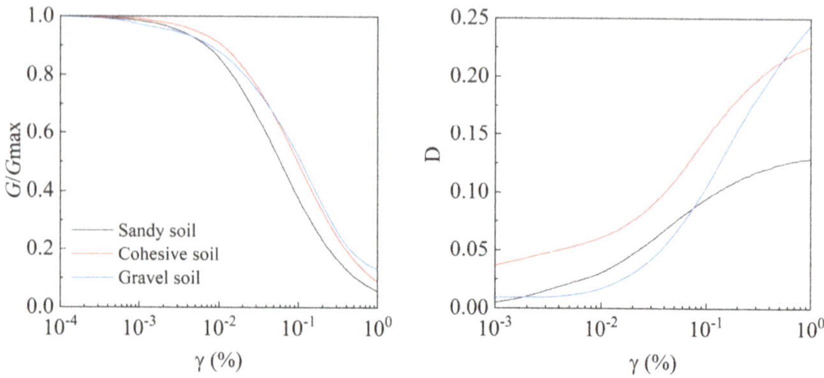

Figure 3. Correlation between damping ratio, shear modulus, and shear strain of materials.

The spatial variability of material is accounted for, and the parameter random field is generated by the above SRM. Then, the parameters are assigned to the well-constructed FEM. In this study, the parameters of each soil layer are presumed as independent of each other. The *COV* of *c* and φ (COV_C and COV_F) are set as 0.1, 0.2, and 0.3. The vertical and horizontal autocorrelation distances (l_h and l_v) are 20 m and 2 m, respectively. The horizontal dimension of the random field unit is 2 m, and the vertical dimension is 0.5 m. The ratios of the vertical and horizontal fluctuation ranges to the vertical and horizontal dimensions of the random field are $\delta_h/l_x = 20\sqrt{\pi}/2 = 17.7$ and $\delta_v/l_y = 2\sqrt{\pi}/0.5 = 7.08$, respectively, which are greater than the accuracy requirements (5.7~7.6) given by Ching and Phoon [48].

5.3. Input of the Mainshock–Aftershock Sequence

One of the randomly generated MAS is employed as the deterministic seismic wave input. The acceleration curve of MAS is presented in Figure 4.

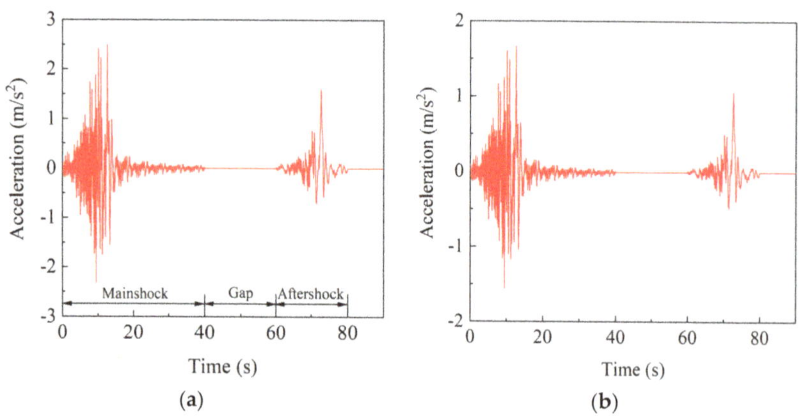

Figure 4. Acceleration change curve of the mainshock–aftershock sequence: (**a**) horizontal; (**b**) vertical.

6. Effect of Coefficient of Variation on Dynamic Reliability of Layered Soil Slope

The D_P is adopted to assess the dynamic stability of slope, so it is necessary to define the critical D_P of the soil slope. According to previous research [6], three D_P thresholds (0.05 m, 0.5 m, and 1 m) were used to assess the dynamic reliability of the layered slope. A total of 86 sets of random material parameters were generated based on spectral representation to explore the impact of spatial variability on slope dynamic reliability.

6.1. Case 1: Clayey Soil–Gravel Soil–Sandy Soil–Foundation Soil

Figure 5 presents the distribution of D_P discrete points for Case 1 when the COV_C values are 0.1, 0.2, and 0.3. It is significant that when the COV_C is small, the D_P is small and relatively concentrated. With the increase in COV_C, the range of variation of soil cohesion increases, the distribution of discrete points of D_P becomes more discrete, and the mean of D_P gradually increases. When the COV_C is 0.1, the mean of D_P caused by the MAS is 0.63 m, while the mean D_P for the slope subjected to the single mainshock is 0.339 m. It is obvious that the mean D_P of the slope induced by MAS is wider than the D_P caused by the single mainshock. Moreover, the mean D_P values of the slope under the MAS are 0.668 m and 0.725 m, respectively, when the COV_C is 0.2 and 0.3. At this time, the mean D_P of the slope due to the single mainshock is 0.368 m and 0.42 m. The mean value of D_P increases continuously along with the increment of COV_C, and the discrepancy of D_P also shows a gradual tendency to increase.

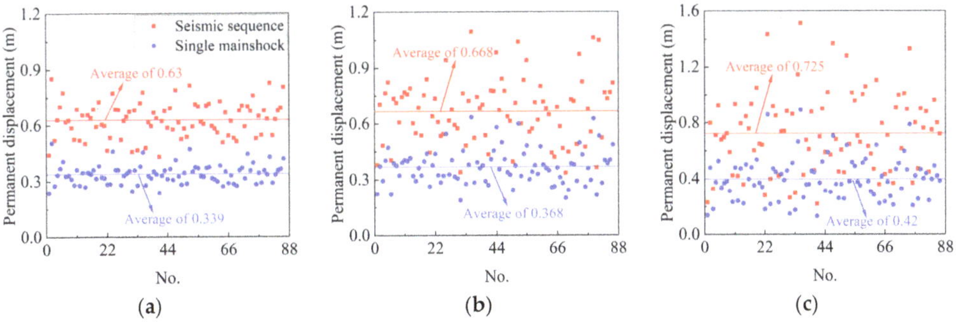

Figure 5. Distribution of discrete points of D_P under different COV_C (Case 1): (**a**) $COV_C = 0.1$; (**b**) $COV_C = 0.2$; (**c**) $COV_C = 0.3$.

Figure 6 displays the distribution of D_P discrete points of slope under various PGA when the COV_C is 0.3. When the PGA values are 0.4 g and 0.6 g, the mean D_P values of the slope caused by the MAS are 0.386 m and 0.924 m, respectively. However, the mean D_P values of the slope induced by the single mainshock are 0.224 m and 0.505 m. The D_P of the slope constantly changed incrementally with the increase in PGA.

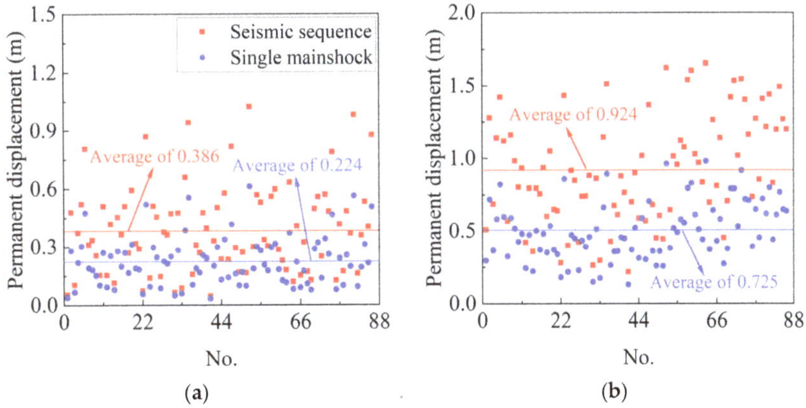

Figure 6. Distribution of discrete points of D_P under different PGA when COV_C=0.3 (Case 1): (**a**) PGA = 0.4 g; (**b**) PGA = 0.6 g.

Figures 7 and 8 show the probability information of D_P of slope under different COV_C. When the COV_C is 0.1, the maximum value of PDF is 3.98, the fluctuation region of D_P

is 0.4–1.0 m, and the D_P is primarily focused around 0.45 m. When the COV_C is 0.3, the maximum value of PDF is 1.2, the fluctuation region of D_P of slope is 0–1.5 m, and the D_P is relatively centralized around 0.8 m. With the growth of the COV_C, the PDF value gradually decreases, the curve gradually shifts to the right, the D_P distribution range is wider, and the failure probability of the slope is higher. Table 2 shows the reliability information of the slope when the cumulative slips are 0.05 m, 0.5 m, and 1 m under different COV_C and PGA. The dynamic reliability of the slope caused by the MAS decreases by 13% with the COV_C increasing from 0.1 to 0.3 when the PGA is 0.5 g, and the displacement threshold is 1 m. When the COV_C is 0.3, the dynamic reliability of the slope under the action of the MAS is also reduced by 13% compared with the single mainshock. In addition, with the PGA increasing from 0.4 g to 0.6 g, the dynamic reliability of the slope induced by MAS decreased by 35%.

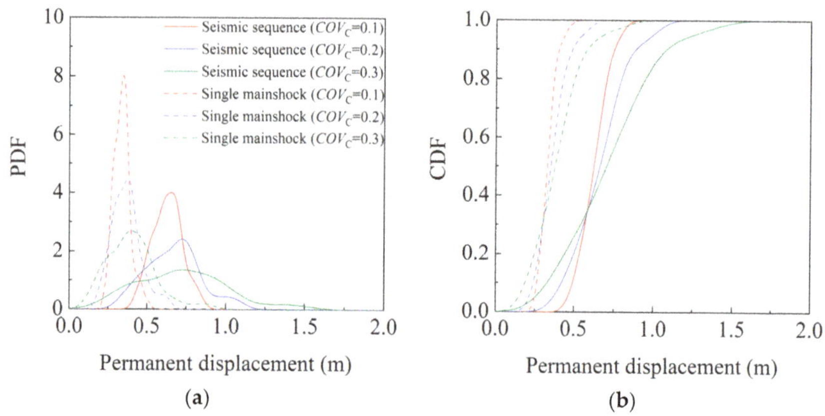

Figure 7. Probability distribution of D_P with different COV_C (Case 1): (**a**) PDF; (**b**) CDF.

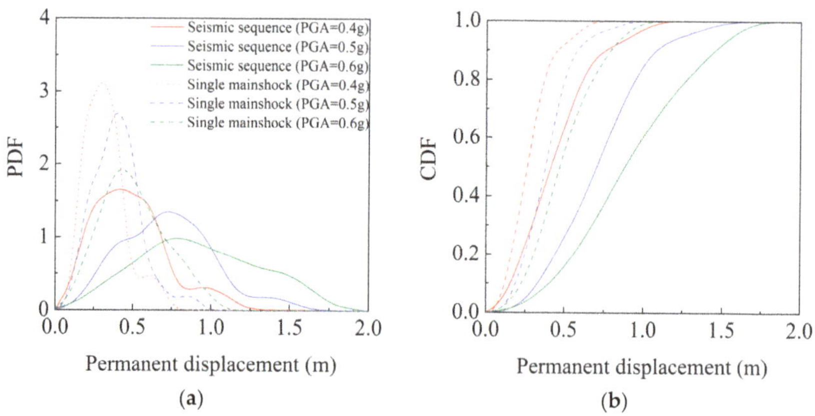

Figure 8. Probability distribution of D_P with different PGA when COV_C = 0.3 (Case 1): (**a**) PDF; (**b**) CDF.

Figure 9 provides the distribution of D_P discrete points for Case 1 when the COV_F values are 0.1, 0.2, and 0.3. As the COV_F is small, the D_P for the slope is low and concentrated. With the increase in COV_F, the fluctuating region increases and the mean value of D_P of slope gradually increases and becomes more discrete. When the COV_F is 0.1, the mean D_P of the slope under the MAS is 0.674 m and the mean D_P of the slope under the single mainshock is 0.372 m. The mean D_P of the slope induced by the MAS is larger than that under the single mainshock. When the COV_F values are 0.2 and 0.3, the mean D_P values of the slope caused by the MAS are 0.743 m and 0.795 m, respectively. At this time, the mean D_P values of the slope due to the single mainshock are 0.409 m and 0.432 m. With increasing COV_F, the mean value of D_P of the slope continuously increases.

Table 2. Dynamic reliability of slope under different COV_C and PGA (Case 1).

Type of Ground Motion	D_P	COV_C			PGA		
		0.1	**0.2**	**0.3**	**0.4 g**	**0.5 g**	**0.6 g**
Mainshock–aftershock sequence	0.05 m	0	0	0.016	0.035	0.016	0.003
	0.5 m	0.09	0.21	0.26	0.77	0.26	0.15
	1 m	1	0.96	0.87	0.98	0.87	0.63
Single mainshock	0.05 m	0	0	0.016	0.035	0.016	0.003
	0.5 m	0.99	0.92	0.84	0.95	0.84	0.58
	1 m	1	1	1	1	1	1

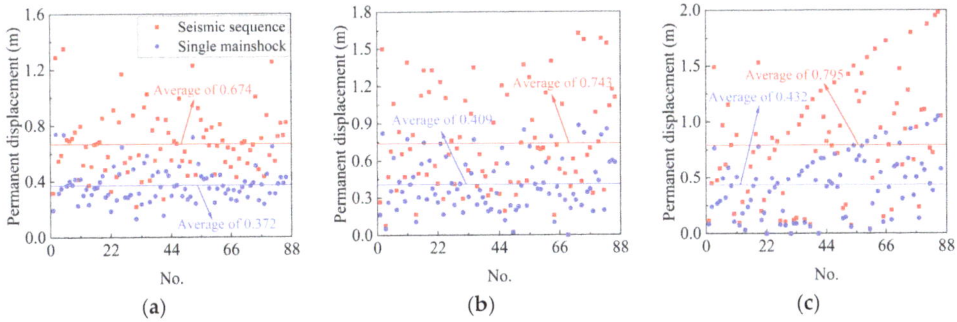

Figure 9. Distribution of discrete points of D_P under different COV_F (Case 1): (**a**) COV_F = 0.1; (**b**) COV_F = 0.2; (**c**) COV_F = 0.3.

Figure 10 illustrates the distribution of D_P discrete points of slope under different PGA when the COV_F is 0.3. When the PGA values are 0.4 g and 0.6 g, the mean D_P values of the slope subjected to the MAS are 0.41 m and 1.165 m, respectively. However, the mean D_P value of for Case 2 caused by the single mainshock are 0.246 m and 0.648 m. Obviously, the D_P of for the slope increases continuously with the growth of PGA.

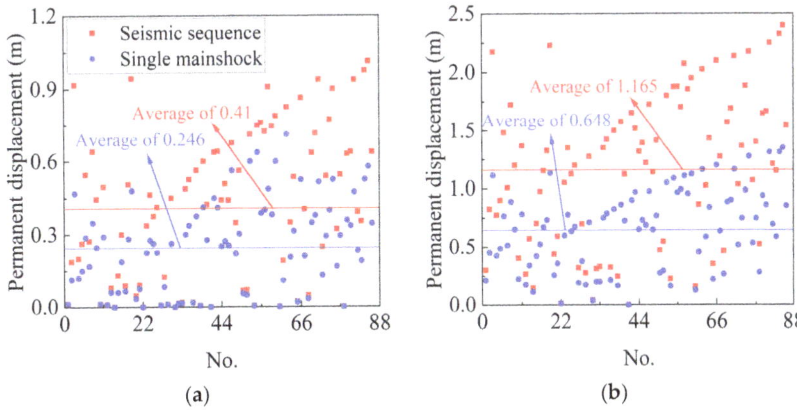

Figure 10. Distribution of discrete points of D_P under different PGA when COV_F = 0.3 (Case 1): (**a**) PGA = 0.4 g; (**b**) PGA = 0.6 g.

Figures 11 and 12 show the probability information of the D_P of the slope under different COV_F values. The maximum PDF value is 1.89 when the COV_F is 0.1, and the D_P of the slope is principally around 0.7 m. With the growth of the COV_F, the PDF value gradually decreases, and the D_P of the slope is more widely distributed. Table 3 shows the information of dynamic reliability for the slope when the cumulative slip is 0.05 m, 0.5 m, and 1 m under different COV_F and PGA. When the PGA is 0.5 g and the threshold is 1 m, the dynamic reliability of the slope induced by the MAS decreases by 25% with the COV_F increasing from 0.1 to 0.3. When the COV_F is 0.3, the dynamic reliability of the slope under the action of the MAS is also reduced by 31% compared with the single mainshock. In addition, with the PGA increasing from 0.4 g to 0.6 g, the dynamic reliability of the slope under the MAS decreased by 17%.

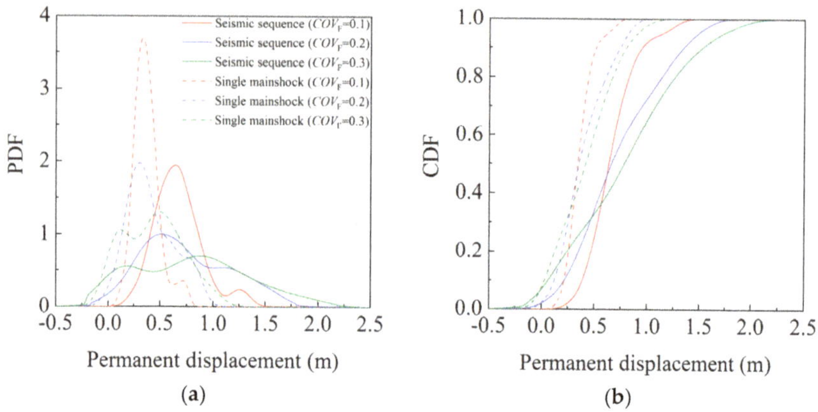

Figure 11. Probability distribution of D_P with different COV_F (Case 1): (**a**) PDF; (**b**) CDF.

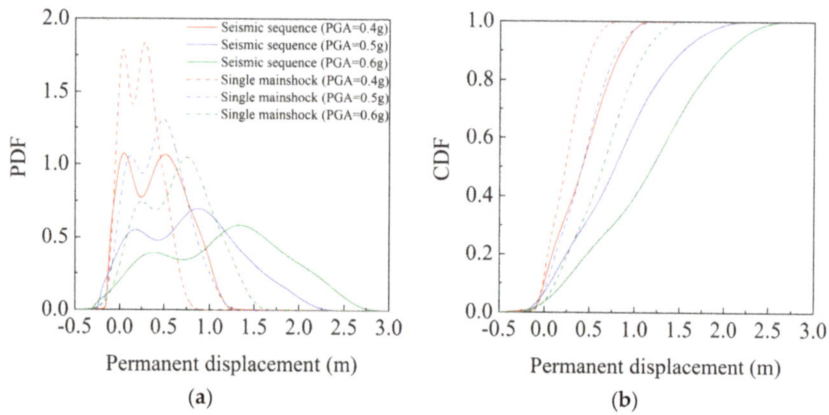

Figure 12. Probability distribution of D_P with different PGA when COV_F = 0.3 (Case 1): (**a**) PDF; (**b**) CDF.

For Case 1, with the increase in COV_C and COV_F, the dynamic reliability gradually decreases, and the failure probability gradually increases under different displacement thresholds. In contrast, the dynamic reliability of slopes is more sensitive to COV_F. Additionally, the dynamic reliability of slopes is more sensitive to the COV_F. Similar conclusions were also obtained by Huang et al. [1], but the impact of aftershocks was not considered in their research.

Table 3. Dynamic reliability of slope under different COV_F and PGA (Case 1).

Type of Ground Motion	D_P	COV_F			PGA		
		0.1	**0.2**	**0.3**	**0.4 g**	**0.5 g**	**0.6 g**
Mainshock–aftershock sequence	0.05 m	0.01	0.03	0.16	0.23	0.16	0.04
	0.5 m	0.23	0.38	0.32	0.61	0.32	0.24
	1 m	0.92	0.71	0.67	0.99	0.67	0.38
Single mainshock	0.05 m	0.01	0.03	0.16	0.26	0.1	0.04
	0.5 m	0.89	0.88	0.61	0.89	0.61	0.38
	1 m	1	1	0.98	1	0.98	0.83

6.2. Case 2: Clayey Soil–Sandy Soil–Gravel Soil–Foundation Soil

Figure 13 illustrates the dispersion of D_P dispersion points for Case 2 when the COV_C values are 0.1, 0.2, and 0.3, respectively. The D_P of the slope is smaller and more concentrated when the COV_C is small. With the increase in the COV_C, the D_P of the slope is gradually increased and became more discrete. When the COV_C is 0.1, the mean of D_P caused by the MAS is 0.293 m, and the mean of D_P of the slope induced by the single mainshock is 0.247 m. When the COV_C values are 0.2 and 0.3, the mean of D_P subjected to the MAS is 0.36 m and 0.458 m, respectively. At this time, the mean of D_P values under a single mainshock are 0.28 m and 0.335 m. The mean of D_P increases continuously, and the difference shows a trend of increasing gradually with the increase in the COV_C.

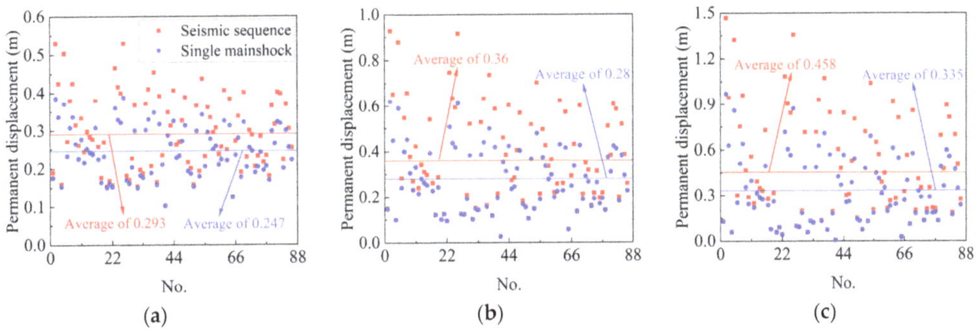

Figure 13. Distribution of discrete points of D_P under different COV_C (Case 2): (**a**) COV_C = 0.1; (**b**) COV_C = 0.2; (**c**) COV_C = 0.3.

Figure 14 summarizes the distribution of D_P discrete points of the slope under different PGA when the COV_C is 0.3. When the PGA is 0.4 g and 0.6 g, the mean of D_P values under the MAS are 0.261 m and 0.881 m, respectively. However, the mean of D_P values under a single mainshock are 0.173 m and 0.577 m. The D_P of the slope is raised step by step with the increase in PGA.

Figures 15 and 16 describe the probability information of D_P of slope under different COV_C. The maximum PDF value is 3.2 when the COV_C is 0.1, and the D_P is mainly concentrated around 0.2–0.4 m. As the COV_C improves, the PDF value gradually decreases, the curve gradually shifts to the right, and the D_P of the slope is more widely distributed. Table 4 shows the information of dynamic reliability of D_P for the slope when the cumulative slip is 0.05 m, 0.5 m, and 1 m under different COV_C and PGA values. When the PGA is 0.5 g and the displacement threshold is 1 m, the reliability of the slope subjected to the MAS decreases by 7% with the COV_C increasing from 0.1 to 0.3. When the COV_C is 0.3, the dynamic reliability of the slope under the action of the MAS is also reduced by 7% compared with the single mainshock. In addition, with the PGA increasing from 0.4 g to 0.6 g, the reliability of D_P of the slope produced by the MAS decreased by 8%.

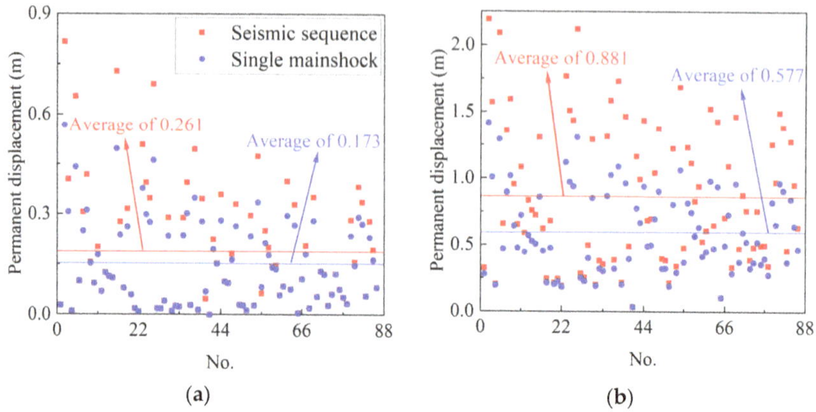

Figure 14. Distribution of discrete points of D_P under different PGA when COV_C = 0.3 (Case 2): (**a**) PGA = 0.4 g; (**b**) PGA = 0.6 g.

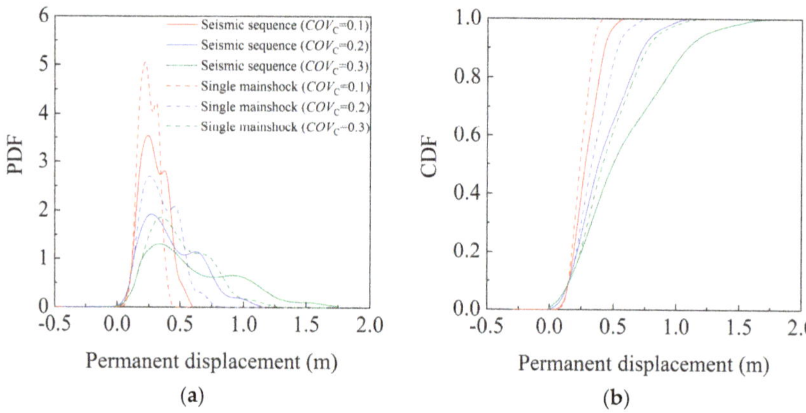

Figure 15. Probability distribution of D_P with different COV_C (Case 2): (**a**) PDF; (**b**) CDF.

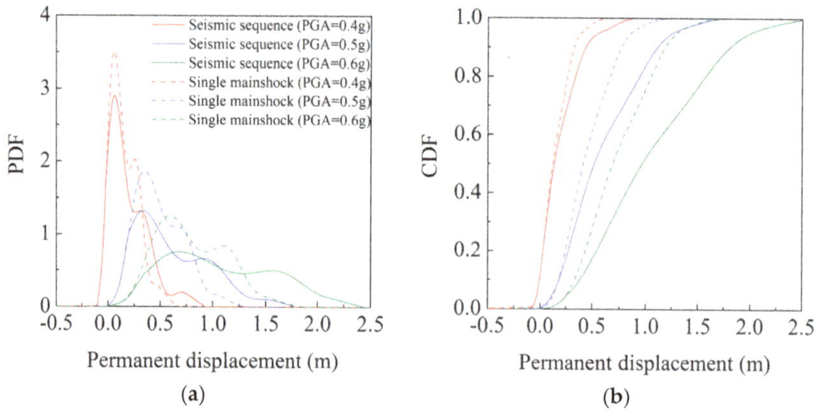

Figure 16. Probability distribution of D_P with different PGA when COV_C = 0.3 (Case 2): (**a**) horizontal; (**b**) vertical.

Table 4. Dynamic reliability of slope under different COV_C and PGA (Case 2).

Type of Ground Motion	D_P	COV_C			PGA		
		0.1	**0.2**	**0.3**	**0.4 g**	**0.5 g**	**0.6 g**
Mainshock–aftershock sequence	0.05 m	0	0.02	0.11	0.3	0.11	0.1
	0.5 m	0.97	0.72	0.63	0.94	0.63	0.37
	1 m	1	1	0.93	1	0.93	0.62
Single mainshock	0.05 m	0	0.03	0.11	0.33	0.11	0.1
	0.5 m	1	0.96	0.78	0.98	0.78	0.52
	1 m	1	1	1	1	1	0.92

Figure 17 presents the distribution of D_P discrete points for Case 2 when the COV_F values are 0.1, 0.2, and 0.3. At a small COV_F value, the D_P of the slope is small and concentrated. The variation range of soil increases with the increase in COV_F, so that the mean of D_P gradually increases and becomes more discrete. When the COV_F is 0.1, the mean of D_P subjected to the MAS is 0.299 m, and the mean of D_P caused by the single mainshock is 0.247 m. The mean of D_P under the MAS is larger than that caused by a single mainshock. When the COV_F values are 0.2 and 0.3, the mean of D_P produced by MAS is 0.404 m and 0.553 m, respectively. At this time, the mean of D_P values due to a single mainshock are 0.293 m and 0.364 m. The mean D_P of the slope keeps increasing with the growth of COV_F, and the discrepancy of D_P also displays a gradual increasing tendency.

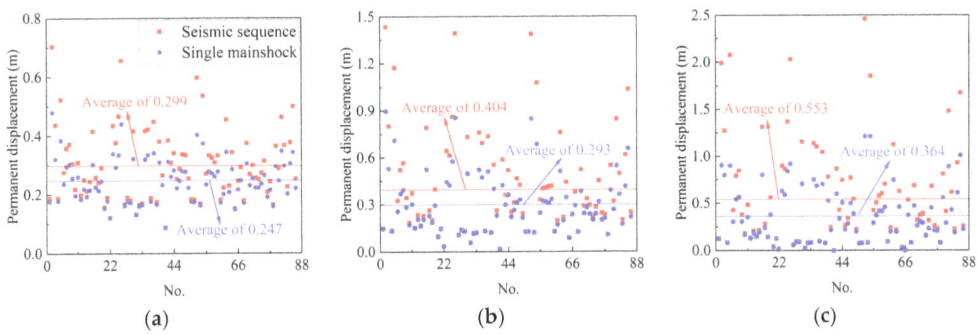

Figure 17. Distribution of discrete points of D_P under different COV_F (Case 2): (**a**) COV_F = 0.1; (**b**) COV_F = 0.2; (**c**) COV_F = 0.3.

Figure 18 shows the distribution of D_P discrete points of slope under the action of different PGA when the COV_F is 0.3. When the PGA values are 0.4 g and 0.6 g, the mean of D_P values of the slope produced by the MAS are 0.261 m and 0.881 m, respectively. However, the mean D_P values of the slope due to the single mainshock are only 0.173 m and 0.577 m. The D_P of the slope increases continuously with the increase in PGA.

Figures 19 and 20 show the probability information of D_P of the slope under different COV_F. The maximum PDF value is 3.2 when the COV_F is 0.1, and the D_P of the slope is mainly focused around 0–0.5 m. The PDF value gradually decreases, the curve gradually shifts to the right with the increase in the COV_F, and the D_P of the slope is more widely distributed. Table 5 lists the dynamic reliability of the slope when the cumulative slips are 0.05 m, 0.5 m, and 1 m under COV_F and PGA. When the PGA is 0.5 g and the threshold is 1 m, the dynamic reliability of the slope induced by the MAS decreases by 17% with the COV_F increasing from 0.1 to 0.3. When the COV_F is 0.3, the dynamic reliability of the slope subjected to the MAS is also reduced by 14% compared with the single mainshock. In addition, the dynamic reliability of the slope under the MAS decreased by 14% with the PGA increasing from 0.4 g to 0.6 g.

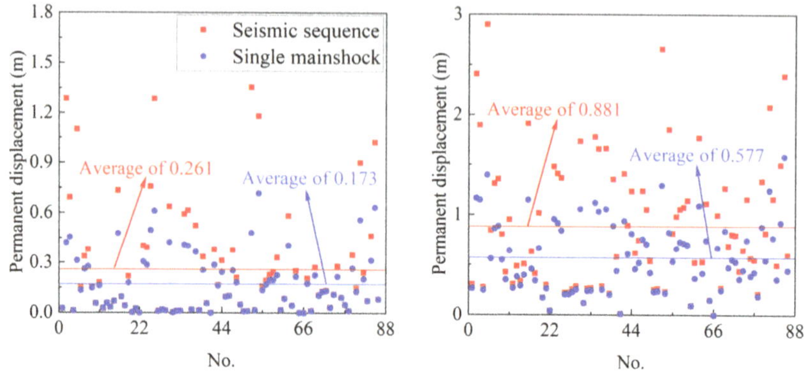

(**a**) (**b**)

Figure 18. Distribution of discrete points of D_P under different PGA when COV_F = 0.3 (Case 2): (**a**) PGA = 0.4 g; (**b**) PGA = 0.6 g.

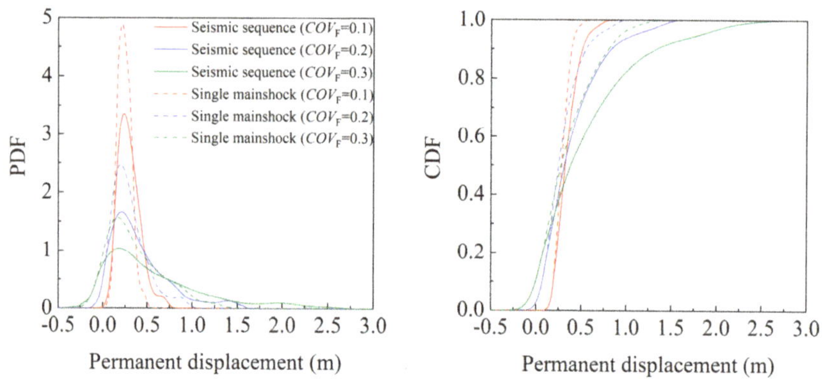

(**a**) (**b**)

Figure 19. Probability distribution of D_P with different COV_F (Case 2): (**a**) PDF; (**b**) CDF.

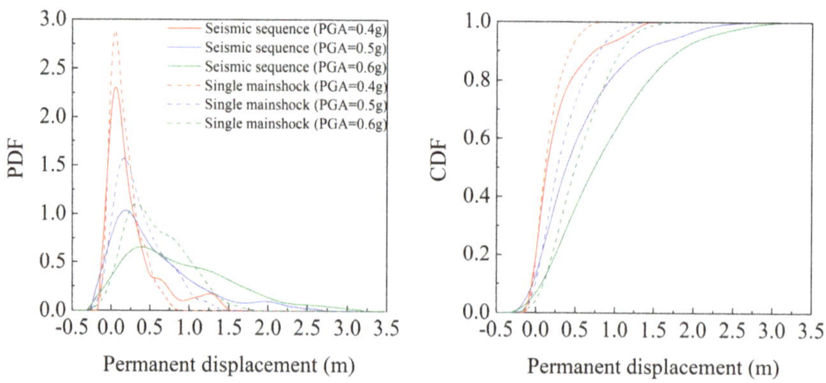

(**a**) (**b**)

Figure 20. Probability distribution of D_P with different PGA when COV_F = 0.3 (Case 2): (**a**) PDF; (**b**) CDF.

Table 5. Dynamic reliability of slope under different COV_F and PGA (Case 2).

Type of Ground Motion	D_P	COV_F			PGA		
		0.1	**0.2**	**0.3**	**0.4 g**	**0.5 g**	**0.6 g**
Mainshock–aftershock sequence	0.05 m	0	0.03	0.18	0.4	0.18	0.05
	0.5 m	0.95	0.74	0.62	0.83	0.62	0.4
	1 m	1	0.94	0.83	0.94	0.83	0.65
Single mainshock	0.05 m	0	0.03	0.19	0.4	0.19	0.06
	0.5 m	1	0.87	0.74	0.95	0.74	0.51
	1 m	1	1	0.97	1	0.97	0.86

On the basis of the above reliability information, it can be seen that the conclusion of Case 1 is comparable to that of Case 2, i.e., the COV_F has a significantly greater influence on the dynamic reliability of the slope. However, due to the different distribution of soil layers in the layered slopes, the influence of the *COV* on the failure probability of a slope is different. Compared with Case 2, the lower layer located on the slope is a sandy soil with poorer properties and its thickness is relatively deep. Therefore, the dynamic reliability of Case 1 is more significantly affected by the *COV*. The influence of different soil layers on the dynamic reliability and sliding surface position of slopes has been discussed in our previous research and detailed content can be found in [6].

7. Conclusions

A slope reliability analysis method based on GPDEM and the Newmark displacement method is proposed to quantify the impact of spatial variability of soil strength parameters on the dynamic reliability. The MAS and parameter random field are generated by the random simulation method of MAS and spectral representation method. Based on the Newmark method, the D_P of layered soil slope is calculated by nonintrusive reliability analysis, and the influence of the COV_C and COV_F on the dynamic reliability of slope is compared. The conclusions of this study are as follows:

(1) A reliability analysis method for D_P of the slope is established based on the GPDEM and Newmark methods. Combined with the noninvasive stochastic analysis method, the failure probability of a slope can be quickly obtained.

(2) According to the stochastic dynamic calculation results of the layered soil slope, COV_C and COV_F have a significant impact on the D_P of the slope induced by the MAS. The mean of D_P of the slope also presents a trend of increasing gradually with an increase in the COV_C and COV_F values. In contrast, the D_P of slope is more sensitive to the COV_F.

(3) Affected by the randomness and nonlinearity of the materials, the PDF curve has nonuniform single or double peaks. As the *COV* increases, the PDF curve becomes lower and wider, and the failure probability of the layered soil slope increases. When the D_P threshold is 1 m and PGA is 0.5 g, the dynamic reliability of the soil slope is continuously reduced, and the failure probability is even increased by about 20% with the *COV* increasing from 0.1 to 0.3.

(4) The impact of aftershocks on the D_P of the soil slope cannot be ignored. The mean of D_P of the slope induced by the MSA is larger than that under a single mainshock. The dynamic reliability of the slope caused by the MAS can even be reduced by 7–30% compared with a single mainshock when the displacement threshold is 1 m and the COV_C is 0.3. Additionally, the impact of aftershocks on the D_P of slope increases with an increase in PGA.

Author Contributions: Conceptualization, R.P.; Methodology, H.Z.; Software, G.W.; Validation, X.Y.; Writing—original draft, G.W.; Writing—review & editing, R.P.; Supervision, R.P. All authors have read and agreed to the published version of the manuscript.

Funding: This work was supported by the National Key Research and Development Program of China (2021YFB2601102), China National Natural Science Foundation (Grant Nos. 52279125, 52279096, and 52009017), and Liaoning Province Science Foundation (2020-BS-06).

Data Availability Statement: Not applicable.

Conflicts of Interest: The authors declare no conflict of interest.

References

1. Huang, Y.; Zhao, L.; Li, X. Slope-Dynamic Reliability Analysis Considering Spatial Variability of Soil Parameters. *Int. J. Geomech.* **2020**, *20*, 04020068. [CrossRef]
2. Huang, J. Chi-Chi earthquake induced landslides in Taiwan. *Earthq. Eng. Eng. Seismol.* **2000**, *2*, 25–33.
3. Xu, C.; Xu, X.; Dai, F. Three (nearly) complete inventories of landslides triggered by the May 12, 2008 Wenchuan Mw 7.9 earthquake of China and their spatial distribution statistical analysis. *Landsides* **2014**, *11*, 441–461. [CrossRef]
4. Yin, Y.; Zheng, X.; Li, P.; Li, S. Catastrophic landslides associated with the M8.0 Wenchuan earthquake. *Bull. Eng. Geol. Environ.* **2011**, *70*, 35–50. [CrossRef]
5. Xu, C.; Xu, X. Statistical analysis of landslides caused by the Mw 6.9 Yushu, China, earthquake of April 14, 2010. *Nat. Hazards* **2014**, *72*, 871–893. [CrossRef]
6. Wang, G.; Pang, R.; Xu, B.; Yu, X. Permanent displacement reliability analysis of soil slopes subjected to mainshock-aftershock sequences. *Comput. Geotech.* **2023**, *153*, 105069. [CrossRef]
7. Zhang, S.; Wang, G.; Sa, W. Damage evaluation of concrete gravity dams under mainshock-aftershock seismic sequences. *Soil Dyn. Earthq. Eng.* **2013**, *50*, 16–27. [CrossRef]
8. Liu, J.; Yi, G.; Zhang, Z.; Guan, Z.; Ruan, X.; Long, F.; Du, F. Introduction to the Lushan, Sichuan M7.0 earthquake on 20 April 2013. *Chin. J. Geophys. Chin. Ed.* **2013**, *56*, 1404–1407. [CrossRef]
9. Kim, B.; Shin, M. A model for estimating horizontal aftershock ground motions for active crustal regions. *Soil Dyn. Earthq. Eng.* **2017**, *92*, 165–175. [CrossRef]
10. Pang, R.; Xu, B.; Zhou, Y.; Zhang, X.; Wang, X. Fragility analysis of high CFRDs subjected to mainshock-aftershock sequences based on plastic failure. *Eng. Struct.* **2020**, *206*, 110152. [CrossRef]
11. Zhou, Z.; Yu, X.; Lu, D. Identifying Optimal Intensity Measures for Predicting Damage Potential of Mainshock-Aftershock Sequences. *Appl. Sci.* **2020**, *10*, 6795. [CrossRef]
12. Shen, J.; Chen, J.; Ding, G. Random field model of sequential ground motions. *Bull. Earthq. Eng.* **2020**, *18*, 5119–5141. [CrossRef]
13. Jibson, R.W. Methods for assessing the stability of slopes during earthquakes—A retrospective. *Eng. Geol.* **2011**, *122*, 43–50. [CrossRef]
14. Li, J.; Chen, J. The principle of preservation of probability and the generalized density evolution equation. *Struct. Saf.* **2008**, *30*, 65–77. [CrossRef]
15. Su, H.; Fu, Z.; Gao, A.; Wen, Z. Numerical simulation of soil levee slope instability using particle-flow code method. *Nat. Hazards Rev.* **2019**, *20*, 04019001. [CrossRef]
16. Su, H.; Hu, J.; Yang, M.J. Evaluation method for slope stability under multianchor support. *Nat. Hazards Rev.* **2014**, *16*, 04014033. [CrossRef]
17. Chousianitis, K.; Del Gaudio, V.; Sabatakakis, N.; Kavoura, K.; Drakatos, G.; Bathrellos, G.D.; Skilodimou, H.D. Assessment of earthquake-induced landslide hazard in Greece; from Arias intensity to spatial distribution of slope resistance demand. *Bull. Seismol. Soc. Am.* **2016**, *106*, 174–188. [CrossRef]
18. Du, W.; Wang, G. A one-step Newmark displacement model for probabilistic seismic slope displacement hazard analysis. *Eng. Geol.* **2016**, *205*, 12–23. [CrossRef]
19. Wang, M.; Li, D.; Du, W. Probabilistic seismic displacement hazard assessment of earth slopes incorporating spatially random soil parameters. *J. Geotech. Geoenvironmental Eng.* **2021**, *147*, 04021119. [CrossRef]
20. Li, Y.; Pang, R.; Xu, B.; Wang, X.; Fan, Q.; Jiang, F. GPDEM-based stochastic seismic response analysis of high concrete-faced rockfill dam with spatial variability of rockfill properties based on plastic deformation. *Comput. Geotech.* **2021**, *139*, 104416. [CrossRef]
21. Pang, R.; Xu, B.; Kong, X.; Zhou, Y.; Zou, D. Seismic performance evaluation of high CFRD slopes subjected to near-fault ground motions based on generalized probability density evolution method. *Eng. Geol.* **2018**, *246*, 391–401. [CrossRef]
22. Pang, R.; Xu, B.; Zou, D.; Kong, X. Stochastic seismic performance assessment of high CFRDs based on generalized probability density evolution method. *Comput. Geotech.* **2018**, *97*, 233–245. [CrossRef]
23. Chen, X.; Gao, R.; Gong, W.; Li, Y.; Qiu, J. Random seismic response and dynamic fuzzy reliability analysis of bedding rock slopes based on Pseudoexcitation method. *Int. J. Geomech.* **2018**, *18*, 04017165. [CrossRef]
24. Liu, L.; Cheng, Y. System reliability analysis of soil slopes using an advanced Kriging metamodel and quasi-Monte Carlo simulation. *Int. J. Geomech.* **2018**, *18*, 06018019. [CrossRef]
25. Hasofer, A.M.; Lind, N.C. Exact and invariant second-moment code format. *J. Eng. Mech. Div.* **1974**, *100*, 111–121. [CrossRef]
26. Metropolis, N.; Ulam, S. The monte carlo method. *J. Am. Stat. Assoc.* **1949**, *44*, 335–341. [CrossRef]
27. Wong, F.S. Slope reliability and response surface method. *J. Geotech. Eng.* **1985**, *111*, 32–53. [CrossRef]

28. Pasculli, A.; Calista, M.; Sciarra, N. Variability of local stress states resulting from the application of Monte Carlo and finite difference methods to the stability study of a selected slope. *Eng. Geol.* **2018**, *245*, 370–389. [CrossRef]
29. Li, J.; Chen, J.B. Probability density evolution method for dynamic response analysis of structures with uncertain parameters. *Comput. Mech.* **2004**, *34*, 400–409. [CrossRef]
30. Li, J.; Chen, J.B. The dimension-reduction strategy via mapping for probability density evolution analysis of nonlinear stochastic systems. *Probabilistic Eng. Mech.* **2006**, *21*, 442–453. [CrossRef]
31. Huang, Y.; Xiong, M. Dynamic reliability analysis of slopes based on the probability density evolution method. *Soil Dyn. Earthq. Eng.* **2017**, *94*, 1–6. [CrossRef]
32. Huang, Y.; Hu, H.; Xiong, M. Probability density evolution method for seismic displacement-based assessment of earth retaining structures. *Eng. Geol.* **2018**, *234*, 167–173. [CrossRef]
33. Pang, R.; Xu, B.; Zhou, Y.; Song, L. Seismic time-history response and system reliability analysis of slopes considering uncertainty of multi-parameters and earthquake excitations. *Comput. Geotech.* **2021**, *136*, 104245. [CrossRef]
34. Xu, B.; Pang, R.; Zhou, Y. Verification of stochastic seismic analysis method and seismic performance evaluation based on multi-indices for high CFRDs. *Eng. Geol.* **2020**, *264*, 105412. [CrossRef]
35. Bai, T.; Hu, X.; Gu, F. Practice of searching a noncircular critical slip surface in a slope with soil variability. *Int. J. Geomech.* **2019**, *19*, 04018199. [CrossRef]
36. Hu, H.; Huang, Y.; Zhao, L. Probabilistic Seismic-Stability Analysis of Slopes Considering the Coupling Effect of Random Ground Motions and Spatially-Variable Soil Properties. *Nat. Hazards Rev.* **2020**, *21*, 04020028. [CrossRef]
37. Li, D.; Qi, X.; Phoon, K.; Zhang, L.; Zhou, C. Effect of spatially variable shear strength parameters with linearly increasing mean trend on reliability of infinite slopes. *Struct. Saf.* **2014**, *49*, 45–55. [CrossRef]
38. Hariri-Ardebili, M.A.; Seyed-Kolbadi, S.M.; Saouma, V.E.; Salamon, J.; Rajagopalan, B. Random finite element method for the seismic analysis of gravity dams. *Eng. Struct.* **2018**, *171*, 405–420. [CrossRef]
39. Zhang, J.; Zhang, M.; Tang, W. New methods for system reliability analysis of soil slopes. *Can. Geotech. J.* **2011**, *48*, 1138–1148. [CrossRef]
40. Cornell, A. *First-Order Uncertainty Analysis of Soil Deformation and Stability*; Publication of University of Hong Kong: Hong Kong, China, 1972.
41. Vanmarcke, E. Reliability of earth slopes. *J. Geotechnical Eng.* **1977**, *103*, 1247–1265. [CrossRef]
42. Der Kiureghian, A.; Ke, J. The stochastic finite element method in structural reliability. *Probabilistic Eng. Mech.* **1988**, *3*, 83–91. [CrossRef]
43. Shinozuka, M.; Deodatis, G. Simulation of multi-dimensional Gaussian stochastic fields by spectral representation. *Appl. Mech. Rev.* **1996**, *49*, 29–53. [CrossRef]
44. Matthies, H.G.; Brenner, C.E.; Bucher, C.G.; Guedes Soares, C. Uncertainties in probabilistic numerical analysis of structures and solids-Stochastic finite elements. *Struct. Saf.* **1997**, *19*, 283–336. [CrossRef]
45. Ghiocel, D.; Ghanem, R. Stochastic finite-element analysis of seismic soil-structure interaction. *J. Eng. Mech.* **2002**, *128*, 66–77. [CrossRef]
46. Pang, R.; Zhou, Y.; Chen, G.; Jing, M. Stochastic mainshock-aftershock simulation and its applications in dynamic reliability of structural systems via DPIM. *J. Eng. Mech.* **2023**, *149*, 04022096. [CrossRef]
47. Huang, L. *Dynamic Stability Analysis of Layered Soil Slope under Earthquake*; Southwest Jiaotong University: Chengdu, China, 2017. (In Chinese)
48. Ching, J.; Phoon, K. Effect of element sizes in random field finite element simulations of soil shear strength. *Comput. Struct.* **2013**, *126*, 120–134. [CrossRef]

water

MDPI

Article

Research on Improvement of Slope Protection Concrete Precast Block Joints Based on Physical Model Experiment

Fang Chen [1,2,3,*], Songtao Hu [1,3,*], Qinghe Fang [4], Liehong Ju [5], Da Liu [1,3] and Zhe Huang [5]

1 Jiangxi Academy of Water Science and Engineering, Nanchang 330029, China
2 College of Civil and Transportation on Engineering, Hohai University, Nanjing 210098, China
3 Jiangxi Engineering Technology Research Center on Hydraulic Structures, Nanchang 330029, China
4 China School of Ocean Engineering, Harbin Institute of Technology at Weihai, Weihai 264209, China
5 Nanjing Hydraulic Research Institute, Nanchang 330029, China
* Correspondence: fangchen@hhu.edu.cn (F.C.); sthu.2000@sohu.com (S.H.)

Abstract: Precast concrete block slope protection is widely used due to its advantages of easy detection and laying, ease of organization, and the limited time required for construction. In order to prevent the soil or gravel bedding of precast concrete from being subjected to wind and wave pressures, the joints between precast concrete blocks are usually filled with mortar. However, the existing standards do not specify the width or material of the joints. Furthermore, excessively wide mortar joints or shrinkage of the mortar can result in loss, a hollowed-out cushion, and damage to the slope, thus compromising the quality of slope protection engineering. To establish standards for controlling the quality of slope protection seams, this paper designed and conducted a physical model test of precast concrete block revetment seams. By embedding pore water pressure sensors in the cushion layer, changes in the pore water pressure were observed under varying conditions, including different water pore pressure sensor locations, water levels in front of the embankment, and different joint widths. Based on the test results, design standards for joint widths and recommendations for the treatment of joint mortar materials were proposed. After adding different amounts of a calcium oxide–calcium sulfoaluminate composite expansion agent (HME) into a joint mortar material, the paper also carried out a shear test on the contact surface between the joint mortar and the slope protection concrete after adding varying amounts of a calcium oxide-calcium sulfoaluminate composite expansion agent (HME) to the joint mortar material. Following a microporous structure test, recommendations for joint mortar construction treatment were proposed. The results indicate that the pore water pressure of the precast concrete slope protection cushion is closely related to the position of the cushion, the water level in front of the embankment, and the width of the paving seam. When the masonry seam width increased from 0.5 mm to 1 mm and from 1 mm to 1.5 mm, the variation ranges of the pore water pressure were 40–80% and 6–20%, respectively, with the latter being significantly lower than the former. Therefore, in practical engineering, joint treatment should take into account the impact of the cushion position, the water level in front of the dike, and the joint mortar width. Mortar shedding within the range of wave climbing height should be addressed promptly, and joint width should be controlled to below 1 cm as much as possible to effectively prevent damage to the cushion surface. The addition of an expansion agent can improve the bond strength of the concrete and mortar to a certain extent. The study found that an 8% content of the expansion agent resulted in the best mortar bond strength and the densest microstructure. These research findings can serve as a basis for the development of quality control standards for precast concrete slope protection.

Keywords: concrete block revetment; paving seam; model pore water pressure; cushion; HME

Citation: Chen, F.; Hu, S.; Fang, Q.; Ju, L.; Liu, D.; Huang, Z. Research on Improvement of Slope Protection Concrete Precast Block Joints Based on Physical Model Experiment. *Water* **2023**, *15*, 1874. https://doi.org/10.3390/w15101874

Academic Editor: Francesco De Paola

Received: 21 December 2022
Revised: 16 April 2023
Accepted: 25 April 2023
Published: 15 May 2023

1. Introduction

Prefabricated slope protection on reservoir banks plays a critical role in maintaining slope stability and reducing the scouring effects of water flow on slope surfaces. This protection is essential to ensure the safety of reservoir dams and river dikes. According to statistics, more than 98,000 reservoirs and 312,000 km of embankments for grade 5 or above rivers have been built in China, with hexagonal precast concrete blocks being the preferred slope protection method. The Chinese standard SL634-2012 [1] outlines the quality requirements and detection methods for appearance dimensions, slope protection thickness, and the quality of paving and filtration. While the standard mandates no gaps in the paving seams, gaps are necessary to maintain the stability of the precast blocks due to construction conditions and temperature deformation [2–4]. Currently, mortar is the most commonly used joint material. However, based on long-term engineering practice, when the gap is too wide, or the joint material falls off, waves can easily dislodge the underlying layer of stone and earth dam soil material used to protect the slope, leading to slope collapse and unstable slope protection.

Regarding slope stability designs, E.I. Mikhnevich [5] proposes determining the diameter of dumped rip-rap material and the thickness of plates for slope protection of earth dams in the zone of active wave impact. Zhao Nenghao [6] has established a coupling model that incorporates both pore water pressure and pore water gravity to realize their mutual influence on slopes. Wang Xiaobing [7] investigated the probabilistic stability of embankment slopes subjected to water level drawdown using the random field finite element method (RFEM) with strength reduction technology. Zhilu Zheng [8] emphasizes the need to control the space between building blocks to within 2–5 mm to ensure their aesthetics and safety. Ji Zheng [9] suggests that horizontal seam widths and seam widths of straight sections of precast block slope protection should be controlled at 8 mm and 10 mm, respectively. Due to the time-consuming nature of three-dimensional analyses of slope stability, Yong Liu [10] proposed a two-dimensional plane strain analysis method that offers an appropriate level of conservativeness, making it useful for slope stability designs. Decheng LIU [11] utilized a two-dimensional limit equilibrium transfer coefficient method to analyze the stability of an entire reservoir bank under various working conditions and proposed comprehensive treatment measures involving foot walls, precast block slope protection, and drainage ditches. In another study, the influence of different slope protection conditions on slope deformation was summarized, and the influence of soil compaction, soil moisture content, and slope ratio on the horizontal displacement restraint capacity was investigated. The stability of slopes was determined through a laboratory model test of eco-concrete slope protection [12].

It is widely known that the most commonly used joint material is cement mortar, which is made up of cement, fine aggregates, and water. However, the fluidity and strength of cement mortar, as well as the construction process, can impact the filling and plastering of joints. Mortar is also susceptible to falling off or resisting cushion loss due to wave-current scour and weight adjustment of precast blocks. Seifan Mostafa [13] conducted a study on the effects of nano and micro Fe_3O_4 on the fresh and mechanical properties of plain mortar and mortar modified with fly ash. Dongyu Niu [14] investigated the mechanical properties of asphalt mortar with different admixtures and found that adding additives improved its water stability. Lv Dan [15] prepared three types of asphalt mortars with different mineral powder fillers and tested their low-temperature cohesive strengths using technology for testing cohesive tensile strength. John Shaise K [16] reviewed the properties of fly ash and fly ash-slag geopolymer mortar, including microstructural properties, fresh properties (such as setting time and workability), hardened properties (such as compressive strength and tensile strength), and durability properties (such as thermal resistance, shrinkage, acid resistance, chloride resistance, and sulfate resistance). Rocha Joaquin Humberto Aquino [17] evaluated the influence of waterproof and adhesive additives on the mechanical properties, such as compressive and flexural strength, of render mortar over a 28-day period. Additionally, some scholars [18–22] have utilized nanotechnology and fibers to enhance the mechanical properties of mortar materials.

The stability of precast block slope protection is directly related to the quality of masonry. Previous research has analyzed the stability of slope protection and provided construction guidance for masonry materials and construction quality control. However, the width of masonry and material improvement has not been adequately researched. Therefore, this paper established a physical model of the masonry joints of precast concrete slope protection. By placing pore water pressure gauges in the slope protection cushion, the changes in the pore water pressure in the cushion were studied under different positions of the gauges, different water levels in front of the dike, and different widths of the joints. The bonding properties of the joint mortar were also improved by adding an expansion agent with a double-expansion source. Appropriate joint widths and expansion agent dosages were proposed for concrete slope protection based on the research findings. These results can serve as a valuable reference for precast concrete slope protection.

2. Experiment Design

2.1. Materials

In this study, ordinary Portland cement (P.O 42.5, GB175-2007, [23]) and a calcium oxide-calcium sulfoaluminate composite expansion agent (HME, produced by Jiangsu Subote New Material Co., Ltd. in China) were used, and the physical properties of the raw materials are presented in Tables 1–3. The materials included natural medium sand with a fineness module of 2.9 and natural gravel with a maximum particle size of 40 mm (following the Chinese standard SL/T 352-2020 [24], Specifications for Hydraulic Concrete Construction), and their grain compositions are given in Tables 4 and 5.

Table 1. Physical and mechanical properties of the cement.

	Specific Surface Area (m^2/kg)	Initial Setting Time (min)	Final Setting Time (min)	Flexural Strength (MPa)		Compressive Strength (MPa)	
				3 d	28 d	3 d	28 d
P.O 42.5	340	168	279	4.8	7.7	27.6	46.4

Table 2. Physical and mechanical properties of HME.

	Density (g/cm^3)	Specific Surface Area (m^2/kg)	Initial Setting Time (min)	Final Setting Time (min)	Compressive Strength (MPa)		Restrained Expansion Rate (%)	
					7 d	28 d	7 d	28 d
HME	2.90	390	170	220	22.9	41.8	0.15	0.10

Table 3. Chemical composition of HME (%).

	SiO_3	CaO	MgO	Fe_2O_3	Al_2O_3	K_2O	Na_2O	SO_3
HME	1.03	52.70	1.81	0.66	13.61	0.40	0.14	28.33

Table 4. Grain composition of the natural medium sand.

Aperture Size (mm)	5.0	2.5	1.25	0.63	0.315	0.16	<0.16	Fineness Module
Cumulative Residue Amount (wt.%)	1.5	5.2	11.9	39.2	90.5	98.0	100.0	2.9

Table 5. Grain composition of the gravel.

Aperture Size (mm)	40	30	20	10	5	<5.0
Cumulative Residue Amount (wt.%)	0	6.7	43.5	73.2	99.2	100.0

2.2. Mix Proportions and Preparation of Samples

2.2.1. Mix Proportions and Preparation of Prefabricated Blocks

The dimensions of hexahedral prefabricated slope protection are regulated by the Hydraulic Design Manual [25] to have a side length of 0.3–0.4 m and a thickness of 0.15–0.20 m. In this study, the prefabricated blocks had a side length of 0.3 m and a thickness of 0.11 m, and a model with a scale of 1:10 was created with a side length of 30 mm and a thickness of 11 mm. The match ratio of the prefabricated blocks was ignored, as it did not affect the test results. The prefabricated blocks were formed with cement paste to have a compressive strength of C15, as detailed in Table 6. After brushing, mixing, pouring, forming, and demolding, all samples were cured and are shown in Figure 1.

Table 6. Mix proportions of cement paste samples (kg/m^3).

	Cement	Sand	Water	Compressive Strength of 28 d/MPa
C15	260	764	173	17.8

(a) (b) (c) (d) (e) (f)

Figure 1. The forming process of prefabricated blocks. (**a**) Brushing. (**b**) Mixing. (**c**) Pouring. (**d**) Forming. (**e**) Removing. (**f**) Curing.

2.2.2. Mix Proportions and Preparation of Joint Mortar

Based on the conditions of precast concrete block laying for slope protection, specific areas were chosen to conduct bond performance tests between the cement mortar and precast concrete blocks intended for slope protection use, as depicted in Figure 2. In order to establish the test model conveniently, a scaled version of the regional model was used with a length scale of 1:2. The test model was carried out using test blocks and molds. During the molding process, concrete ① and concrete ② were formed and maintained first, with both having a strength of C15. Following the curing process, cement mortar was poured into the paving seam. Figure 3 illustrates the pouring situation of the test model.

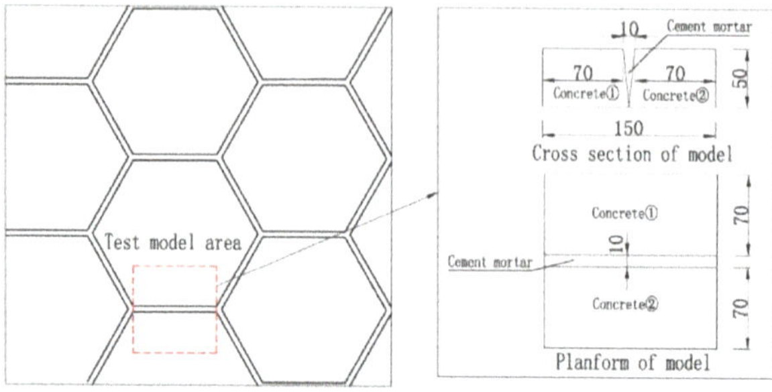

Figure 2. Bond property test model.

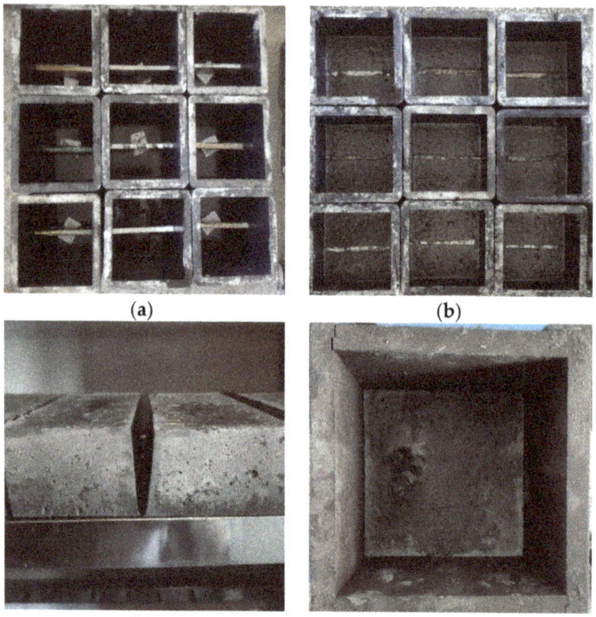

Figure 3. Pouring conditions of the test model. (**a**) Fix the position and shape of the paving seams. (**b**) Pour concrete on both sides of the paving seams. (**c**) Cure after demolding. (**d**) Pour cement mortar on paving seams.

To investigate the effect of different admixture strengths and dosages on the bonding property between cement mortar and precast concrete blocks used for slope protection, 3 groups of mixing ratios were designed using the mixing ratio design method of hydraulic concrete and hydraulic mortar, in accordance with the SL/T 352-2020 Hydraulic Concrete Test Specification. These were M15 cement mortar, M20 cement mortar, and M20+ cement mortar (with 8% HME added to consider the mortar shrinkage cracking effect of the seam falling off). The mix proportions of the cement mortar are detailed in Table 7.

Table 7. Mix proportions of paving seam treatment mortar (kg/m^3).

	Cement	Water	Sand	Expanding Agent	W/b
M15	400	350	1350	/	0.875
M20	300	210	1350	/	0.70
M20$^+$	300	210	1350	24	0.70

Figure 4 displays the specimen model after molding, and all specimens were cured for 24 h in a standard curing chamber (20 °C/95% RH) before demolding. Subsequently, the specimens were further cured for 28 days in the same curing chamber. After curing, the strengths of the cement paste, cement mortar, and concrete were tested.

Figure 4. Specimen model after molding.

2.3. *Width Model of Paving Seam*

2.3.1. Observation and Monitoring Equipment

The strain-type micropore water pressure gauge sensor has a measuring range of 0–10 kPa, an accuracy of less than 0.03 kPa, a sensitivity coefficient of 0.5 mV/kPa, and a conversion coefficient of 0.002 kPa/με after acquisition. Figure 5 shows the strain-type micropore water pressure gauge sensor.

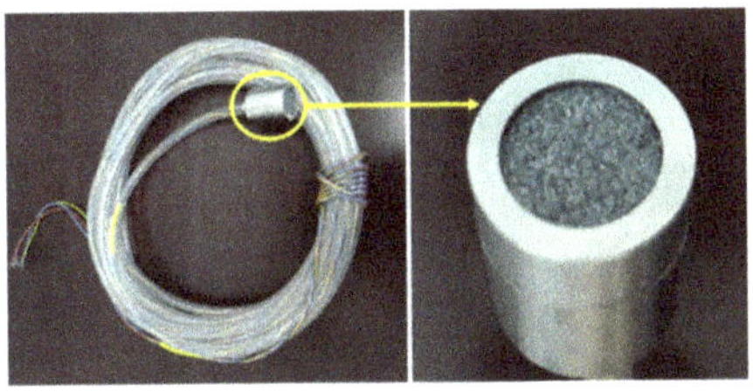

Figure 5. Strain micropore water pressure gauge sensor.

In order to realize the collection of stress in the strain-type micropore water pressure sensor, a static resistance strain tester was used to measure the variation in the physical parameters of the pore water pressure through the transformation of a strain full-bridge circuit so as to study the influence of the variation in the pore water pressure in the cushion layer under wave action.

Assuming that the value of the pore water pressure gauge collected through the static resistance strain gauge is F_0, then the real-time stress value was calculated according to Formula (1):

$$F = K(F_i - F_0) \tag{1}$$

where K is the sensor sensitivity coefficient, F is the real-time stress value of the sensor, F_0 is the 0 output strain value of the sensor, and F_i is the output strain value corresponding to P_i.

2.3.2. General Layout of Model

The gravity similarity criterion was used to design the physical model of the paving seam with a geometric scale of 1:10 for the slope protection section of the soil embankment. The slope model had a height of 0.5 m, a ratio of 1:2.5, and a width of 0.4 m. A hexagonal precast block with a side length of 300 mm, made at a geometric scale of 1:10, was used for the model test to truly restore the soil embankment slope protection. The bedding material for slope protection was arranged using the coarse-grained material grading scale method proposed by Guo Wanli [26]. A wave generator was placed at the end of the model water tank to simulate different wave heights. To observe the influence of different locations, water levels in front of the embankment, and paving seam widths on the change in the pore water pressure in the cushion layer and to provide a reference value for the width of paving seams in prefabricated slope protection blocks, 4 strain-type micropore water pressure sensors were arranged on the same dam axis with a vertical spacing of 60 mm. The pore water pressure gauges 14 corresponded to measuring points 14, and their arrangement is shown in Table 8 and Figure 6, while Figure 7 shows a profile view of the precast concrete block slope.

Table 8. Arrangement of pore water pressure gauge.

Sensor Number	1	2	3	4
Vertical distance from the axis of the embankment top/mm	360	510	660	810
Vertical distance from the embankment bottom/mm	330	270	210	150

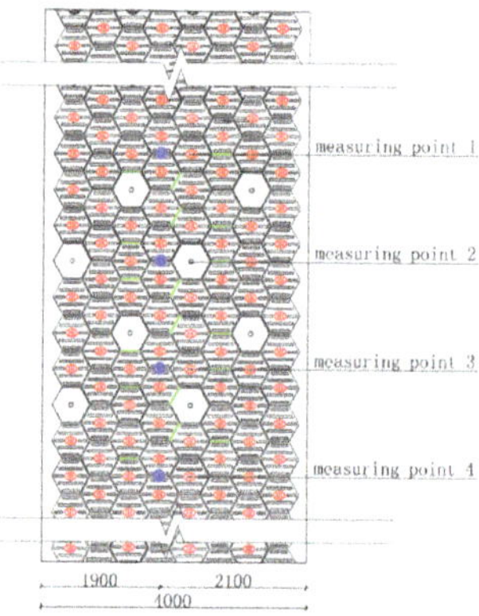

Figure 6. Plan of pore water pressure gauge.

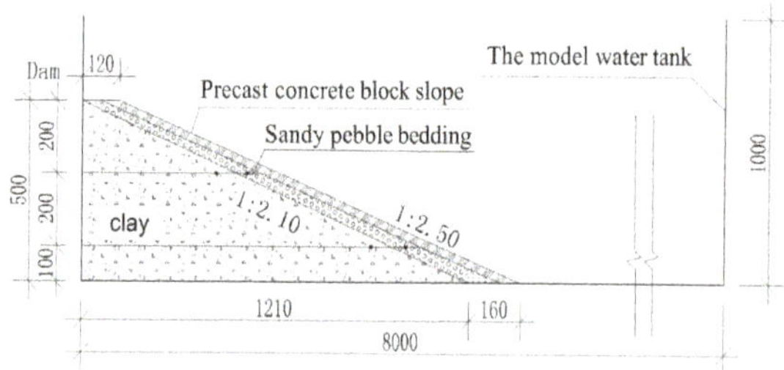

Figure 7. Profile view of precast concrete block slope.

2.3.3. Laying the Precast Block and Setting the Paving Seam

Considering that the test mainly studies the falling off of prefabricated block joint material caused by long-term erosion attributable to waves, and the subsequent failure of the cushion layer, a reference value for the width of the joint was proposed. Therefore, in order to simulate the changes in the cushion layer under the action of waves under different shedding conditions, prefabricated blocks with different paving seam widths were laid by inserting acrylic plates with different thicknesses in combination with the actual possible different widths of paving seams. The laying process of the precast blocks is shown in Figure 8.

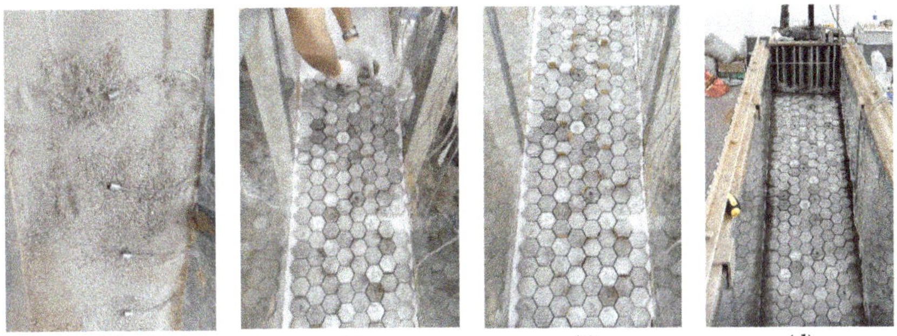

(a) (b) (c) (d)

Figure 8. The laying process of precast concrete blocks. (**a**) Placing the sensors. (**b**) Laying the precast blocks. (**c**) Paving seam arrangement. (**d**) Finished model.

2.4. Test Program Design

2.4.1. Paving seam Width

To understand the paving seam widths of existing slope protection materials, a key survey was conducted on a precast block slope protection project on site, and a total of 60 paving seams were measured at 20 points. The measurement method consisted of measuring the maximum seam width at 1 place every 50 m^2 of slope protection.

For the convenience of analyses and comparisons, the layout pattern of the paving seams (mainly referring to the numbers of transverse, oblique, and vertical joints) between the 2 measurement points was consistent. Three different water levels were set in front of the embankment: 22.5 cm, 27.6 cm, and 30.6 cm. The wave-making period was set to 800 RPM, and the conversion formula between the speed of the generator and the wave period is shown in Equation (2). The wave-making period was calculated to be 1.875 s. According to the wave height calculated by Mingwei Wei et al. [27] using the water depth and the wave-making period, the wave heights simulated by the model were 0.12 m, 0.15 m, and 0.17 m. The specific working conditions are shown in Table 9.

$$T = 60/(V/25) \tag{2}$$

where T is the wave-making period of the wave-making machine. V is the speed of the wave generator (turns per minute).

Table 9. Working conditions.

Working Condition	The Width of Paving Seam/mm	Water Depth in Front of the Dam/cm	Actual Simulation of Wave Height/m	Wave Period/s
1	0.5	27.6	0.15	1.875
2	1.0	27.6	0.15	1.875
3	1.5	27.6	0.15	1.875
4	1.5	22.5	0.12	1.875
5	1.5	30.6	0.17	1.875

The test operation procedure is as follows: After the slope protection is laid, water is injected into the model groove. When the water is injected into the set water level, the generator is turned on, and the strain gauge is turned on at the same time to record the sensor data. After the completion of the first group of working conditions, the water in the tank is drained, and the precast concrete blocks are removed; at the same time, the clay and the sand are added to the cushion, the pore water pressure gauge is buried in the same position, and the precast concrete blocks are laid again. In the test, the laying conditions of the cushion, pore water pressure gauge, and precast concrete blocks should be as consistent as possible under multiple working conditions. For specific collection, see Figure 9: Acquisition system layout.

Figure 9. Acquisition system layout.

2.4.2. Paving Seam Materials

(a) Compressive/Shear strength of paving seam materials

Material model samples of 70.7 mm × 70.7 mm × 70.7 mm were used for a compressive strength test, and material model samples of 150 mm × 150 mm × 50 mm with the paving seam materials were used for a shear strength test. The compressive and shear strengths were tested using a universal machine and a microcomputer-controlled electric stress-type rock shear instrument. Considering that no axial stress was present in the actual situation, the shear strength of the concrete and joint mortar was examined during the test process in order to analyze the bond performance between them.

During a shear test, normal stress is first applied to a predetermined level, and then the normal deformation is measured. When the difference between 2 normal deformation readings is less than 1%, a shear load is applied. During the shear process, the normal stress of the sample should be kept constant, the initial shear load rate should be 0.4 MPa/min, and the load and shear deformation values should be measured and recorded at appropriate loading intervals. When the horizontal deformation caused by the increased shear load is 1.5 times greater than that of the previous load, the shear load rate is halved, and the measurement interval is shortened until shear. The shear failure loads were recorded. The loading process of the specimens is shown in Figure 10.

(a) (b)

Figure 10. Specimen loading process. (**a**) Compression test. (**b**) Shear strength test.

(b) Microstructure analysis of paving seam materials

To further analyze the improvements in the properties of paving seams treated with an expansion agent, the internal structures of M20 and M20 with an 8% expansion agent were analyzed using a YG-97A capacitive mercury injection apparatus. The main measurements taken were pore volume and pore size distribution after treatment.

The capacitive mercury injection apparatus operates based on the principle of capillary pressure and the non-wettability of mercury on solid samples. Under the action of applied pressure, mercury is pressed into solid pores. The volume of mercury injected under different pressures can be measured to obtain the pore volume with the corresponding pore radius, according to Washburn's [28] cylindrical pore model (as shown in Formula (3)). By measuring the pore volume, the pore size distribution curve (porosity) of the material can be obtained.

Cylindrical pore model:

$$\pi r^2 \rho = -4\pi d\gamma \cos\theta \tag{3}$$

where p is the pressure of the mercury pressed into the pores (MPa), d is the equivalent opening size (mm), r is the surface tension coefficient of the mercury (mN/m), and θ is the contact angle between the mercury and the porous concrete material.

3. Analysis of Model Test Results

3.1. Statistics of Paving Seam Width

The survey statistics provided the paving seam widths of the precast concrete block slopes, as shown in Table 10. According to the data, the maximum widths of the level-direction, vertical-direction, and oblique-direction paving seams are 11.3 mm, 10.6 mm, and 15.8 mm, respectively. All of the seam widths are irregular, with the oblique-direction paving seam being slightly wider. Seam widths under 6 mm account for 11.7% of the total, seam widths between 6 mm and 15 mm account for 11.7% of the total, and seam widths between 15 mm and 20 mm account for 3.3% of the total. Based on these findings, the model test used three paving seam widths, namely 15 mm, 10 mm, and 5 mm, with corresponding model paving seam layouts of 1.5 mm, 1 mm, and 0.5 mm, respectively.

Table 10. The paving seam widths of precast concrete block slope from survey statistics.

No.	The Width of Paving Seam/mm			No.	The Width of Paving Seam/mm		
	Level Direction	Vertical Direction	Oblique Direction		level Direction	Vertical Direction	Oblique Direction
1	7.5	6.7	11.2	11	8.3	7.6	13.5
2	5.8	6.9	13.2	12	10.2	6.7	10.3
3	8.4	9.3	10.7	13	9.6	6.9	11.3
4	9.6	8.2	15.8	14	7.9	7.9	15.2
5	7.6	10.5	5.2	15	11.3	8.2	9.5
6	7.7	9.5	5.5	16	8.7	9.3	9.6
7	5.5	8.6	10.6	17	6.8	10.1	9.8
8	6.9	6.9	5.9	18	10.6	10.6	5.7
9	10.9	8.3	11.3	19	5.9	8.8	7.6
10	7.3	7.8	12.5	20	6.9	7.6	9.8

3.2. Variation Law of Pore Water Pressure

3.2.1. Variation Law of Pore Water Pressure at Different Positions

In total, five working conditions were tested. Overall, the pore water pressure changed in a manner similar to the wave period under different working conditions, with the difference mainly reflected in the water level in front of the embankment. For example, taking the strain value of the static resistance strain gauge in the first 30 acquisition cycles of working condition 4, Formula (1) was used to convert the pore water pressure, and the variation trend of the pore water pressure is shown in Table 11 and Figure 11. Based on the results, the following conclusions can be drawn:

Table 11. Pore water pressure at different measuring points in working condition 4.

Acquisition Cycles	Pore Water Pressure of Measuring Point/kPa				Acquisition Cycles	Pore Water Pressure of Measuring Point/kPa			
	1	2	3	4		1	2	3	4
1	0.008	0.004	0.005	0.056	16	0.014	0.089	0.238	0.570
2	0.014	0.060	0.102	0.162	17	0.002	0.098	0.236	0.582
3	0.026	0.102	0.156	0.358	18	0.020	0.128	0.368	0.430
4	0.038	0.116	0.200	0.454	19	0.034	0.114	0.208	0.566
5	0.007	0.120	0.210	0.508	20	0.003	0.076	0.264	0.436
6	0.014	0.096	0.190	0.468	21	0.011	0.100	0.278	0.582
7	0.032	0.070	0.260	0.394	22	0.025	0.080	0.326	0.444
8	0.005	0.130	0.270	0.524	23	0.034	0.112	0.226	0.594
9	0.017	0.144	0.372	0.590	24	0.004	0.137	0.224	0.588
10	0.024	0.080	0.262	0.400	25	0.002	0.086	0.328	0.416
11	0.040	0.180	0.356	0.554	26	0.033	0.112	0.358	0.594
12	0.006	0.108	0.306	0.470	27	0.002	0.114	0.224	0.508
13	0.018	0.122	0.230	0.440	28	0.015	0.134	0.340	0.426
14	0.008	0.152	0.350	0.558	29	0.035	0.086	0.346	0.420
15	0.038	0.132	0.208	0.450	30	0.008	0.114	0.242	0.578

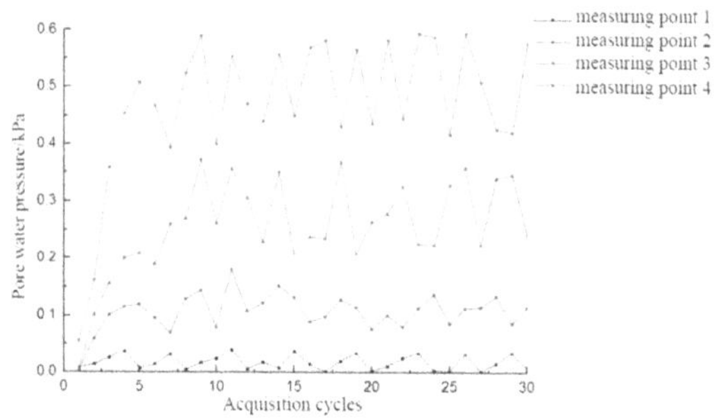

Figure 11. Variation curves of pore water pressure at different measuring points under working condition 4.

(1) Under the condition of the same seam width and water level in front of the embankment, the variation pattern of pore water pressure at different positions of the cushion varied during the wave process. At measuring point 1, located at a vertical height of 330 mm from the dam, the pore water pressure changed during the wave process due to a maximum wave height of 0.12 m and a water level in front of the dam of 22.5 cm;

(2) At measuring point 2, located at a vertical height of 270 mm from the bottom of the embankment, the arrangement of the paving seams in the slope led to water seepage through the seams. The pore water pressure increased linearly from the first to the fifth wave periods, reaching 0.12 kPa in the fifth wave period and then fluctuating around 0.10 kPa;

(3) The pore water pressure at measuring point 3 at a vertical height of 210 mm from the bottom of the embankment and at measuring point 4 at a vertical height of 150 mm fluctuated greatly, and the pore water pressure at measuring point 4 was obviously greater than that at measuring point 3. The pore water pressure at measuring points 3 and 4 increased linearly from the first to the fifth wave periods, and it fluctuated from the sixth wave period. The pore water pressure at measuring point 3 fluctuated within the range of 0.29–0.46 kPa, and that at measuring point 4 fluctuated within the range of 0.49–0.68 kPa.

In conclusion, when the masonry joint width and the water level in front of the embankment were kept constant, the pore water pressure in the cushion was found to be influenced by the location of the pore water pressure gauge, which was consistent with the impact of wave climbing height on the cushion. However, since pore water penetrates upward slowly, the pore water pressure in the cushion above the water level in front of the embankment changed only slightly. For the cushion below the water level in front of the embankment, the pore water pressure differed at different measuring points. This was because the water pressure at different measuring points was influenced by the water level in front of the embankment, while the reciprocating cycle of waves affected the seepage time of water in the joint, resulting in an unsynchronized pore water pressure between different measuring points.

3.2.2. Variation Law of Pore Water Pressure at Different Levee Water Levels

The changes in the pore water pressure under different levee water levels in working conditions 3 (crack width is 1.5 mm), 4, and 5 are shown in Table 12 and Figure 12. The pore water pressure at the three different water level measuring points fluctuates with the wave period. When the water level is 22.5 cm, the fluctuation range of the pore water pressure at measuring point 4 is 0.49–0.69 kPa. When the water level in front of the embankment is 27.6 cm, the fluctuation range of the pore water pressure at measuring point 4 is 1.1–1.5 kPa. When the water level in front of the embankment is 30.6 cm, the fluctuation range of the pore water pressure at measurement point 4 is 1.30–2.10 kPa.

Table 12. Pore water pressure at different levee levels.

Acquisition Cycles	Pore Water Pressure of Different Levee Levels/kPa			Acquisition Cycles	Pore Water Pressure of Different Levee Levels/kPa		
	20.5 cm	25.8 cm	30.6 cm		20.5 cm	25.8 cm	30.6 cm
1	0.006	0.216	0.432	16	0.570	0.976	1.529
2	0.162	1.386	1.320	17	0.582	1.007	1.551
3	0.358	1.137	2.084	18	0.430	1.284	1.310
4	0.454	1.456	1.576	19	0.566	1.015	2.019
5	0.508	1.182	2.090	20	0.436	1.352	1.338
6	0.468	1.382	1.891	21	0.582	1.043	1.295
7	0.394	1.415	1.966	22	0.444	0.996	2.084
8	0.524	1.023	2.106	23	0.594	1.335	1.507
9	0.590	1.239	1.485	24	0.588	1.090	1.504
10	0.400	1.325	2.094	25	0.416	1.049	1.828
11	0.554	1.131	2.097	26	0.594	1.178	1.616
12	0.470	1.284	1.298	27	0.508	1.080	1.994
13	0.440	1.107	1.944	28	0.426	1.348	2.072
14	0.558	1.013	2.025	29	0.420	1.241	1.367
15	0.450	1.350	1.373	30	0.578	1.043	2.109

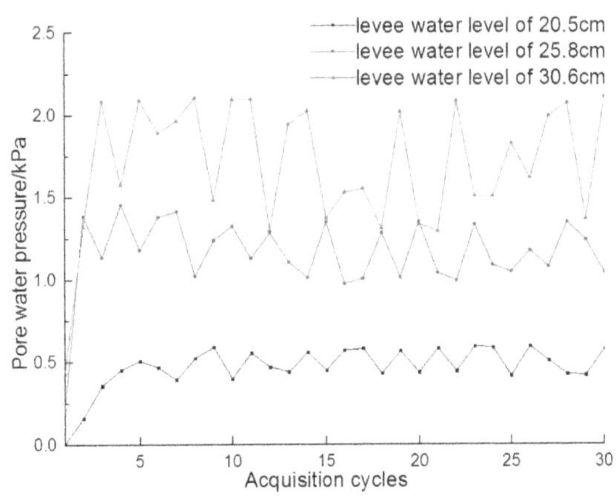

Figure 12. Variation curves of pore water pressure at different levee levels.

Finally, when the seam width remains constant, the pore water pressure of the cushion at different water levels in front of the dike varies with the wave cycle. The larger the water level in front of the dike, the higher the measured pore water pressure and the greater the fluctuation range. This is because the wave water penetrates into the cushion through gaps, influenced by the static water pressure in front of the dike and the seepage pressure of waves, leading to a gradual increase in the original water pressure in the cushion. When the increasing water pressure cannot dissipate in time, the buried pore water pressure sensor reads a higher pore water pressure value. This explains why a higher water level in front of the embankment increases the likelihood of the cushion being affected by water pressure, causing sand to run and resulting in cushion hollowing over time. Therefore, when the water level in front of the dike is high, it is recommended to repair the fallen masonry joints as soon as possible to minimize the risk of cushion hollowing.

3.2.3. Variation Law of Pore Water Pressure at Different Paving Seam Widths

The changes in the average and maximum pore water pressures under different levee water levels under working conditions 1, 2, and 3 are shown in Figure 13. The following conclusions can be drawn:

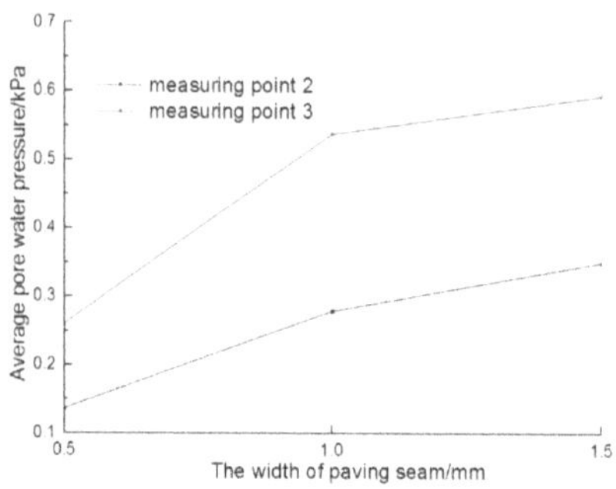

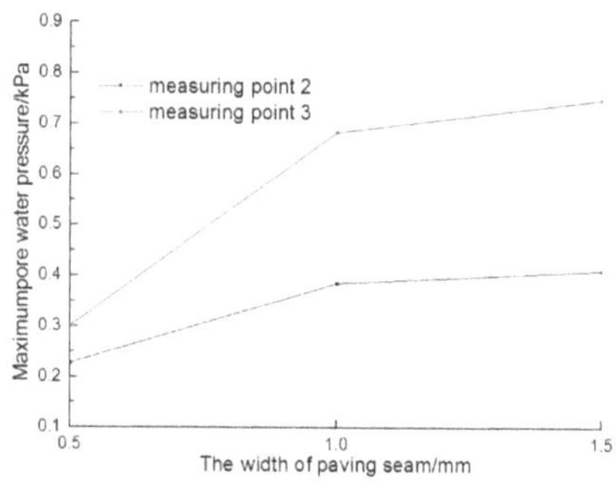

Figure 13. Graph of pore water pressure variation. (**a**) Average pore water pressure. (**b**) Maximum pore water pressure.

(1) As the width of the paving seam increases, the pore water pressure values at measuring points 2 and 3 also increase. In addition, the maximum and average pore water pressure during the wave period range increase;

(2) The pore water pressure at measuring point 3 is consistently greater than that at measuring point 2 under various paving seam widths, as observed by comparing the average and maximum pore water pressure values at both points. This variation in pore water pressure is also consistent with the differences in pore water pressure observed at different positions;

(3) Comparing the maximum pore water pressure values at measuring points 2 and 3 with the same paving seam widths shows that the values at measuring point 2 are 0.397 kPa, 0.545 kPa, and 0.577 kPa, while those at measuring point 3 are 0.472 kPa, 0.844 kPa, and 0.913 kPa for paving seam widths of 0.5 mm, 1.0 mm, and 1.5 mm, respectively. The maximum pore water pressure at measuring point 2 increases by 37.3% from 0.5 mm to 1.0 mm and by 5.9% from 1.0 mm to 1.5 mm, while the maximum pore water pressure at measuring point 3 increases by 78.8% and 8.2%, respectively. The increase in pore water pressure at measuring points 2 and 3 is much more significant when the width of the paving seam changes from 0.5 mm to 1.0 mm compared to the changes observed in the range of 1.0–1.5 mm;

(4) When the paving seam width is the same, the average pore water pressure at measuring point 2 is compared to that at measuring point 3. The average pore water pressure values at measuring point 2 are 0.295 kPa, 0.427 kPa, and 0.508 kPa, while the

average pore water pressure values at measuring point 3 are 0.420 kPa, 0.678 kPa, and 0.743 kPa for paving seam widths of 0.5 mm, 1.0 mm, and 1.5 mm, respectively. The average pore water pressure at measuring point 2 increases by 44.7% from 0.5 mm to 1.0 mm and by 19.0% from 1.0 mm to 1.5 mm, while the average pore water pressure at measuring point 3 increases by 61.4% and 9.6% for the same paving seam widths. The increase in average pore water pressure at measuring points 2 and 3 is much greater in the range of 0.5 mm to 1.0 mm than in the range of 1.0 mm to 1.5 mm.

From the above, it can be observed that under the same water level and paving seam width conditions, the average and maximum pore water pressures at the low position are higher than those at the high position. When the paving seam width increases from 0.5 mm to 1 mm, there is a significant difference in the increased amplitude of the maximum and average pore water pressures at measuring points, even though the width only increases by 0.5 mm. When the paving seam width is less than 1 mm, the maximum and average pore water pressures at the measuring points increase by about 50%. However, when the width is greater than 1 mm, the maximum and average pore water pressures increase by at least 10%. Therefore, it is recommended to control the joint width to be less than 1 cm during the process of laying precast concrete blocks, as this can help reduce the risk of cushion loss even if the cushion falls off during operation.

3.3. Compressive/Shear Strength of Paving Seam Materials

The results of the measured shear strength of the cemented surface are presented in Table 13. The experimental results show that the destructive development situation of mortar paving seams of different strength grades is almost the same, with destruction occurring in the paving seam. Upon examining the destroyed part of the mortar, it can be observed that the cement surface of the paving seam materials is well-poured, and the holes are relatively small, with no obvious impact on the strength. The findings reveal that the shear strength of the mortar paving seam materials increases with the mortar strength.

Table 13. Shear strength of the cemented surface.

Grade	Serial Number	Mortar Strength/MPa	Measured Value of Mortar Strength/MPa	Average Shear Strength/MPa
M15	1	17.2	0.86	0.82
	2		0.79	
	3		0.82	
M20	4	22.5	1.25	1.23
	5		1.24	
	6		1.21	
$M20^{+}$	7	23.0	1.46	1.49
	8		1.48	
	9		1.52	

Comparing the M15 and M20 mortar paving seam materials indicates that the greater the mortar strength, the higher the shear strength. The results show that the mortar compressive strength increases by 23.5%, and the mortar shear strength increases by 32.5%. Furthermore, comparing M20 with M20 containing an 8% expansion agent, the compressive strength of the latter increases by 2.2%, and the mortar shear strength increases by 17.4%. Therefore, a reasonable amount of an expansion agent can lead to better pore distribution of the mortar of a cemented surface, improve the strength of mortar to a certain extent, and significantly enhance the shear strength of mortar.

3.4. Microstructure Analysis of Paving Seam Materials

In order to accurately reflect the field conditions, the experiment used samples obtained after the mortar paving seam shear test. However, the irregular shape of the mortar material after the test made it difficult to measure its apparent volume. As a result, the experiment only analyzed the pore size distribution of M20 and M20+. Pores were classified into four grades based on pore size: pores with a size of less than 20 nm were deemed harmless, pores with a size between 20 and 50 nm were considered less harmful, pores between 50 and 200 nm were regarded as harmful, and pores larger than 200 nm were labeled as very harmful [29].

Table 14 indicates that the addition of an expansion agent has a significant effect on the pore size distribution of the mortar, particularly on the percentages of multiple damaged holes and less damaged holes. The percentage of pore size distribution with harmful holes decreases from 57.4% to 41.6%, while that of less damaged holes increases from 12.7% to 30.5%, and that of harmful holes decreases from 27.9% to 25.9%. Overall, a reasonable amount of admixture makes the structure of the mortar more compact. This is due to the expansion agent promoting the growth of needle-stick ettringite Aft crystals in the mortar. As the content of the expansion agent increases, the number of needle-stick ettringite Aft crystals also increases, which results in a more compact contact surface between the concrete and the mortar due to the expansion effect. This is conducive to improving the shear strength of cemented surfaces.

Table 14. Pore size distribution of M20 and M20$^+$ mortars.

Classes of Mortar	Materials of Paving Seam	Pore Size Distribution/%		
		Less Damaged Holes 20–50 μm	**Harmful Holes/ 50–200 μm**	**Very Harmful Holes/over 200 μm**
M20		12.70%	27.90%	57.40%
M20$^+$		30.50%	25.90%	41.60%

4. Conclusions

In this study, a physical model of precast concrete blocks with different paving seam widths and joint materials was established to analyze the changes in pore water pressure under various positions, water levels, and paving seam widths through the use of embedded pore water pressure sensors. The aim was to provide suggestions for the width and material treatment of paving seams. Based on the results, the following conclusions can be drawn:

(1) Pore water pressure is significantly affected by the position of the slope cushion, the water level in front of the embankment, and the width of the paving seam;
(2) Control of the paving seam width to less than 1 cm is an effective measure for preventing cushion loss considering the influence of construction and other factors.
(3) Paving seam plumpness and mortar strength are both crucial factors that affect the shear strength of the cemented surface of paving seam materials, and they have a mutual influence on each other;
(4) Properly adding an expansion agent can improve the properties of mortar paving seam materials and significantly enhance the bond property between mortar and slope protection concrete.

Author Contributions: Conceptualization, F.C. and S.H.; methodology, F.C.; software, D.L.; validation, F.C., S.H. and Q.F.; formal analysis, Q.F.; investigation, L.J.; resources, F.C.; data curation, Z.H.; writing—original draft preparation, F.C.; writing—review and editing, F.C. and S.H.; visualization, D.L.; supervision, Z.H.; project administration, F.C.; funding acquisition, F.C. and S.H. All authors have read and agreed to the published version of the manuscript.

Funding: This work was supported by the Open Research Fund Projects of the Changjiang River Scientific Research Institute (CKWV20221027/KY), Open Research Fund Projects of Jiangxi Academy of Water Science and Engineering (2020GCZX05, 2021SKSG07), Water conservancy major science and technology project of Jiangxi Province (202325ZDKT23).

Data Availability Statement: Data will be made available upon request.

Conflicts of Interest: The authors declare no conflict of interest.

References

1. *SL634-2012*; Standard for Acceptance and Evaluation of Construction Quality of Embankment Projects. CWP (China Water Power Press): Beijing, China, 2012.
2. Kang, J. Deformation analysis and treatment suggestions of precast concrete block slope protection. *Hebei Prov. Water Conserv.* **2018**, *10*, 38.
3. Li, B. Key points of construction technology of precast concrete slope protection for Henan Keji Dam. *Heilongjiang Sci. Technol. Inf.* **2016**, *6*, 196.
4. Zhou, A. Damage reason, harm and prevention of slope protection of earth-rock dam. *Henan Sci. Technol.* **2014**, *19*, 176–177.
5. Mikhnevich, E.I. Methodology for calculation of of rigidity in earth dam slope protection under conditions of wave impact. *Sci. Tech.* **2018**, *12*, 100–105. [CrossRef]
6. Zhao, N.; Lu, H.; Zhang, R. The Coupling Effect of Pore Water Pressure and Pore Water Gravity in Unsaturated Soils under Rainfall Condition and Its Influence on Slope Stability. *Geofluids* **2022**, *2022*, 9492514. [CrossRef]
7. Wang, X.; Xia, X.; Zhang, X.; Gu, X.; Zhang, Q. Probabilistic Risk Assessment of Soil Slope Stability Subjected to Water Drawdown by Finite Element Limit Analysis. *Appl. Sci.* **2022**, *12*, 10282. [CrossRef]
8. Zheng, Z. Application of concrete block slope protection in hydraulic engineering. *China's Water Transp.* **2011**, *11*, 140–141.
9. Zheng, J.; Zhang, C.; Liu, C. Discussion on quality acceptance of laying joint width in precast concrete block slope protection project. *China Water Transp.* **2015**, *15*, 305–306.
10. Liu, Y.; Zhang, W.; Zhang, L.; Zhu, Z.; Hu, J.; Wei, H. Probabilistic stability analyses of undrained slopes by 3D random fields and finite element methods. *Geosci. Front.* **2018**, *9*, 1657–1664. [CrossRef]
11. Niu, D.; Han, S.; Li, X.; Li, B.; Ren, W. Mechanical property of asphalt with additive. *J. Traffic Transp. Eng.* **2016**, *16*, 8–16.
12. Liu, D.; Chen, J.; Yue, W.; Zhang, Y. Stability Analysis of Reservoir Bank and Study on Landslide Control Measures—Taking Baiheqiao Reservoir in Kaixian County, Chongqing as an example. In Proceedings of the 8th International Symposium on Project Management, Beijing, China, 4–5 July 2020; pp. 223–236. [CrossRef]
13. Seifan, M.; Mendoza, S.; Aydin, B. Effect of nano and micro iron oxide particles on the workability, strength and absorption rate of cement mortar containing fly ash. *Eur. J. Environ. Civ. Eng.* **2022**, *26*, 3898–3912. [CrossRef]
14. Chen, H.; Wang, F.; Xu, G.; Guo, L. Laboratory Model Test of Eco-Concrete Slab Slope Protection. In Proceedings of the 8th International Conference on Civil Engineering, Nanchang, China, 4–5 December 2021; pp. 372–381. [CrossRef]
15. Dan, L.; Zheng, C.; Qin, Y.; Bi, H.; Li, K.; Huang, J. Analysing the effects of the mesoscopic characteristics of mineral powder fillers on the cohesive strength of asphalt mortars at low temperatures. *Constr. Build. Mater.* **2014**, *65*, 330–337. [CrossRef]
16. John, S.K.; Nadir, Y.; Girija, K. Effect of source materials, additives on the mechanical properties and durability of fly ash and fly ash-slag geopolymer mortar: A review. *Constr. Build. Mater.* **2021**, *280*, 122443. [CrossRef]
17. Rocha, J.H.A.; Galarza, F.P. Mechanical Properties Evaluation of Render Mortars with Different Waterproof Additive Compositions. *Key Eng. Mater.* **2021**, *6202*, 161–167. [CrossRef]
18. Savchenko, S.; Antoniuk, N. Assessment of the impact of modifier and filler on the physical and mechanical and operational properties of plastering mortars. *IOP Conf. Ser. Mater. Sci. Eng.* **2021**, *1164*, 012004. [CrossRef]
19. Corinaldesi, V.; Nardinocchi, A.; Donnini, J. The influence of expansive agent on the performance of fibre reinforced cement-based composites. *Constr. Build. Mater.* **2015**, *91*, 171–179. [CrossRef]
20. Feng, Y.; Yang, Q.; Chen, Q.; Kero, J.; Andersson, A.; Ahmed, H.; Engström, F.; Samuelsson, C. Characterization and evaluation of the pozzolanic activity of granulated copper slag modified with CaO. *J. Cleaner Prod.* **2019**, *232*, 1112–1120. [CrossRef]
21. Polat, R.; Demirbog, R.; Khushefati, W.H. Effects of nano and micro size of CaO and MgO, nano-clay and expanded perlite aggregate on the autogenous shrinkage of mortar. *Constr. Build. Mater.* **2015**, *81*, 268–275. [CrossRef]
22. Jia, Z.; Yang, Y.; Yang, L.; Zhang, Y.; Sun, Z. Hydration products, internal relative humidity and drying shrinkage of alkali activated slag mortar with expansion agents. *Constr. Build. Mater.* **2018**, *158*, 198–207. [CrossRef]
23. *GB175-2007*; Common Portland Cement. CNSMC (China National Standardization Management Committee): Beijing, China, 2007.
24. *DL/T 5144-2015*; Specifications for Hydraulic Concrete Construction. CNEA (China National Energy Administration): Beijing, China, 2015.
25. CWP. *Hydraulic Design Manual*, 2nd ed.; China Water Power Press: Beijing, China, 2014.
26. Guo, W.; Zhu, J.; Wen, Y. Unified description for four grading scale methods for coarse aggregate. *Chin. J. Geotech. Eng.* **2016**, *38*, 1473–1480.

27. Wei, M. System Parameters Calculation and Software Design for Port Engineering Model Text. Ph.D. Thesis, Tianjin University of Technology, Tianjin, China, 2012.
28. Brue, F.; Davy, C.A.; Skoczylas, F.; Burlion, N.; Bourbon, X. Effect of temperature on the water retention properties of two high performance concretes. *Cem. Concr. Res.* **2012**, *42*, 384–396. [CrossRef]
29. Stovall, T.; de Larrard, F.; Buil, M. Linear packing density model of grain mixtures. *Powder Technol.* **1986**, *48*, 1–12. [CrossRef]

water

MDPI

Article

Geotechnical Assessment of Rock Slope Stability Using Kinematic and Limit Equilibrium Analysis for Safety Evaluation

Aftab Ur Rahman [1], Guangcheng Zhang [1,*], Salman A. AlQahtani [2,*], Hammad Tariq Janjuhah [3], Irshad Hussain [4,*], Habib Ur Rehman [5] and Liaqat Ali Shah [6]

1 Faculty of Engineering, China University of Geosciences, Wuhan 430074, China
2 Department of Computer Engineering, College of Computer and Information Sciences, King Saud University, Riyadh 11543, Saudi Arabia
3 Department of Geology, Shaheed Benazir Bhutto University, Sheringal, Dir Upper 18050, Pakistan
4 Faculty of Electrical and Computer Engineering, University of Engineering and Technology Peshawar, Peshawar 25000, Pakistan
5 Gems & Jewellery Centre of Excellence, University of Engineering & Technology, Peshawar 25120, Pakistan
6 School of Architecture and Civil Engineering, Loughborough University Epinal Way, Loughborough LE11 3TU, UK
* Correspondence: zhangguangc@cug.edu.cn (G.Z.); salmanq@ksu.edu.sa (S.A.A.); ee.irshad@gmail.com (I.H.)

Abstract: The purpose of this study is to identify the leading causes of slope instability along a local highway in Anhui, People's Republic of China. As part of the east expansion project, the mountain range will be excavated to create a two-way, nearly 30 m wide highway. The site's topography consists of a hill with palm-shaped faces carved from limestone running along its sides. The geological characteristics and slope stability of the research area highlight the possibility of slope failure along both sides of the roadway. Slope stability analysis was performed in order to determine the failure mechanism and create a stable slope. Initial slope characterization and shear properties of the rock were determined by means of fieldwork and laboratory analysis. By causing wedging failure and toppling collapse, the bedding joints and discontinuity orientations increase instability, as determined by a kinematic analysis performed with DIP.6 software. The Limit Equilibrium Method (LEM) of analysis is presented in the software Slide 6.020 to illustrate the instability of the slope. The unstable condition of the slopes was determined using empirical methods that were validated and enhanced by limit equilibrium analysis.

Keywords: rock slope stability; failure mechanism; kinematic analysis; limit equilibrium method finite; DIPS 6.0; Slide 6.020

Citation: Rahman, A.U.; Zhang, G.; AlQahtani, S.A.; Janjuhah, H.T.; Hussain, I.; Rehman, H.U.; Shah, L.A. Geotechnical Assessment of Rock Slope Stability Using Kinematic and Limit Equilibrium Analysis for Safety Evaluation. *Water* **2023**, *15*, 1924. https://doi.org/10.3390/w15101924

Academic Editors: Helena M. Ramos and Laura Bulgariu

Received: 19 March 2023
Revised: 4 May 2023
Accepted: 7 May 2023
Published: 19 May 2023

1. Introduction

Analysis of rock slope stability is required to plan safe excavations of slopes such as road cuts, mines, and railways to evaluate the stability of natural slopes. Especially in steep and mountainous regions, rock slopes are one of the most critical concerns in road construction. Distinct rock mass conditions, different discontinuities, shear zones, faults, thrusts, an unprofessional method of slope cutting, excessive rainfall, seismic activity, and tectonic activities make slopes more susceptible to instability problems [1–4]. Nevertheless, the most frequent types of slope failure are planar, wedge, rock topples, and rockfalls [5]. A complete geological analysis of the slope necessitates a detailed survey and research to assist the engineer in designing the appropriate support system and mitigating slope failures. A well-designed slope improves slope stability and safety, reduces expenses, increases mine life, and minimizes the stripping ratio [3,6].

Due to advancements in geology, we can understand such events, eventually enabling us to predict them so that the damage can be reduced. Modern technology and advanced techniques regarding slope stability have improved in recent years, but rock slope stability is still challenging for engineers. This is mainly due to the use of primitive techniques and the complexity of the rock slope considering the discontinuities influenced by joints, bedding, folds, faults, etc. Geological characteristics of rock mass include location and the number of joint sets, joint spacing, joint orientations, joint material, slope geometry, slope material strength, and seepage [7–10]. A rock slope can fail due to one or a combination of the different failure mechanisms. When the presence of pre-existing discontinuities dictates the instability, the failure will be in the form of plane sliding, wedge sliding, or toppling [11,12].

Many researchers around the globe have used various approaches for slope stability evaluation. Unlike deterministic analysis, which provides a single measure of the factor of safety (FoS), the kinematic analysis and limit equilibrium method can determine the probability of failure [8,13–15]. Several earlier studies have adopted probabilistic approaches to assess the stability of slopes. Probabilistic analysis is a technique for modeling the uncertainty and variation in parameters. The deterministic method minimizes the safety factor to determine the critical failure surface. In contrast, the probabilistic method allows measuring slope reliability based on the probability of failure and reliability index. The deterministic method is characterized by the high uncertainty of the variables considered in the analysis [16–18]. As a result, the probabilistic method allows for a new perspective on risk and reliability beyond the scope of traditional deterministic methods [13,18]. The factor of safety is normally expressed as a random variable. It can be substituted by the probability of failure in the assessment of slope stability. In contrast, the sensitivity analysis utilizes the maximum and minimum values of the parameter to determine the slope's critical condition. Therefore, the relationships among the parameters and their effects on FoS can be evaluated [19,20].

With the development of computer technology, the finite element method has recently been a concern at home and abroad. When the slope reaches the failure state through the finite element strength reduction, the displacement on the sliding surface will change suddenly and produce a large and unlimited plastic flow. Discontinuities play a key contributing factor in slope stability. In addition, rock type, joint properties, geometry, water condition, and failure-plane characteristics are the key parameters that affect slope instability [9,14,16,21,22]. Moreover, assessing rock slopes is challenging to select the appropriate parameters for the stability assessment. Therefore, combining field data with a laboratory analysis approach leads to suitable consideration of the appropriate parameters [23].

Kinematics analysis is the most important method to investigate the slope instability and failure mechanism associated with landsides. The height of the slope also plays a role in the mode of failure; as the height increases, complex failures may occur. Kinematic analysis, the Limit Equilibrium Method (LEM), and numerical modeling are modern techniques widely used for slope stability analysis. To perform kinematic analysis and understand the failure mode, discontinuity orientation, joint sets, and bedding planes are key parameters that can be considered. Kinematic analysis is a very useful method to understand types of failure by using discontinuities and joint orientations [22,24,25]. Various techniques, including kinematic analysis, limit equilibrium analysis, rock mass classification system (SSPC system), and probabilistic analysis, among others, are available for the investigation of rock slope stability [26–29]. Kinematic analysis using the stereographic projection method is typically carried out before performing a detailed study in nearly all slope stability analyses [30–32]. Kinematic analysis is a geometric method that employs angular interactions between discontinuity planes to determine the possibility and failure types in a jointed rock mass [33].

In engineering analysis, numerous limit equilibrium techniques have been developed to analyze and design slopes in both soil mechanics and rock mechanics. It is founded on the idea of safety factors. To compute the global failure analysis, the factor of safety, and the failure surface, limit equilibrium analyses were carried out. Several approaches, including LEM, were used to assess the slope stability for the safety factor. The discrepancies in the benefits and limitations of each strategy must be emphasized. The LEM technique basically segments a soil/rock mass above the slip's surface into a finite number of slices that can be segmented either vertically or horizontally [34,35]. Despite the widespread usage of traditional 2D modeling, it is generally believed that 2D slope stability analysis is inadequate when compared to 3D [20,36]. However, many researchers have conducted 2D slope stability analysis [8,9,13]. When using the limit equilibrium approach to calculate the forces pushing the rock mass and the opposing forces, the ratio of opposing forces to driving forces at equilibrium is known as the factor of safety (FOS) [11,37,38]. Numerous controlling parameters, including slope geometry, failure plane features, water forces, and external triggering factors, will have an impact on the FOS. Slope height has an inverse relationship with the FOS. Shear strains will increase as slope height (h) increases; as a result, FOS will decline [39,40].

The current study investigated the possible mode and potential of failure along the left and right slopes of a highway. Kinematic analysis tools were used to evaluate the potential failure of rock masses moving along geologic structures on the slope face. The analysis was conducted through DIPS software using graphical stereographic projection to examine the failure potential in both locations. In this study, the main factors affecting the mode of instabilities observed in the field are discontinuity orientation and bedding planes. The geometry of the slope, joints, and bedding plane was considered for the possible design and mode of failure.

Location and Geology of the Study Area

The study area is located in northeastern China in a moderately mountainous area (longitude 30.928194702590133, latitude 117.88073690223622) with an elevation of 35 m in Tongling city, Anhui province, P.R. China, as shown in Figure 1.

The province has different climates in the north and south, associated with geography. The seasons are more pronounced and more moderate towards the north. In January, temperatures are often between 1 °C and 3 °C, whereas in July, they are typically 27 °C or higher (Anhui Meteorological Department). The vegetation on the mountain is not developed, the overall terrain in the upper middle is high, and the two ends are low. The hillsides slope toward the inlet and outlet ends. The natural slope is 18°–26°. The Tongling area is a tectonically active zone with a low thickness of about 32 km. The Tongling district is located in the center of the Middle-Lower Yangtze River Block (MLYB) [41,43]. The dominant lithologies in this region are Silurian Middle Triassic marine sedimentary rocks. NE-trending fold structures are present and are accompanied by a series of NE-, NNE- NW-, and NNW-trending fault systems that control the emplacement of Mesozoic intrusions. The structures in this area include the NE-trending Qingshan anticline [42,44,45]. The sedimentary rocks exposed in the area range from Middle-Upper Silurian to Lower Triassic in age, but no Lower to Middle Devonian rocks are recognized. The main host rocks of the deposit are sandstone, limestone, and dolomitic limestone [46].

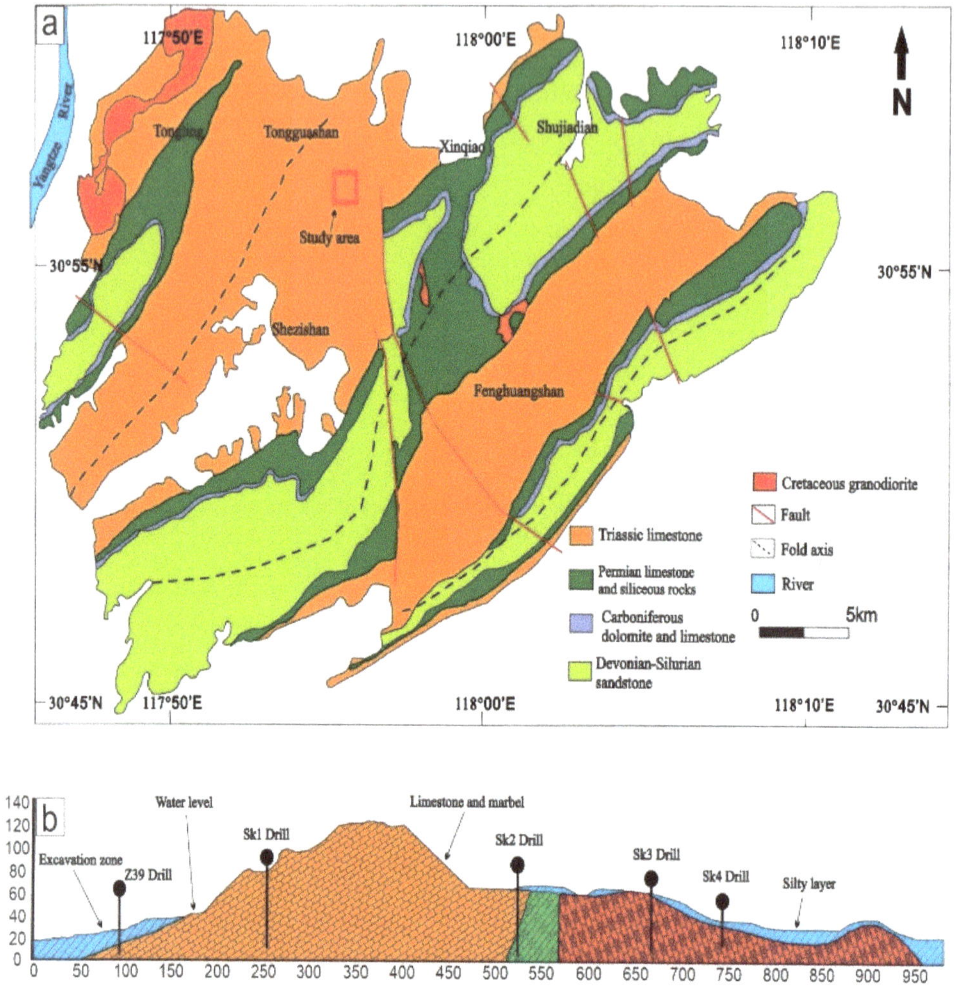

Figure 1. Maps of the study area. (**a**) Geological map of Tongling district, Anhui province, modified after [41,42]; (**b**) cross-sectional area of the highway to be constructed.

2. Materials and Methods

2.1. Field Investigation

The east extension project plan is to excavate the mountain body as a two-way lane, and the width of the road is nearly 30 m. The site terrain consists of a hill with limestone palm faces along both sides and a spherical peak. The planned route joins Tongjing Middle Road and Qifeng Road at their intersection in the west, crosses the northern foot of Qingshan Mountain in the east, travels past the deserted quarrying pond on the mountain's eastern side, and finally joins Tongjing East Road. The total length of the road is about 2.2 km. It is 18.9~25.6 m from west to east. The project extends east along Tongjing Middle Road. The excavation section of the low mountain is about 600 m, the highest point of the terrain is 144 m, the relative height difference of the mountain is 130 m, and the north side cutting slope is 110 m. The 70 m south slope is shown in Figure 1b. Field investigation shows that the rock formation and structural surface in the northern foothills of Qingshan are well outcropped, the rock formation level is flat, and the joint surface is developed. The local rock mass is severely cut.

2.2. Laboratory Testing and Geotechnical Assessment

In order to assess the rock mass metrics and geotechnical features, fresh bulk samples were extracted and transported to the rock mechanics lab at the China University of Geosciences Wuhan. The cylindrical samples were prepared with dimensions of 100 mm in length and 50 mm in diameter. Both sides of the cylinders were flattened to get a uniform outcome and characteristics across all samples. All samples were drilled perpendicular to the depositional layers in a pattern in which no apparent fractures were observed (Figure 2a,b). All samples were dipped in the water tank before being put in a vacuum saturator to absorb more water and become thoroughly saturated (Figure 2d). Each sample mass was evaluated every 24 h until it reached complete saturation. Throughout every measurement, the saturated mass of each specimen remained unchanged. Therefore, reaching a completely saturated condition in rock is challenging. After calculating the saturated mass, samples were baked for 24 h at 120 °C (Figure 2c). To assess water absorption, the difference between dry and wet mass is divided by the rock dry mass, according to the Chinese rock mass testing standard. The calculated data for physical properties are listed in Table 1.

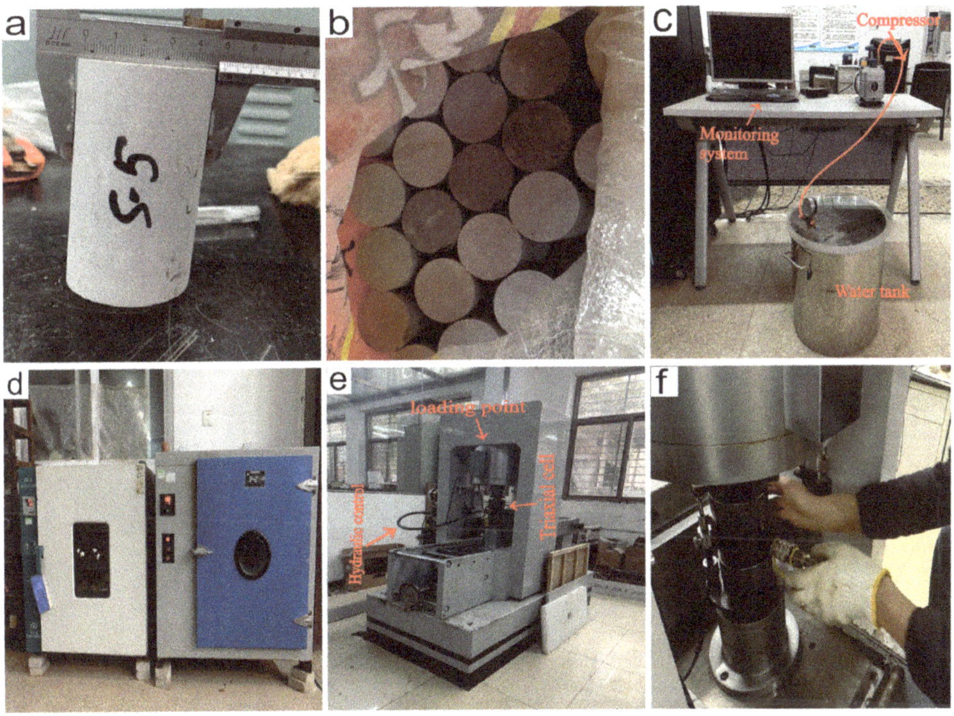

Figure 2. Different materials and instruments. (**a**) Core sample extracted from core drilling machine. (**b**) All prepared core samples from limestone and dolomite. (**c**) Oven for dying the saturated samples. (**d**) Vacuum saturator. (**e**) RMT-150. (**f**) Triaxial cell.

Table 1. Physical Properties of Limestone.

Sample	Water Absorption (%)	Error (±)	Porosity (%)	Error (±)	Unit Weight (γ)	Error (±)
Limestone dry	0.12	0.042	0.36	0.112	2.71	0.010
Dolomite wet	0.23	0.020	0.66	0.055	2.82	0.011

Before the triaxial test, cylindrical samples were subjected to a drying and wetting procedure. All specimens were triaxially tested to determine the result (Figure 2e). Each sample was placed in a triaxial cell and subjected to 0.20 KN/s of steady vertical pressure until failure (see Figure 2f). ISRM-recommended laboratory testing methods were followed [40]. Figure 2e depicts an RMT-150 rock mechanics testing machine used to perform triaxial tests. These experiments were done at the Chinese Academy of Sciences, Rock and Soil Mechanics Laboratory in Wuhan, China.

2.3. Strength Parameters

According to the Mohr-Coulomb technique, one of the most crucial factors affecting the stability of rock slopes is the shear strength of the discontinuities, which may be computed using the cohesion and friction angle parameters [47]. After metamorphism, the rock is a marbleized limestone and dolomite unit with fine grain structure, a grey-black hue, and thin layers. For limestone, porosity, unit weight, and water absorption were estimated. The test was conducted in the rock mechanics lab at the China University of Geosciences in Wuhan. To assess water absorption, the difference between dry and wet mass is divided by the rock dry mass, according to Chinese rock mass testing standard. The interpretation of stress-strain characteristics (maximum (σ_1), minimum (σ_3) normal stresses in MPa, and elastic modulus Ei (GPa)) and shear strength parameters (cohesion (c), internal friction angle (φ), and Poisson ratio) were based on the difference between the dry and saturated states and the confining stress (Table 2).

Table 2. Shear Strength Parameters of Limestone.

Serial No:		Unit Weight (kN/m^3)	Cohesion (kPa)	Internal Friction Angle (°)	Deformation Modulus (MPa)	Poisson Ratio (MPa)
Strong, weathered limestone	Natural	26.8	350	38	25,000	0.28
	Saturated	27.1	300	35	23,000	0.31

2.4. Failure Mechanisms and Slope Stability

Kinematic analysis was conducted to highlight the possibility for several types of rock slope failures that arise due to adversely oriented discontinuities [5]. The rock slope moments and likely failure direction were calculated and anticipated using kinematic analysis along both slopes. Three frequent failure modes (planar, wedge, and toppling) that result from adversely oriented discontinuities and bedding planes were examined [11]. The required parameters obtained from the field through geological survey and geotechnical investigation are summarized.

In order to evaluate the slope kinematics, the discontinuity parameters acquired from the scan-line survey that was conducted for each of the analyzed slopes were entered into the DIPS program, which is shown in Figure 3 [11,24,48]. The failure potential and cause, impacted by the numerous slope angles, exiting discontinuities, and bedding planes, were calculated using an arbitrary slope dip angle of 40° to 70°. For this purpose, the slope was modeled to the right and left banks with different slope angle.

To compute the global failure analysis, the factor of safety, and the failure surface, limit equilibrium analysis was carried out. A Slide 6.020 numerical model was utilized to simulate how the material would react to stress, strain, and shearing as the road was built. To better understand the mechanism of weak rock instability brought on by highway excavation, two-dimensional cutting slope models were created. The slopes have an excavation ratio of 130 m long and 130 m high. The model still requires the creation of borders, mesh generation, boundary conditions, addition of traction, field stress, normal condition, seismic condition, and simulation at the end. These studies employed the Bishop, Janbu, and Spencer four-slice methodologies.

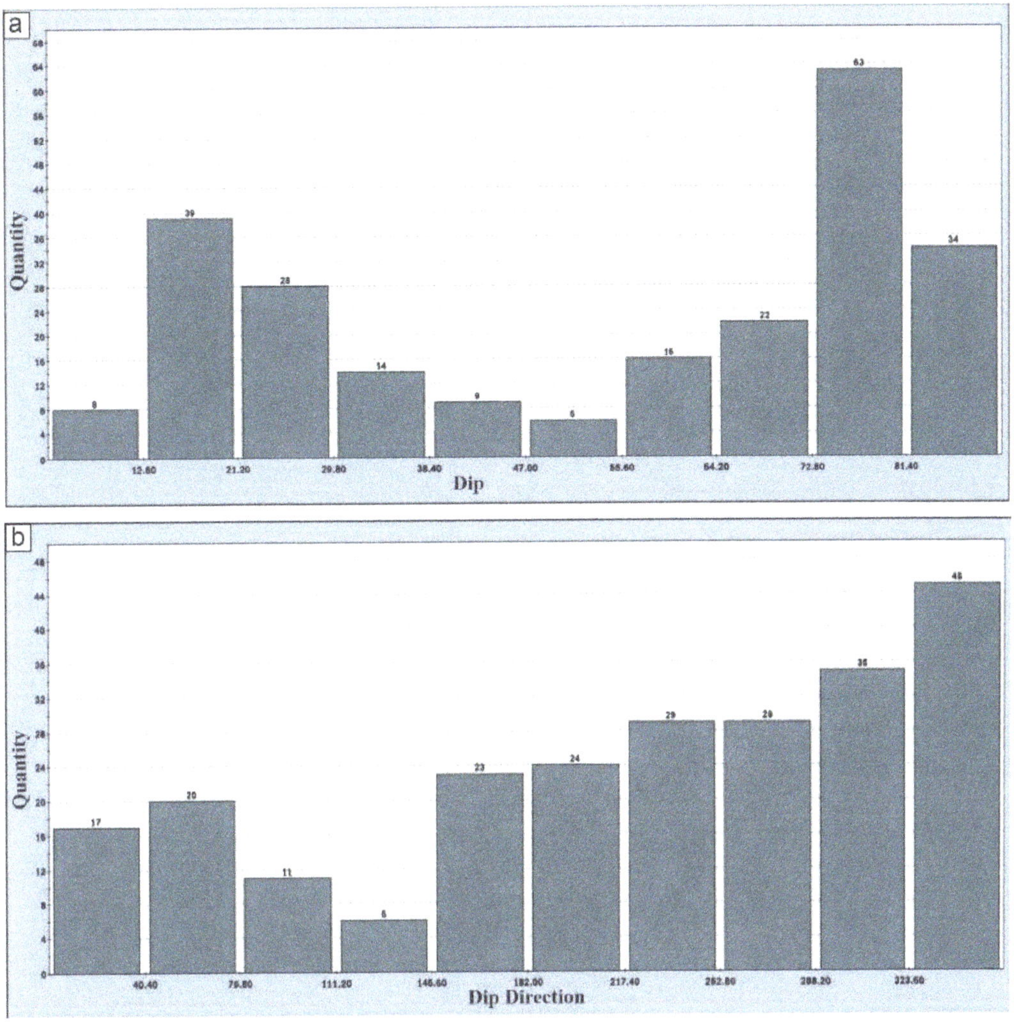

Figure 3. Quantitative data of discontinuities and bedding joints: (**a**) dip (**b**) dip direction.

3. Results

Plane failure, wedge failure, and toppling failure are the three basic types of failure that occur along rock discontinuities in hard rock slopes [5]. This study sought to determine how geological structures affected the frequency of rock failure along a road in a hilly area with several geological characteristics crossing it. Geological and geotechnical approaches were used to investigate the issue. Kinematic analysis (Stereo-net plots) is one of the geotechnical techniques to determine the failure mechanism (wedge, planar, and toppling analysis) and critical slope failure.

Three main failure mechanisms were examined on both sides of the road using the orientation data and kinematic analysis generated by Dips 6.0 software. We tabulated the rock discontinuity data collected from the sloped site and grouped the data sets based on their similarities. The kinematic analysis findings and the actual field situation agree rather well. Field measurements and observations revealed a joint-controlled toppling and wedge-type failure. The data also show that failure is more critical and likely to occur along the left bank than the right. The result shows that when the slope angle is low, the critical intersection points in the critical zone are lower, while with an increase in slope dip angle, the number of critical intersections also increases (Table 3). The kinematic study of the left bank makes it evident that the toppling mode has been tested with a higher failure probability than the planar and wedge failure.

Table 3. Summary of critical interaction points, left bank (LB), right bank (RB), planar (P), wedge (W), toppling (T), critical intersection (CI), total critical interaction (TCI).

Site	Failure Type	45°	50°	55°	60°	65°	70°	TCI
LB	P	2	3	4	5	6	8	239
	W	593	907	1267	1571	1967	2488	28,426
	T	2190	2366	2500	2667	2989	5773	28,426
RB	P	0	0	0	0	0	0	239
	W	599	780	1113	1509	1915	28,426	28,426
	T	484	628	725	915	1059	1247	28,426

Overall, from the right bank, it is clear that three forms of failure display rising failure possibilities with increasing dip slope angle, with toppling and wedge failure exhibiting higher failure probabilities than planar (Figure 4a). On the other hand, it can be seen from the left bank that toppling failure and wedge failure both exhibit a pattern of increasing the likelihood of failure and are more problematic, although toppling failure is more prevalent than wedge failure. Additionally, planar failure exhibits very low odds (Figure 4b).

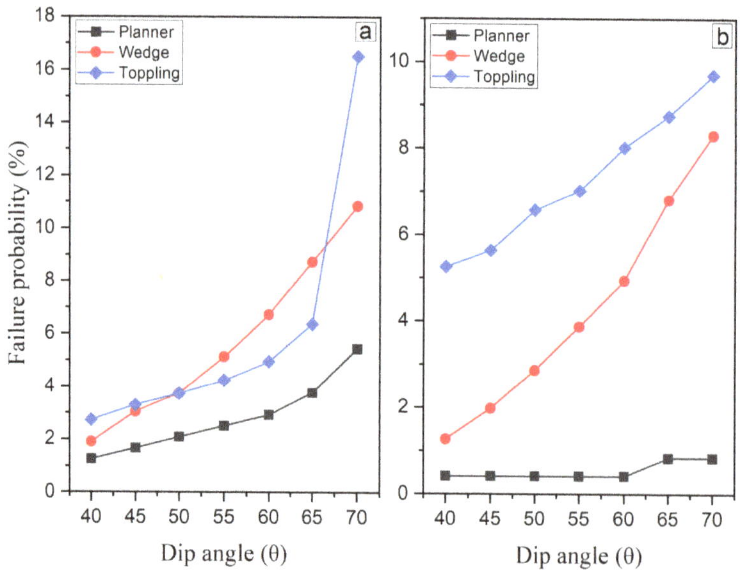

Figure 4. Failure probabilities of different failure mechanism on the (**a**) left bank and (**b**) right bank.

The toppling mode was tested with a higher failure probability than the planar and wedge failure, as shown by the kinematic analysis of the left bank (Figure 5a). According to the results, the likelihood of a planar failing is slightly higher than 1% when the slope dip angle is 40°, but it increases to 1.6% when the angle is 5°. The number of critical failures is 2 (Figure 5a). Also, when the slope dip angle is 55°, 60°, and 70°, failure chances rise to 2.5%, 2.9%, and 5%, respectively. The chance of wedge failure increases to 10% at a slope angle of 55°, as shown in Figure 5d, from 5% at 45° (Figure 5e). The failure probability is highest (17%) when the slope's peak orientation has a dip angle of 70° (Figure 5f). The majority of the joints have a dip angle of 20° or more, as shown by the data. In total,

34 joints have a dip angle of 80°, whereas 63 joints dipped with an angle greater than 70° (Figure 2a). The toppling failure is the rotation of a block or stone column located on a sloping surface, around a point [49]. As demonstrated in Figure 4a, toppling failure has a higher failure probability. Failure probabilities are just below 3% when the slope dip angle is 40°, but they rise somewhat with slope dip angle and can reach up to 4.5% when the dip angle is 55° (Figure 5g,h).

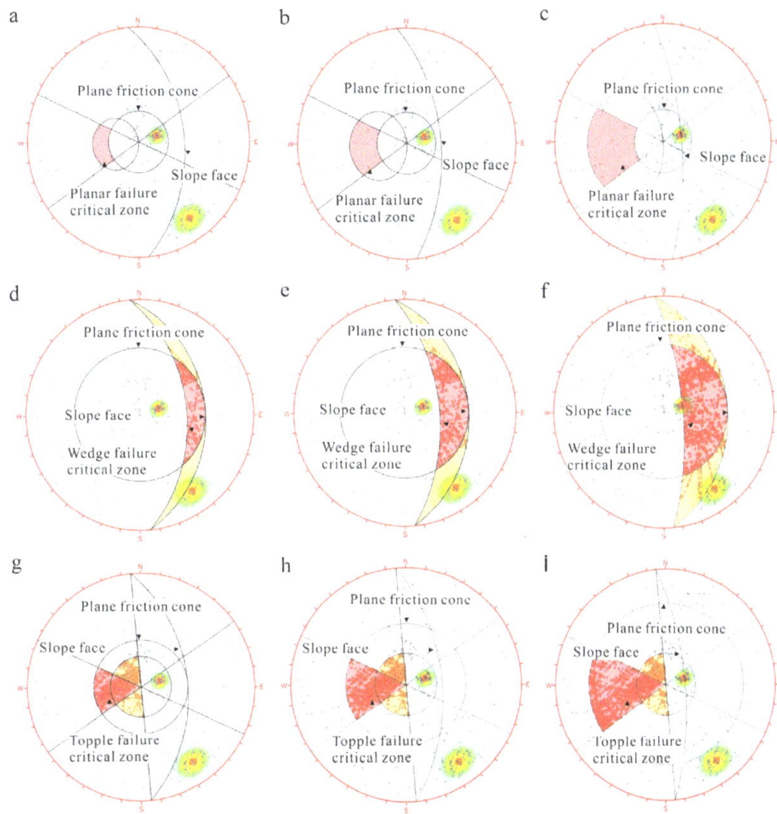

Figure 5. Dip analysis for the left bank of the highway: planar failure at (**a**) 45°, (**b**) 55°, and (**c**) 70°; wedge failure at (**d**) 45°, (**e**) 55°, and (**f**) 70°; and toppling failure at (**g**) 45°, (**h**) 55°, and (**i**) 70°.

According to the findings from the right bank, the likelihood of a planar failing is 0% at a slope angle of 40°, while the same percentages are repeated at a slope angle of 60°. When the slope dip angle is 65° and 70°, the failure chances increase slightly with values of 1%, which is very low (Figure 4b). The right bank wedge failure study yields nearly identical results as the left bank. Although the odds of failure were lower for a modest dip angle, they rose as the slope dip increased (Figure 4b). The outcome shows that the slope dip angle of 45° had a very low failure chance of 2% (Figure 6d). According to the right bank kinematic analysis results, toppling failure probability is somewhat more than 5% at a slope dip angle of 40°, but gradually increases as the slope dip angle increases, obtaining a maximum of 5.5% at a dip angle of 55°. At 65° and 70°, toppling exhibits the greatest values of 8.5% and 9.5%, respectively (Figure 4b).

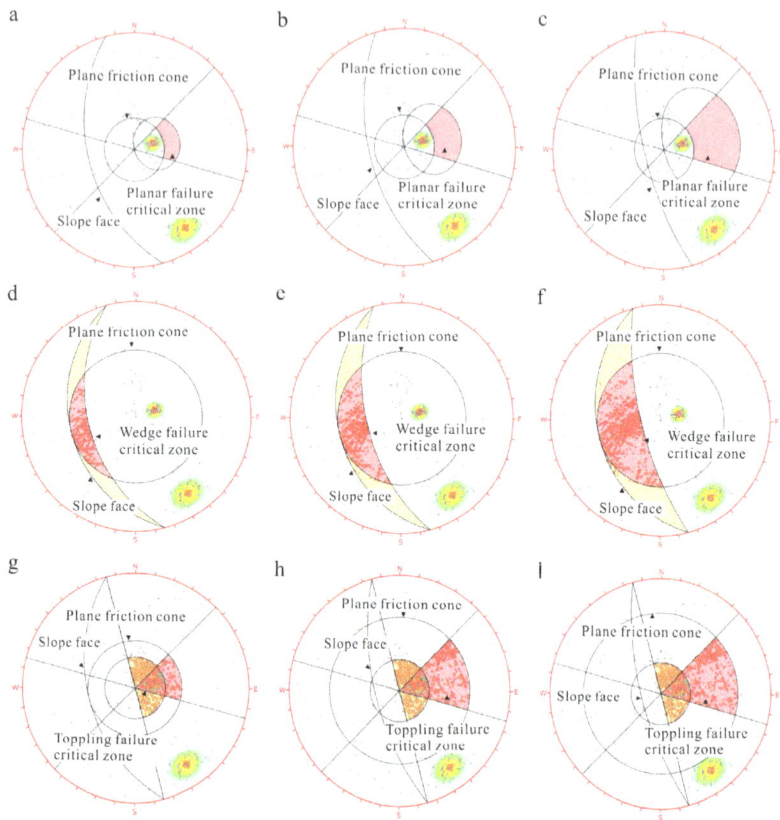

Figure 6. Dip analysis for the right bank of the highway: planar failure at (**a**) 45°, (**b**) 65°, and (**c**) 70°; wedge failure at (**d**) 45°, (**e**) 65°, and (**f**) 70°; and toppling failure at (**g**) 45°, (**h**) 65°, and (**i**) 70°.

To simulate the impact of road building and geological characteristics concerning rock mass displacement, the Finite Element Method (Slide 6.020 software) was utilized. Because of the highly jointed and varied joint spacing of the rocks along the research area road, different-sized blocks can develop. Using a Brunton compass, more than 150 bedding joints and discontinuities with dip/dip directions were measured at each location. The results show that the calculated factor of safety in normal conditions for the Bishop, Janbu, and Spencer method is 1.128, 1.105, and 1.126, respectively (Figure 7a–c). This result of the left bank indicates that the slope is stable when the slope angle is 60° or less, but the slope becomes vulnerable to failure when the overall slope angle increases to 65° and 70°. However, the slope becomes more unstable when the ground acceleration is applied. Along with this increasing slope angle, the FoS value decreases, which ultimately increases the failure probability (Figure 9).

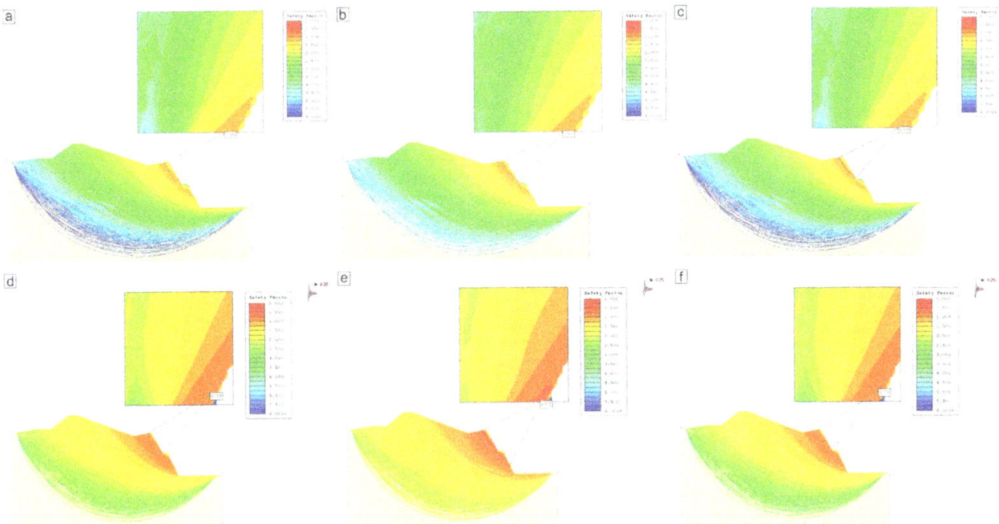

Figure 7. Slope stability analysis for the left bank of the highway: (**a**) Bishop method under normal conditions, (**b**) Janbu method under normal conditions, (**c**) Spencer method under normal conditions, (**d**) Bishop method under seismic conditions, (**e**) Janbu method under seismic conditions, (**f**) Spencer method under seismic conditions.

4. Discussion

4.1. Kinematic Analysis

4.1.1. Left Slope

(a) *Planar failure*: It is clear from the kinematic analysis of the left bank that the toppling mode tested with a greater failure probability than the planar and wedge failure (Figure 5a). The results shows that when the slope dip angle is 40°, planar failure is slightly more than 1%, but with an increase in 5°, the failure chances reach 1.6%. The number of critical failures is 2 (Figure 5a). Also, when the slope dip angle is 55°, 60°, and 70°, failure probability increases to 2.5%, 2.9%, and 5%, respectively. The key critical zone for planar failure is located on the left slope, as seen in the highlighted red area. The danger of the developing planar slide is represented by intersections in the crucial zone (Figure 5b,c). The condition that must be met for a plane to fail is that the slope-face dip must be larger than the slide-plane dip, and both must be higher than the slide-surface friction angle, i.e., $\psi f > \psi p > \phi$ [12]. According to the pole points, none of the major joint sets on the slopes is critical for planar slope failure. As a result of the discontinuities of pole points not being inside the planar daylight envelope, planar failure does not occur on the slopes of the examined area (Figure 5a–c). Because the dip angle of the majority of joints is lower and extremely small, the pole of the junction point of the joints falls within the friction cone or safe region. The number of critical failures likewise rises as the slope dip angle increases (Table 3). Thus, the kinematic analysis indicates that planar failure chances are very low, approaching zero.

(b) *Wedge failure*: The failure probability of a wedge is 5% at a slope angle of 45° as shown in Figure 5d, and increases to 10% at a slope angle of 55° (Figure 5e). The failure probability is highest (17%) when the slope's peak orientation has a dip angle of 70° (Figure 5f). The majority of the joints have a dip angle of 20° or more, as shown by the data; moreover, 34 joints have a dip angle of 80°, whereas 63 joints dipped with an angle greater than 70° (Figure 2a). This indicates that wedge failure is more likely to occur along the left bank. The number of critical failures also increases as the slope dip angle increases even though more joints are in the unsafe zone, which raises the probability of wedge failure (Table 3). According to [50], the prior landslide in construction is mostly caused by steep slope angles. Additionally, [51] reported that most of the intersection points of the joints fall in the safe zone because of the low dip angle, which decreases the probability of wedge failure. Azimuth-based discontinuities with NE and SE orientations produce intersection points in the critical zone that lead to wedge failure. The number of crucial locations likewise grows as the dip slope angle rises. Therefore, the probability of wedge failure increases and rises up to 11% when the slope dip angle is 70° (Figure 5f).

(c) *Toppling failure*: The toppling failure is the rotation of a block or stone column located on a sloping surface, around a point [49]. From Figure 4a, it can be seen that toppling failure shows a higher failure probability. When the slope dip angle is 40°, failure chances are slightly less than 3%, while a small increase in slope dip angle increases the failure chances up to 4.5% when the dip angle is 55° (Figure 5g,h). There is a lower chance of a toppling failure than a wedge failure (Figure 4a). Failure probabilities spike dramatically at a slope dip angle of 70°, where they reach 17%, indicating a dangerously high risk for toppling failure (Figure 5i). By observing the pole orientation of joints, one can determine the most vulnerable areas that could topple. Ning et al. [52] found that the change in slope angle and stratum dip angle could impact the stability of the slope. From Figure 5a, it can be seen that 109 discontinuities are oriented in the NW and SW where the joints and slope dip in opposite directions. This suggests that toppling failure shows a higher probability. From Figure 5g, it can be seen that with a slope angle of 45°, critical failure is comparatively lower, but when the slope angle increases, the critical intersection points increase. With slope angles of 60° and 70°, the toppling failure chances are high and more critical. All results of the critical interactions are listed in Table 3.

4.1.2. Right Slope

(a) *Planar failure*: Results from the right bank show that when the slope angle is 40°–60°, planar failure chances are 0%. When the slope dip angle is 65° and 70°, the failure chances increase slightly to 1%, which is very low (Figure 4b). The failure probability is very low when the slope dip is lower than the joint dip. On the other hand, there will be no chance for planar failure if the pole points of the joints are not in the critical zone [51,53]. The number of critical failures is 0 when the slope dip angle is kept lower or higher. All results are shown in Table 3. The failure plane created by the friction angle circle and the daylight envelope demonstrates that any junction point is susceptible to failure. According to the results of quantitative data on dip direction, several joints are orientated in the same direction as the slope. As a result, no joint is in danger of failing from the right bank (Figure 6a–c). The azimuth of most discontinuities is oriented in NW and NE where the criteria for planar failure are not met. Additionally, when the slope angle is 45°, the failure chance is zero, but with an increase in slope dip angle up to 70°, the critical intersection points and failure chances are zero and 1%, respectively. This indicates that an increase in slope dip angle has no effect on planar failure in this condition (Figure 6a,c).

(b) *Wedge failure*: The right bank wedge failure study yields nearly identical results as the left bank. A low dip angle has lower odds, but as the slope dip increases, the likelihood of failure increases (Figure 4b). From the result, it can be seen that the slope dip angle of 45° has a failure probability of 2%, which is very low (Figure 6d). Additionally, the number of intersection points in the critical intersection zone is comparatively lower (Table 3). Additionally, the failure probability exhibits a regular increase with increasing slope dip angle. From the quantitative analysis of discontinuities, it can be seen that 67 discontinuities are oriented in a SW direction. In contrast, 94 are oriented NW (Figure 2b). The kinematic measurements show that wedge-type failure is controlled by azimuth and slope dip and dip direction. In the critical zone, azimuth-based discontinuities with NW and SW orientations result in intersection spots that cause wedge failure. The failure probability becomes very high when the slope angle is 65° and 70°; then, failure chances rise to 7% and 8%, respectively (Figure 6e,f). This indicates that as crucial intersection sites rise, they depend on the slope dip angle, which raises the chance of failure.

(c) *Toppling failure*: From the results, it can be seen that toppling failure is more crucial in the right bank at any dip angle of the slope. When rock masses have a dominating discontinuity set (often bedding or foliation) with a strike almost parallel to the sloping surface and inward dip, rock slopes are more likely to topple [43]. Landslides mainly occur on dip slopes and sporadically happen elsewhere [14]. In addition, it is possible to decrease the slope angle because doing so lowers the weight of the material, enhancing the slopes' stability [1]. From the kinematic analysis result of the right bank, it can be seen that toppling failure probabilities are slightly more than 5% when the slope dip angle is 40°, but they gradually increase as the slope dip angle increases, obtaining a maximum of 5.5% at a dip angle of 55°. At 65° and 70°, toppling exhibits the greatest values of 8.5% and 9.5%, respectively (Figure 4b). From Figure 3a,b, 54 discontinuities are oriented NE and SE, where the joints and slope dip in the opposing directions. In addition, most of the discontinuities dip with a high dip angle. This implies that toppling failure is a possibility. In comparison, when the slope angle is 45° and the failure chance is 5.5%, then crucial junction locations are relatively lower (Figure 6g). Critical interaction sites and failure probability both rise as the slope dip angle rises (Table 3). When the slope dip angle rises up to high value of 65° and 70°, then failure probability is slightly more than 8.5% and 9%. This result concurs with kinematic analysis, which indicates that NE–SW dipping discontinuities have a high potential for toppling failure. Moreover, the number of critical points also increases with the increase in slope dip angle (Figure 6h,i). Based on the azimuth and slope geometry of the right bank, it is evident that high failure probability of toppling exists with a high dip slope angle. In this condition, the slope is unstable and more likely to fail.

4.2. Limit Equilibrium Analysis

The limit equilibrium method (LEM) was used to determine the factor of safety (FoS) and evaluate the surface failure along both sides of the highway slopes. The primary method used with the Slide 6.020 application to assess stability is this analytical methodology for failure surfaces and factor of safety in normal and seismic conditions is shown in Figure 8. For a variety of limit equilibrium approaches, including the simplified Bishop method, the simplified Janbu method, and the simplified Spencer method, the FoS values using the Mohr-Coulomb criteria were computed. Using this technique, forces that lead to rock mass instability, as well as resistant forces are examined, and the resistance to driving force ratio (FoS) is calculated [11,24,42,45]. In a 2-dimensional analysis, a slice that is one unit thick is thought to represent a typical piece of the slope. Previous studies have demonstrated that the limit equilibrium approach provides acceptable results for engineering applications [5,26].

4.2.1. Left Slope

In a deterministic investigation of the left bank slope using LEM under typical circumstances, several FoS values were obtained. Two rock slope stability scenarios were examined: one under typical conditions and the other under peak ground acceleration (PGA). The results show that the calculated factor of safety in normal conditions for the Bishop, Janbu, and Spencer methods is 1.128, 1.105, and 1.126, respectively (Figure 7a–c). This result of the left bank indicates that the slope is stable when the slope angle is 60° or less, but the slope becomes vulnerable to failure when the overall slope angle increases to 65° and 70°. [52] revealed that movement of the anti-dip slope can be influenced by the slope angle when it increased from 60° to 70° while the dip angle of the strata is kept fixed. According to [50], the prior collapse at the quarry was mostly caused by steep slope angles. The Bishop and Spencer technique values are close to 1.7, whereas the Janbu FoS value with its lowest slope angle of 45° is slightly less than 1.6. Higher safety factor values appear with a decreasing slope angle, indicating a reduced chance of failure (Figure 9a). When the slope angle is kept lower, according to LEM analysis of these methods, the left bank is stable. When the slope angle rises to 60° or 70°, however, the slope becomes highly unstable and has a low value of FoS.

However, for the seismic condition with ground acceleration of 0.25, the values of safety for Bishop, Janbu, and Spencer decrease by 0.759, 0.715, and 0.758, respectively (Figure 7d–f). In addition to this, a small change in slope angle can also change the factor of safety, which ultimately affects the slope stability. The FoS of the slope with various slope angles and a peak ground acceleration (PGA) of 0.25 are shown in Figure 9b. The lowest FoS values are 0.715 for the Janbu approach, and 0.759 and 0.758 for the Bishop and Spencer methods, respectively. These results lead to slope vulnerability, which makes it more likely to collapse at any time under these circumstances. In contrast, it can be seen that, due to a decrease in slope angle with PGA of 0.25, the FoS value rises. The FoS value for Janbu, with the lowest slope angle of 45°, is slightly higher than 1, while the values for the Bishop and Spencer methods are somewhat higher than 1.1. This indicates that failure probability increases, and the slope is more unstable (Figure 9b).

However, for the seismic condition with ground acceleration of 0.25, the values of safety for Bishop, Janbu, and Spencer decrease by 0.759, 0.715, and 0.758, respectively (Figure 7d–f). In addition, a small change in slope angle can also change the factor of safety, which ultimately affects the slope stability. The FoS of the slope with various slope angles and a peak ground acceleration (PGA) of 0.25 are shown in Figure 9b. The lowest FoS values are 0.715 for the Janbu approach, and 0.759 and 0.758 for the Bishop and Spencer methods, respectively. These results indicate the slope vulnerability, which makes it more likely to collapse at any time under these circumstances. In contrast, it can be seen that, due to a decrease in slope angle with PGA of 0.25, the FoS value rises. The FoS value for Janbu, with the lowest slope angle of 45°, is slightly higher than 1, while the values for the Bishop and Spencer methods are somewhat higher than 1.1. This indicates that failure probability increases and the slope is more unstable (Figure 9b).

4.2.2. Right Slope

The LEM analysis for right bank resulted in very low factor of safety (FoS) values. In normal conditions, the FoS values for the Bishop and Spencer methods are 1.179 and 1.168, respectively (Figure 8a,c), whereas the FoS for the Janbu method is 1.092, which is comparatively very low (Figure 8b). The results demonstrate that according to the Bishop and Spencer methods, the right slope is stable, while the Spencer method indicates that it is fairly stable when the slope angle is 60°. From Figure 9c, it can be seen that when the slope dip angle decreases, FoS rises. When the slope angle is kept at 45° and 50°, then the values of FoS are 1.3 for all three methods. This indicates that the rock slope is stable when the dip angle is kept lower. However, the slope becomes extremely unstable and has a low value of FoS as the slope angle increases to 60° or 70°.

The computed safety factor for Bishop, Janbu, and Spencer methods decreases with 0.25 ground acceleration in seismic circumstances. The right bank is unstable and at risk of failure at any slope dip angle, while the Bishop and Janbu methods calculate the FoS as 1 when the slope dip angle is 45°; however, the slope is not stable and can fail at any time. Figure 8b shows that the lowest value of FoS for the Janbu method is 0.706, which is very low. On the other hand, the FoS values for the Bishop and Spencer methods are 0.808 and 0.183, which are comparatively higher than the Janbu result, but lower than 1 which indicates a very unstable condition of the slope (Figure 8a,c). In seismic conditions, the FoS of the slope with various slope dip angles and a peak ground acceleration (PGA) of 0.25 is shown in Figure 9d. In this condition, it is observed that the slope is still unstable when the slope angle decreases and shows chances of failure. In addition, the FoS value is slightly less than 0.7 and 0.6 when the slope dip angle increases up to 60° and 70° with PGA of 0.25. This value indicates that the slope becomes very unstable and the chances of failure are higher.

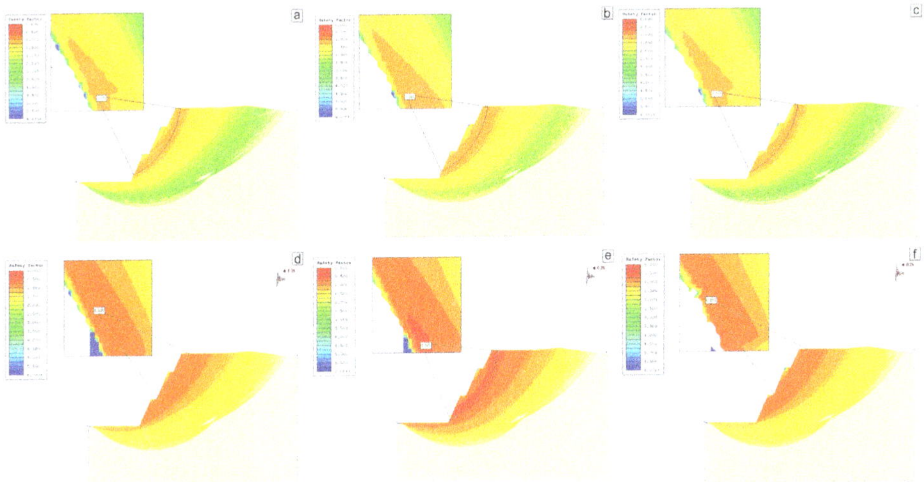

Figure 8. Slope stability analysis for the right bank of the highway. (**a**) Bishop method under normal conditions, (**b**) Janbu method under normal conditions, (**c**) Spencer method under normal conditions, (**d**) Bishop method under seismic conditions, (**e**) Janbu method under seismic conditions, (**f**) Spencer method under seismic conditions.

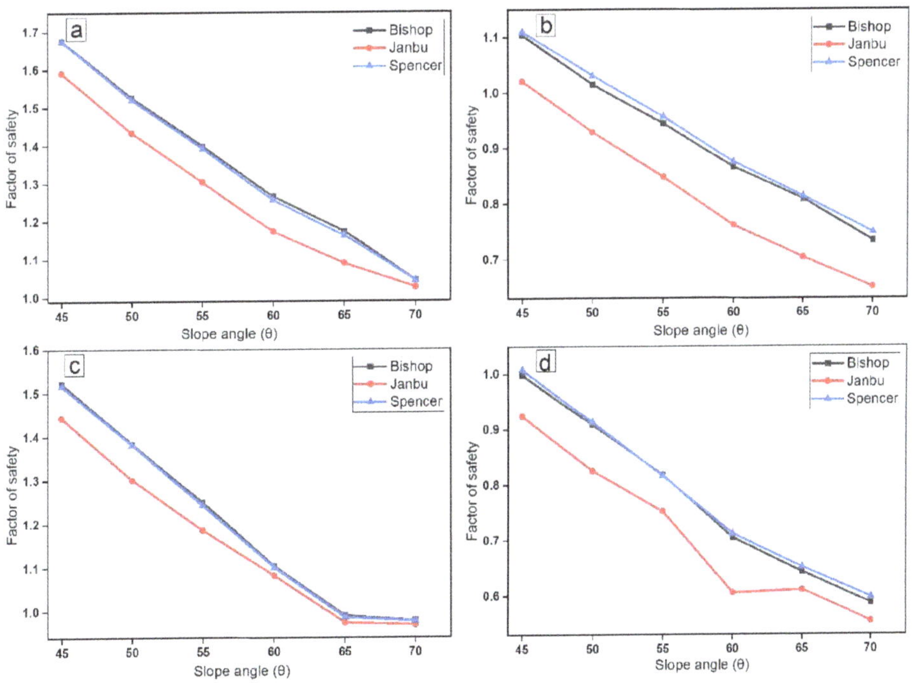

Figure 9. Factors of safety for different slope angles. (**a**) Left bank slope under normal conditions, (**b**) left bank slope under seismic conditions, (**c**) right bank slope under normal conditions, (**d**) right bank slope under seismic conditions.

5. Conclusions

The purpose of this study was to investigate potential failure and mode along the left and right slopes of an embankment. Kinematic analysis techniques were used to evaluate the potential for failure of rock masses moving along geologic structures on the slope face. The geometry of the bedding plane, joints, and slope were considered for both design and failure scenarios. The failure potential at both sites was investigated utilizing the graphical stereographic projection of the DIPS software. The two main factors in this study that affect the type of instabilities seen in the field are bedding planes and discontinuity orientation. The results are drawn from the field visit and the slope stability calculations conducted using kinematic and finite element methods. Based on kinematic analysis, the rock slope is stable when the slope angle is lower than 60°, while the failure probabilities are enhanced with the rise in slope dip angle up to 70°.

In the same way, wedging and toppling failures exist even when there is no expected planar collapse. According to the analysis, on the right and left sides, toppling failure has a higher probability than wedge failure. On the other hand, the Limit Equilibrium Method (LEM) output shows that the slope is critical to failure with a large slope dip angle and stable when the slope dip angle is reduced. In normal conditions, the slope is stable on both sides with factors of safety (FoS) more than 1. When the peak ground acceleration (PGA) reaches 0.25, the slope is extremely unstable, and the factor of safety (FoS) is less than 1.

Author Contributions: Conceptualization, L.A.S.; Data curation, H.U.R.; Formal analysis, H.T.J. and L.A.S.; Funding acquisition, S.A.A. and I.H.; Investigation, A.U.R., G.Z. and H.U.R.; Methodology, A.U.R., G.Z., H.T.J. and H.U.R.; Project administration, G.Z., S.A.A., I.H. and L.A.S.; Resources, S.A.A. and I.H.; Software, A.U.R., S.A.A., H.T.J. and I.H.; Supervision, G.Z., S.A.A. and I.H.; Validation, G.Z. and H.T.J.; Visualization, A.U.R. and I.H.; Writing—original draft, A.U.R.; Writing—review & editing, A.U.R. and G.Z. All authors have read and agreed to the published version of the manuscript.

Funding: This research was funded by Research Supporting Project Number (RSPD2023R585), King Saud University, Riyadh, Saudi Arabia.

Data Availability Statement: The dataset and the samples for this study are available from the corresponding author upon reasonable request.

Conflicts of Interest: The authors declare no conflict of interest.

References

1. Agyei, G.; Osei, J.B.; Adoko, A.C. Slope stability analysis of Nkran Pit at Asanko Gold Mine, Ghana. *J. Appl. Sci. Environ. Manag.* **2018**, *22*, 1673. [CrossRef]
2. Ahmad, I.; Jan, M.Q.; DiPietro, J.A. Age and Tectonic implications of granitoid rocks from the Indian plate of Northern Pakistan. *J. Virt. Explor. Electron. Ed.* **2003**, *11*, 21–28.
3. Akram, M.S.; Ullah, M.F.; Rehman, F.; Ali, M.; Ahmed, L.; Gillani, A.A. Stability Evaluation of Slopes Using Kinematic and Limit Equilibrium Analyses in Seismically Active Balakot, KPK, Pakistan. *Open J. Geol.* **2019**, *9*, 795–808. [CrossRef]
4. Akram, M.S.; Zeeshan, M.; Mirza, K.; Ahmed, L.; Noor, A.; Ali, M. Slope stability analyses using classification systems and numerical methods: Case study from Lower Dir, Khyber Pakhtunkhwa, Pakistan. *Himal. Geol.* **2019**, *40*, 67–77.
5. Aksoy, C.O. Review of rock mass rating classification: Historical developments, applications, and restrictions. *J. Min. Sci.* **2008**, *44*, 51–63. [CrossRef]
6. Alejano, L.R.; Ferrero, A.M.; Ramírez-Oyanguren, P.; Fernández, M.I. Comparison of limit-equilibrium, numerical and physical models of wall slope stability. *Int. J. Rock Mech. Min. Sci.* **2011**, *48*, 16–26. [CrossRef]
7. Bao, Y.; Chen, J.; Su, L.; Zhang, W.; Zhan, J. A novel numerical approach for rock slide blocking river based on the CEFDEM model: A case study from the Samaoding paleolandslide blocking river event. *Eng. Geol.* **2023**, *312*, 106949. [CrossRef]
8. Bao, Y.; Han, X.; Chen, J.; Zhang, W.; Zhan, J.; Sun, X.; Chen, M. Numerical assessment of failure potential of a large mine waste dump in Panzhihua City, China. *Eng. Geol.* **2019**, *253*, 171–183. [CrossRef]
9. Bieniawski, Z.T. Rock Mechanics Design in Mining and Tunneling. *Int. J. Rock Mech. Min. Sci. Geomech. Abstr.* **1984**, *84*, 193–194. [CrossRef]
10. Bieniawski, Z.T. Engineering rock mass classifications: A complete manual for engineers and geologists in mining, civil, and petroleum engineering. In *Engineering Rock Mass Classifications: A Complete Manual for Engineers and Geologists in Mining, Civil, and Petroleum Engineering*; John Wiley & Sons: Hoboken, NJ, USA, 1989.

11. Bieniawski, Z.T. The geomechanics classification in rock engineering applications. In Proceedings of the 4th ISRM Congress, Montreux, Switzerland, 2–7 September 1979.
12. Bowa, V.M.; Xia, Y. Stability Analyses of Jointed Rock Slopes with Counter-tilted Failure Surface Subjected to Block Toppling Failure Mechanisms. *Arab. J. Sci. Eng.* **2018**, *43*, 5315–5331. [CrossRef]
13. Bye, A.R.; Bell, F.G. Stability assessment and slope design at Sandsloot open pit, South Africa. *Int. J. Rock Mech. Min. Sci.* **2001**, *38*, 449–466. [CrossRef]
14. Cerri, R.I.; Reis, F.A.G.V.; Gramani, M.F.; Rosolen, V.; Luvizotto, G.L.; Giordano, L.D.C.; Gabelini, B.M. Assessment of landslide occurrences in Serra do Mar mountain range using kinematic analyses. *Environ. Earth Sci.* **2018**, *77*, 1–16. [CrossRef]
15. Chakraborty, R.; Dey, A. Random Finite Element and Limit Equilibrium Methods-Based Probabilistic Stability Analyses of a Cut Slope. *Indian Geotech. J.* **2022**, *52*, 969–978. [CrossRef]
16. Chang, Y.F.; Liu, X.P.; Wu, Y.-C. *The Copper-Iron Belt of the Lower and Middle Reaches of the Changjiang River*; Geology Publication House, Beijing: Beijing, China, 1991; pp. 1–239.
17. Khan, Z.; Khan, F.A.; Khan, A.U.; Hussain, I.; Khan, A.; Shah, L.A.; Khan, J.; Badrashi, Y.I.; Kamiński, P.; Dyczko, A.; et al. Climate-Streamflow Relationship and Consequences of Its Instability in Large Rivers of Pakistan: An Elasticity Perspective. *Water* **2022**, *14*, 2033. [CrossRef]
18. Chowdhury, R.; Rao, B.N. Probabilistic stability assessment of slopes using high dimensional model representation. *Comput. Geotech.* **2010**, *37*, 876–884. [CrossRef]
19. Deng, J. Experimental remolding on the caprock's 3D strain field of the Indosinian-Yanshanian epoch in Tongling deposit concentrating area. *Sci. China Ser. D Earth Sci.* **2005**, *48*, 863–874. [CrossRef]
20. Deng, J.; Wang, Q.; Xiao, C.; Yang, L.; Liu, H.; Gong, Q.; Zhang, J. Tectonic-magmatic-metallogenic system, Tongling ore cluster region, Anhui Province, China. *Int. Geol. Rev.* **2011**, *53*, 449–476. [CrossRef]
21. Fereidooni, D.; Khanlari, G.R.; Heidari, M. Evaluación del sistema Q modificado de clasificación del macizo rocoso para el análisis de estabilidad de pendiente de roca. *Earth Sci. Res. J.* **2015**, *19*, 147–152. [CrossRef]
22. Goodman, R.E. *Introduction to Rock Mechanics*; Wiley: New York, NY, USA, 1989; Volume 2.
23. Gurocak, Z.; Alemdag, S.; Zaman, M.M. Rock slope stability and excavatability assessment of rocks at the Kapikaya dam site, Turkey. *Eng. Geol.* **2008**, *96*, 17–27. [CrossRef]
24. Hoek, E. Practical Rock Engineering 2007 (2007 ed.) Retrieved June 30, 2010, 2007 from Rocscience. PracticalRockEngineering. Asp. Available online: http://www.rocscience.com/hoek (accessed on 1 January 2023).
25. Hoek, E.; Bray, J.D. *Rock Slope Engineering*; Wyllie, D.C., Mah, C.W., Eds.; Taylor and Francis: London, UK, 2004.
26. Hoek, E.; Bray, J.D. *Rock Slope Engineering*; CRC Press: Boca Raton, FL, USA, 1981.
27. Karaman, K.; Ercikdi, B.; Kesimal, A. The assessment of slope stability and rock excavatability in a limestone quarry. *Earth Sci. Res. J.* **2013**, *17*, 169–181.
28. Kesimal, A.; Ercikdi, B.; Cihangir, F. Environmental impacts of blast-induced acceleration on slope instability at a limestone quarry. *Environ. Geol.* **2008**, *54*, 381–389. [CrossRef]
29. Kumar, N.; Ismail, M.A.M. 3D limit equilibrium method for rock slope stability analysis using generalised anisotropic material model. In Proceedings of the 2020 International Symposium on Slope Stability in Open Pit Mining and Civil Engineering, Australian Centre for Geomechanics, Perth, WA, USA, 12–14 May 2020; pp. 715–730. [CrossRef]
30. Lee, S.-H.; Kim, H.-B. A study to incorporate engineering probability techniques into deterministic slope stability methods. *KSCE J. Civ. Eng.* **2000**, *4*, 153–160. [CrossRef]
31. Meybodi, E.E.; Golestani, M.; Hussain, S.K. Kinematic Analysis of Rock Falls (Case Study: Southeast Wall of Sadat Siriz Iron Ore Mine). Researchgate.Net, March. 2022. Available online: https://www.researchgate.net/profile/Khaliq-Hussain/publication/358982221_Kinematic_analysis_of_rock_falls_Case_study_Southeast_wall_of_Sadat_Siriz_iron_ore_mine/links/6220cd2019d1945aced2e174/Kinematic-analysis-of-rock-falls-Case-study-Southeast-wall-of-Sadat-Siriz-iron-ore-mine.pdf (accessed on 1 December 2022).
32. Nagendran, S.K.; Ismail, M.A.M. Probabilistic and Sensitivity Analysis of Rock Slope Using Anisotropic Material Models for Planar Failures. *Geotech. Geol. Eng.* **2021**, *39*, 1979–1995. [CrossRef]
33. Yang, Y.; Sun, G.; Zheng, H.; Qi, Y. Investigation of the sequential excavation of a soil-rock-mixture slope using the numerical manifold method. *Eng. Geol.* **2022**, *256*, 93–109. [CrossRef]
34. Naghadehi, M.Z.; Jimenez, R.; KhaloKakaie, R.; Jalali, S.-M.E. A probabilistic systems methodology to analyze the importance of factors affecting the stability of rock slopes. *Eng. Geol.* **2011**, *118*, 82–92. [CrossRef]
35. Ing, Y.; Zhang, G.; Tang, H.; Shen, W.; Shen, P. Process Analysis of Toppling Failure on Anti-dip Rock Slopes under Seismic Load in Southwest China. *Rock Mech. Rock Eng.* **2019**, *52*, 4439–4455. [CrossRef]
36. Pantelidis, L. Rock slope stability assessment through rock mass classification systems. *Int. J. Rock Mech. Min. Sci.* **2009**, *46*, 315–325. [CrossRef]
37. Pantelidis, L. An alternative rock mass classification system for rock slopes. *Bull. Eng. Geol. Environ.* **2010**, *69*, 29–39. [CrossRef]
38. Park, H.-J.; Lee, J.-H.; Kim, K.-M.; Um, J.-G. Assessment of rock slope stability using GIS-based probabilistic kinematic analysis. *Eng. Geol.* **2016**, *203*, 56–69. [CrossRef]
39. Phillips, F.C. *The Use of Stereographic Projection in Structural Geology*; Taylor & Francis Group: London, UK, 1972.

40. Rahman, A.U.; Guangcheng, Z.; Khan, A.; Puniya, K.; Rahman, S.U.; Xin, Z.; Kumar Puniya, M. Rock mass classification systems and kinematic analysis of slates from dir group, nw, himalaya, pakistan; implication for slope stability. *Geol. Behav.* **2022**, *6*, 39–45. [CrossRef]
41. Rahman, A.U.; Zhang, G.; Khan, A.; Muhammad, J.; Sohail, A.; Rehman, H.U.; Puniya, M.K.; Ashraf, B. The Effect of Mineral Composition, Microstructure on Physical Properties of Aggregate from River Punjkora and Kunai, Dir (Lower), Khyber Pakhtunkhwa, Pakistan. *J. Geosci. Environ. Prot.* **2022**, *10*, 127–144. [CrossRef]
42. Robani, A. Analysis of Slope Failure at Maran Highway Using Slope/W Software. Ph.D. Thesis, Universiti Malaysia Pahang, Kuantan, Malaysia, November 2009.
43. Sagaseta, C.; Sánchez, J.; Cañizal, J. A general analytical solution for the required anchor force in rock slopes with toppling failure. *Int. J. Rock Mech. Min. Sci.* **2001**, *38*, 421–435. [CrossRef]
44. Sharma, R.K.; Mehta, B.S.; Jamwal, C.S. Cut slope stability evaluation of NH-21 along Nalayan-Gambhrola section, Bilaspur district, Himachal Pradesh, India. *Nat. Hazards* **2013**, *66*, 249–270. [CrossRef]
45. Sharma, S.; Raghuvanshi, T.K.; Anbalagan, R. Plane failure analysis of rock slopes. *Geotech. Geol. Eng.* **1995**, *13*, 105–111. [CrossRef]
46. Turner, A.K.; Schuster, R.L. *Landslides: Investigation and Mitigation*; Special Report 247; Transportation Research Board, National Academy Press: Washington, DC, USA, 1996.
47. Ukritchon, B.; Ouch, R.; Pipatpongsa, T.; Khosravi, M.H. Investigation of Stability and Failure Mechanism of Undercut Slopes by Three-Dimensional Finite Element Analysis. *KSCE J. Civ. Eng.* **2018**, *22*, 1730–1741. [CrossRef]
48. Vatanpour, N.; Ghafoori, M.; Talouki, H.H. Probabilistic and sensitivity analyses of effective geotechnical parameters on rock slope stability: A case study of an urban area in northeast Iran. *Nat. Hazards* **2014**, *71*, 1659–1678. [CrossRef]
49. Wang, Q.; Deng, J.; Huang, D.; Xiao, C.; Yang, L.; Wang, Y. Deformation model for the Tongling ore cluster region, east-central China. *Int. Geol. Rev.* **2011**, *53*, 562–579. [CrossRef]
50. Wu, F.Y. Collapsed Himalayan-Tibetan orogen and the rising Tibetan Plateau. *Acta Petrol. Sin.* **2008**, *24*, 1–30.
51. Wyllie, D.C.; Mah, C.W. *Rock Slope Engineering—Civil and Mining*, 4th ed.; Taylor and Francis: London, UK, 2005.
52. Xiao, X.; Zhou, T.; White, N.C.; Zhang, L.; Fan, Y.; Chen, X. Multiple generations of titanites and their geochemical characteristics record the magmatic-hydrothermal processes and timing of the Dongguashan porphyry-skarn Cu-Au system, Tongling district, Eastern China. *Miner. Depos.* **2021**, *56*, 363–380. [CrossRef]
53. Yoon, W.; Jeong, U.; Kim, J. Kinematic analysis for sliding failure of multi-faced rock slopes. *Eng. Geol.* **2002**, *67*, 51–61. [CrossRef]

water

MDPI

Article

Analysis of Structure Stability of Underwater Shield Tunnel under Different Temperatures Based on Finite Element Method

Lei Zhu [1], Qianwen Wu [2], Yuke Jiang [1], Zhenyu Li [3] and Yuke Wang [3,*]

1 CCCC Tunnel and Bridge (Nanjing) Technology Co., Ltd., Nanjing 211899, China; a13298106805@163.com (L.Z.); a17726406394@163.com (Y.J.)
2 Productivity Centre of Jiangsu Province, Nanjing 210042, China; a18790233360@163.com
3 College of Water Conservancy and Civil Engineering, Zhengzhou University, Zhengzhou 450001, China; li337240542@163.com
* Correspondence: ykewang@163.com

Abstract: The structural stability of the underwater shield tunnel during operations is affected by temperature variations. The effect of different structure temperatures on the underwater shield tunnel during the operation period was studied. By numerical simulation, the variation in the underwater shield tunnel temperature circle was analyzed. The variation patterns of the top arch, bottom arch, waist arch temperature, maximum principal stress, and settlement of the soil under different temperatures were obtained. The results showed that: (1) The early excavation time of the tunnel was short, and the temperature circle was small. The temperature circle expanded rapidly after 50 days of operating. The diffusion range increased from 1.5 m to 5.35 m: an increase of 256.7%. With the increase in time, the expansion rate of the temperature circle gradually slowed down. (2) The higher the temperature of the soil, the more complex the temperature transfer between the soil and the lining was while generating greater temperature stresses and reducing the safety of the tunnel. (3) When the tunnel was just excavated, the compression settlement of the top arch and the waist arch increased rapidly, reaching 5.43 mm and 0.24 mm, respectively. The bottom arch was squeezed by the soil on both sides, resulting in an uplift and rapid increase, reaching 4.94 mm. The settlement rate increased with the increase in the tunnel structure's temperature. After the excavation, with the decrease in temperature, the strength of the soil and lining increased. The settlement of the top arch, bottom arch, and waist arch increased slowly with time, and the growth rate decreased gradually.

Keywords: underwater shield tunnel; numerical simulation; temperature circle; maximum principal stress; tunnel subsidence

Citation: Zhu, L.; Wu, Q.; Jiang, Y.; Li, Z.; Wang, Y. Analysis of Structure Stability of Underwater Shield Tunnel under Different Temperatures Based on Finite Element Method. *Water* **2023**, *15*, 2577. https://doi.org/10.3390/w15142577

Academic Editors: Giuseppe Oliveto and Giuseppe Pezzinga

Received: 6 May 2023
Revised: 25 June 2023
Accepted: 3 July 2023
Published: 14 July 2023

1. Introduction

With the continuous development of water conservancy in China over recent years, the technology of underwater shield tunnels has been developed rapidly. Some safety accidents in shield tunnels are caused by high temperatures in the tunnel's structure [1–3]. The environmental control of the underwater shield tunnel operation represented by structure temperature has become a problem that must be considered in the underwater shield tunnel design [4,5]. This phenomenon could lead to safety hazards in the tunnel. The life safety of workers could also be affected [6–8]. Therefore, problems relating to structure temperature in underwater shield tunnels need to be studied in depth.

At present, there have been a large number of related studies to temperature [9–14]. The influence of the temperature disturbance caused by underwater shield tunnel construction on the temperature distribution of the soil and lining structure was studied, and a new finite difference temperature prediction model was proposed [13]. Based on the k-ε turbulence equation, a two-dimensional axisymmetric model coupling the convective-conduction heat transfer was established, and the airflow temperature field in a high-temperature underwater shield tunnel was investigated [15]. Zhou et al. [16] used the finite difference method to study a transient heat transfer model during the tunnel operation and investigated the effects of mechanical ventilation and the train piston wind on the tunnel's temperature field. The effect of high temperatures from fires on tunnel structures has also been studied by many scholars [17–19]. The flow route of the fire-induced high-temperature airflow through the tunnel was proposed [20,21]. However, numerical simulation studies on the effects of high temperatures on underwater shield tunnel structures during operation are scarce. The high temperature, stress, and settlement changes in the tunnel cannot be better simulated by constructing a two-dimensional model. It is necessary to analyze the mechanical properties of tunnels under high-temperature conditions in the context of actual working conditions.

In this paper, a 3D model of an underwater shield tunnel in a different temperature environment was established. According to the temperature conditions of the underwater shield tunnel, the corresponding model parameters were given. Different thermodynamic parameters were set for the tunnel lining according to different temperatures, and temperature changes in the tunnel and lining were observed. The temperature, maximum principal stresses, and settlement changes to the top arch, bottom arch, and waist arch of the tunnel structure under different temperatures were analyzed. The results of this study provide a reference for similar working conditions.

2. Model Working Conditions

Based on the geological conditions of a tunnel, ABAQUS was used to model the soil and lining parameters, as shown in Tables 1 and 2. The 3D underwater shield tunnel model was 33 m × 33 m in height and width, 50 m in longitudinal length, with a tunnel diameter of 15.0 m and a lining thickness of 0.5 m. The initial temperatures of the soil and lining were taken to be 100 °C and 20 °C, respectively. The forced convection heat transfer coefficient between the soil and the air was 30 W/(m^2·°C). The forced convection heat transfer coefficient between concrete and air was 45 W/(m^2·°C). The ambient temperature of the soil and lining was 20 °C. The densities of the soil and lining were 2630 kg/m^3 and 2400 kg/m^3, respectively. Transient temperature-displacement coupled analysis steps were used. The model had a hexahedral C3D8T unit. To ensure the accuracy and reliability of the calculations, the tunnels and their surroundings were subdivided into meshes to make the results more accurate.

Table 1. Soil layer parameters.

Elastic Modulus (GPa)	Poisson's Ratio	Angle of Internal Friction (°)	Cohesion (MPa)	Thermal Conductivity (W/m·°C)	Coefficient of Linear Expansion (°C^{-5})	Specific Heat Capacity (J/Kg·°C)	Temperature (°C)
6.5	0.25	42	1.1	7.6	8.3	1285	100
6.7	0.25	42	1.1	8.0	7.6	1240	80
6.8	0.25	42	1.1	8.4	6.9	1195	65
6.9	0.25	42	1.1	8.9	6.2	1150	50
7.0	0.25	42	1.1	9.4	5.6	1105	35
7.1	0.25	42	1.1	10	5.0	1060	20

Table 2. Lining parameters.

Elastic Modulus (GPa)	Poisson's Ratio	Angle of Internal Friction (°)	Cohesion (MPa)	Thermal Conductivity (W/m·°C)	Coefficient of Linear Expansion (°C^{-5})	Specific Heat Capacity (J/Kg·°C)	Temperature (°C)
30.0	0.17	54	2.42	1.69	1.00	913	20
29.6	0.17	54	2.42	1.68	1.01	916	35
29.1	0.17	54	2.42	1.67	1.02	919	50
28.9	0.17	54	2.42	1.66	1.03	923	65
28.7	0.17	54	2.42	1.65	1.04	926	80
28.4	0.17	54	2.42	1.64	1.05	929	100

The underwater shield tunnel model was divided into two solid parts, the soil, and the lining, as shown in Figure 1. Earth stresses were balanced in the soil prior to the excavation of the tunnel. The stiffness migration method was used, i.e., by killing the soil during excavation and activating the lining during assembly through the ABAQUS life and death unit function to achieve a change in stiffness from low to high. The transfer in temperature and stress in the tunnel was achieved by setting up a transient temperature-displacement coupled analysis step. Displacements in the X and Y directions were constrained by the horizontal X and Y axes, respectively, and these displacements at the bottom were constrained by the Z direction. No constraints were added to the upper part.

Figure 1. Tunnel model.

3. Analysis of Results

3.1. Analysis of the Temperature Field in Underwater Shield Tunnels

As can be seen from Figure 2, after the excavation of the diversion tunnel was completed and after 100 days of operating, the temperature of the part of the soil in contact with the lining changed, and a temperature circle appeared and expanded. As can be seen from Figure 2a, when the tunnel was freshly excavated, the soil was just in contact with the lining, and there was no temperature transfer. From Figure 2a,b, after 2 days of excavation, the temperature transfer between the soil and the liner occurred, and a temperature circle appeared. Due to the short contact time, the temperature circle was small, with a radius of 4.5 m, excluding the lining radius, and the temperature circle spread was 1.5 m. Because of the heat transfer between the liner and the air, the temperature on the inside of the liner showed a slight increase to 20.02 °C. From Figure 2b,c, the temperature circle expanded rapidly after 50 days of operation. The diffusion range increased from 1.5 m to 5.35 m at an increase of 256.7%. The temperature on the inside of the liner showed an increase compared to the completion of the excavation, reaching 21.16 °C. From Figure 2c,d, after 100 days of operating, the growth rate of the temperature circle slowed down, and the diffusion range increased from 5.35 m to 7.1 m: an increase of 32.7%.

Figure 3 shows the lining temperature variation diagram. As can be seen from Figure 3a, after the excavation of the tunnel was complete, the temperature of the contact surface between the outer side of the lining and the soil was 92 °C. The modulus of elasticity of the liner decreased at high temperatures, and there was a difference of 71.98 °C from 20.02 °C on the inside of the liner. Large temperature gradients tend to produce large temperature stresses, and care should be taken during construction. Figure 3b shows that after 50 days of operating, the temperature of the outer liner contact surface with the soil was 60.84 °C: a significant reduction of 33.87% compared to when the excavation was completed. Figure 3c shows that after 100 days of operating, the temperature at the contact surface between the outer liner and the soil was 55.82 °C: a decrease of 8.25% compared to the temperature at 50 days of operating. This indicates that the temperature drop at the contact between the liner and the soil was greater in the early stage when the temperature difference between the liner and the soil was larger and slowed down in the later stage when the temperature difference decreased.

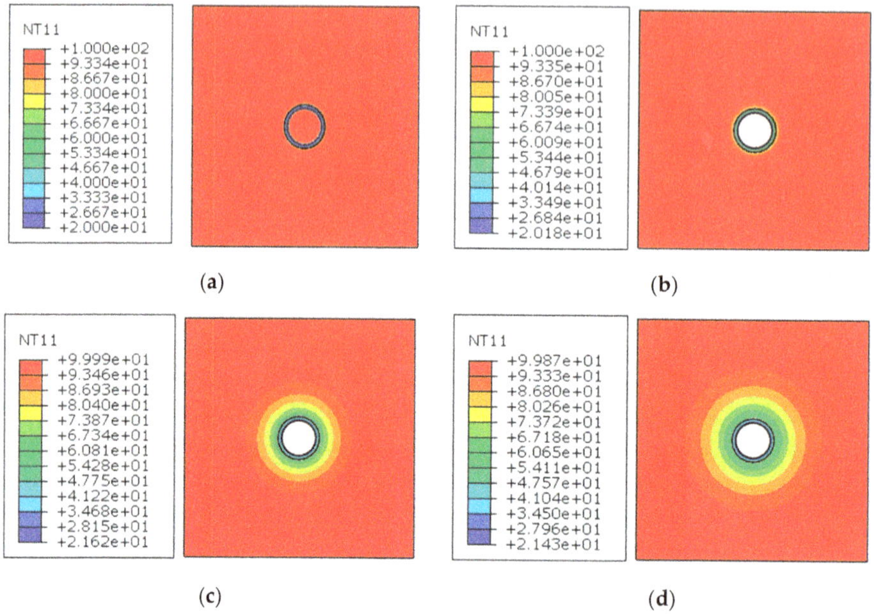

Figure 2. Temperature field distribution of the tunnel: (**a**) Freshly excavated; (**b**) Excavation completed; (**c**) Operated for 50 days; (**d**) Operated for 100 days.

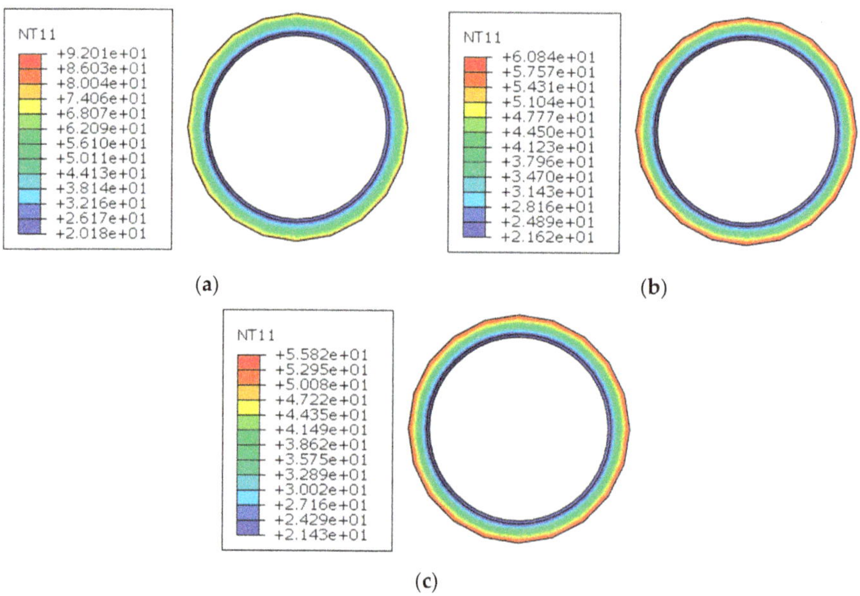

Figure 3. Temperature field distribution of lining: (**a**) Freshly excavated; (**b**) Operated for 50 days; (**c**) Operated for 100 days.

3.2. Analysis of Temperature Transfer between Envelope and Lining

In order to analyze the temperature variation rule at the contact point between the soil and the lining with high temperature, based on the numerical simulation results, the temperature variation curves of the top arch, bottom arch, and waist arch of the tunnel were selected for analysis. As can be seen from Figure 4, the temperature curves of the top and bottom arches of the tunnel were the same over time. The change in temperature at the waist arch was slower than that of the top and bottom arches in the early part of the curve, and the same was true of the later part. The overall temperature of the three curves showed a rapid decrease with time, followed by a slow decrease, which was consistent with the numerical simulation results of Xu et al. [22], indicating the feasibility of the model developed in this paper. At the completion of the tunnel excavation, the temperature of the top, bottom, and waist arches was around 79 °C. After 50 days of operating, the temperature dropped to 60 °C, which was 24% lower than when the excavation was completed. After 100 days of operating, the temperature was 55 °C, which was 8.3% lower than after 50 days of operating. In the early stages, the temperature dropped faster because of the larger temperature difference between the soil and the lining and the large heat transfer coefficient, while in the later stages, the temperature difference between the soil and the lining decreased; as the heat transfer coefficient decreased the temperature dropped slower.

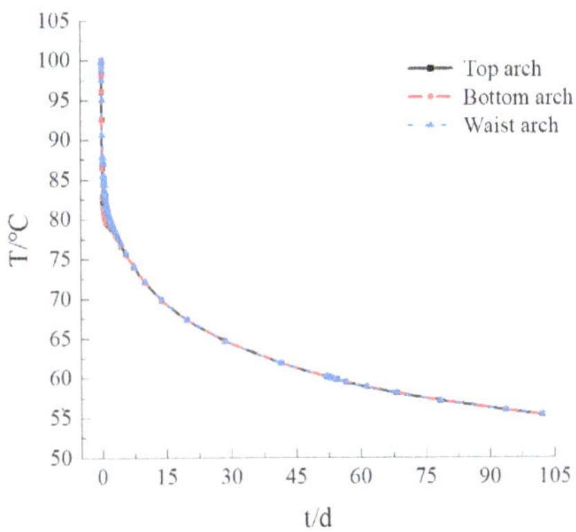

Figure 4. Temperature variation for top arch, bottom arch and waist arch of the tunnel.

3.3. Analysis of the Maximum Principal Stress between the Soil and the Lining

In order to analyze the law of change in the stress field at the contact between the high-temperature soil and lining, based on the numerical simulation results, the law of change in maximum principal stresses at the top, bottom, and waist arches of soil with time was selected for analysis. As shown in Figure 5, the positive values in the diagram are tensile stresses. For the sake of analysis, this analysis was carried out uniformly with positive tensile stresses. As can be seen from Figure 5, the top arch, bottom arch, and waist arch changes were divided into three stages: the sudden stress change period, stress fluctuation period, and stress stabilization period. The first stage was the sudden stress change period. At this stage, the maximum principal stresses in the top, bottom, and waist arches reached a minimum value of −3.12 MPa, −3.17 MPa, and −3.15 MPa, respectively, due to the fresh excavation of the tunnel soil and the influence of the self-weight of the soil. The soil was excavated and came into contact with air for heat exchange. The temperature at the contact between the soil and the air decreased abruptly under a large temperature difference, generating a large temperature stress and a rapid rise in the maximum principal stress occurred. The maximum principal stress in the girdle arch rose to −1.67 MPa, while the maximum principal stresses in the top and bottom arches rose to 0.70 MPa and 0.64 MPa, respectively. With the assembly of the first ring lining, the top arch, bottom arch, and waist arch were extruded by the lining, and the maximum principal stresses on the top and bottom arches decreased rapidly to 0.53 MPa and 0.52 MPa, respectively. The waist arch was subject to a large drop in the maximum principal stress of −2.54 MPa due to the combined extrusion of the upper and lower parts of the lining. The second stage was the period of stress fluctuation. With the disturbance in the tunnel excavation and lining assembly, the maximum principal stresses in the top, bottom, and waist arches fluctuated. The maximum principal stress in the top and bottom arches first decreased to 0.47 MPa and then gradually increased, while the maximum principal stress in the girdle arch first decreased to −3.02 MPa and then gradually increased. The third stage was the stress stabilization period. The maximum principal stresses in the top, bottom, and waist arches increased slowly with time in the early stages and reached stable values of 0.79 MPa and 0.81 MPa in the later stages as the temperature difference became smaller. The maximum principal stress in the waist arch decreased slowly with time, with a greater amount of variation compared to that in the top and bottom arches, which also gradually stabilized at a later stage, reaching a stable value of −1.40 MPa.

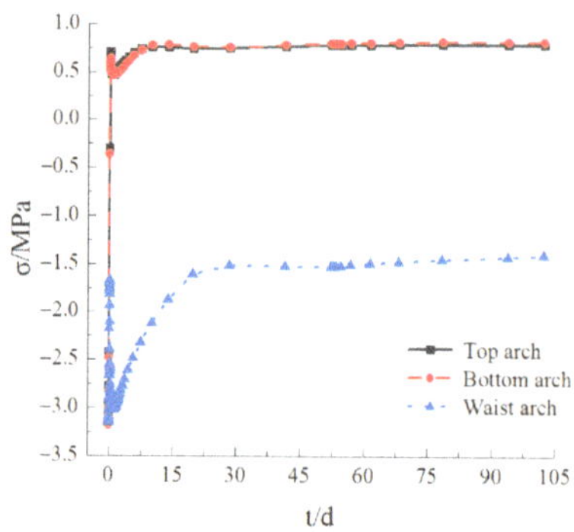

Figure 5. The maximum stress variation for top arch, bottom arch and waist arch of the tunnel.

3.4. Settlement Analysis between Enclosure and Lining

In order to analyze the variation rule of the settlement at the contact point between the high-temperature soil and lining and based on the numerical simulation results, the variation rule of the settlement of the soil vault, bottom arch, and waist arch over time was selected for analysis. As can be seen from Figure 6, the top arch, bottom arch, and waist arch could be divided into three stages: the abrupt settlement period, settlement fluctuation period, and settlement creep period. The settlement change curve was consistent with the experimental results of Wang et al. [23] and showed that the model established in this paper had a certain degree of reasonableness. In the first settlement of the abrupt change period, when the underwater shield tunnel was just excavated, under the influence of the dead weight of the soil, the compression settlement of the top arch and the waist arch increased rapidly, reaching 5.43 mm and 0.24 mm, respectively. The bottom arch was squeezed by the soil on both sides, resulting in an uplift that rapidly increased, reaching 4.94 mm. In the second stage of settlement fluctuation, with the assembly of the lining, the settlement of the top arch and waist arch slowed down in the early stage of excavation, reaching a maximum value of 7.98 mm and 1.48 mm, respectively, and the settlement decreased gradually in the later stage. The settlement increased slowly with the application during the third stage of settlement creep.

After the completion of the excavation, the settlement of the top arch and the waist arch were 7.90 mm and 1.43 mm, respectively, and the uplift of the bottom arch was 4.97 mm. After 50 days of operating, the settlement of the top arch and the waist arch was 9.55 mm and 2.42 mm, which increased by 1.65 m and 0.99 mm, respectively. The uplift of the bottom arch was 4.53 mm, which decreased by 0.44 mm. After 100 days of operating, the settlement of the top arch and waist arch was 10.85 mm and 3.22 mm, respectively, which increased by 1.3 mm and 0.8 mm compared to standing for 50 days. The uplift of the bottom arch was 4.14 mm, which decreased by 0.39 mm. After the excavation, with the decrease in the temperature, the strength of the soil and lining increased, and the settlement of the top arch, bottom arch, and waist arch increased slowly with time while the growth rate decreased gradually.

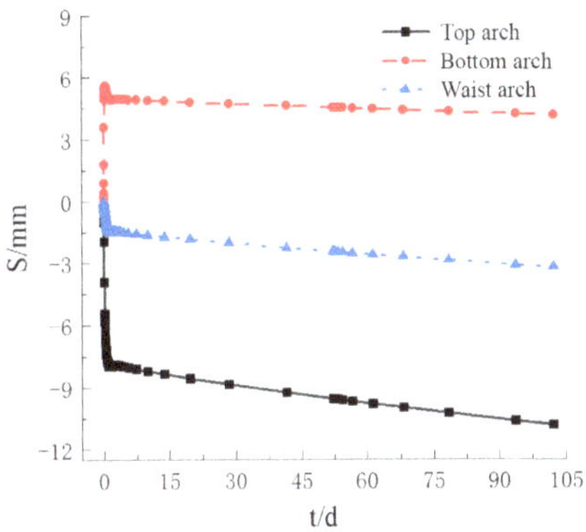

Figure 6. The settlement variation for top arch, bottom arch and waist arch of the tunnel.

3.5. *Temperature Transfer between Soil and Lining at Different High Structure Temperatures*

The temperature change curves of the top and waist arches of the tunnel at 40 °C, 70 °C, and 100 °C were extracted for further analysis of the temperature transfer between the soil and the lining at different high temperatures. As can be seen from Figure 7, the top, bottom, and waist arches followed the same trend at all three temperatures, with a rapid drop in temperature in the early stages and a slow and gradual stabilization in the later stages. At 40 °C of the soil temperature, the temperature of the top, bottom, and waist arches was around 35 °C after 2 days of excavation completion in the early stages: a reduction of 5.0 °C. With the later operating period of 50 days, the top and waist arch temperature was 30.3 °C, which was 9.7 °C lower than the initial temperature. After 100 days of operation, the temperature of the top and waist arches was 29.1 °C, which was 10.9 °C lower than the initial temperature. At 70 °C of the soil temperature, the temperature of the top, bottom, and waist arches was about 57 °C after 2 days of excavation completion in the early stages: a reduction of 13 °C. With the later operating period of 50 days, the top and waist arch temperatures were the same at 45.4 °C: a reduction of 14.6 °C from the initial temperature. After 100 days of operating, the temperature of the top and waist arches was 42.5 °C: a reduction of 27.5 °C from the initial temperature. At 100 °C of the soil temperature, the top, bottom, and waist arch temperatures were at 79.0 °C after 2 days of excavation completion in the early stages: a reduction of 21 °C. With the later operating period of 50 days, the top and waist arch temperatures were the same at 60.1 °C: a decrease of 39.9 °C from the initial temperature. After 100 days of operation, the top and waist arch temperature was 55.4 °C: a reduction of 44.6 °C from the initial temperature.

As the temperature of the soil increased, the greater the temperature drop in the tunnel, the greater the susceptibility to temperature stresses and the longer it took to reach stability. The difference in temperature between the top, bottom, and waist arches of the previous excavation increased as the temperature of the soil increased. Additional temperature stresses could easily be generated between the two. This stress was detrimental to the stability of the underwater shield tunnel.

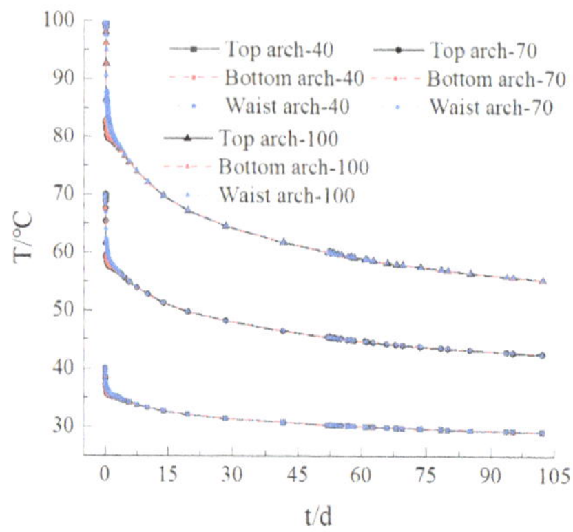

Figure 7. Curves of vault and waist temperature in tunnels with different high structure temperature.

3.6. Analysis of the Maximum Principal Stress between the Soil and the Lining at Different High Temperatures

The maximum principal stress variation curves for the top, bottom, and waist arches of the tunnel at 40 °C, 70 °C, and 100 °C were extracted for further analysis of the stress variation pattern at the contact between the tunnel envelope and the lining at different high temperatures. As can be seen from Figure 8, the maximum principal stresses in the top, bottom, and waist arches at 40 °C, 70 °C, and 100 °C followed the same trend, all showing a surge, then a decrease, and finally a slow increase to a gradual stabilization. When the underwater shield tunnel was first excavated, the initial maximum principal stresses at 40 °C, 70 °C, and 100 °C were the same, with −3.12 MPa, −3.17 MPa, and −3.15 MPa for the top arch, bottom arch, and waist arch, respectively. As the soil was excavated and came into contact with air, the maximum principal stresses in the top, bottom, and waist arches all rose significantly. The maximum principal stresses in the top, bottom, and waist arches at 40 °C were 0.2 MPa, 0.15 MPa, and −1.72 MPa, respectively. The maximum principal stresses in the top, bottom, and waist arches at 70 °C were 0.20 MPa, 0.15 MPa, and −1.68 MPa, respectively. The maximum principal stresses in the top, bottom, and waist arches at 100 °C were 0.70 MPa, 0.64 MPa, and −1.67 MPa, respectively. As the tunnel was excavated and the lining was assembled, the maximum principal stresses in the top, bottom, and waist arches decreased first. The maximum principal stresses in the top, bottom, and waist arches at 40 °C were 0.01 MPa, −0.05 MPa, and −3.06 MPa, respectively. The maximum principal stresses in the top, bottom, and waist arches at 70 °C were 0.35 MPa, 0.40 MPa, and −3.16 MPa, respectively. The maximum principal stresses at 100 °C were 0.53 MPa, 0.52 MPa, and −2.54 MPa for the top, bottom, and waist arches, respectively, which then gradually increased. After the tunnel excavation was complete, the maximum principal stress increased slowly with time and finally stabilized. The maximum principal stresses in the top, bottom, and waist arches at 40 °C were 0.46 MPa, 0.34 MPa, and −2.94 MPa, respectively. The maximum principal stresses in the top, bottom, and waist arches at 70 °C were 0.65 MPa, 0.67 MPa, and −1.95 MPa, respectively. The maximum principal stresses in the top, bottom, and waist arches at 100 °C were 0.79 MPa, 0.81 MPa, and −1.40 MPa, respectively. Under the action of the self-weight of the soil, lining support, and temperature, the maximum principal stresses in the top, and bottom arches were predominantly in tension, while the maximum principal stresses in the waist arch were predominantly in compression. The higher the temperature of the soil surrounding the underwater shield tunnel, the more complex the temperature transfer between the soil and the lining, the higher the temperature stresses generated, and the greater the fluctuation of maximum principal stresses, which reduced the safety of the tunnel.

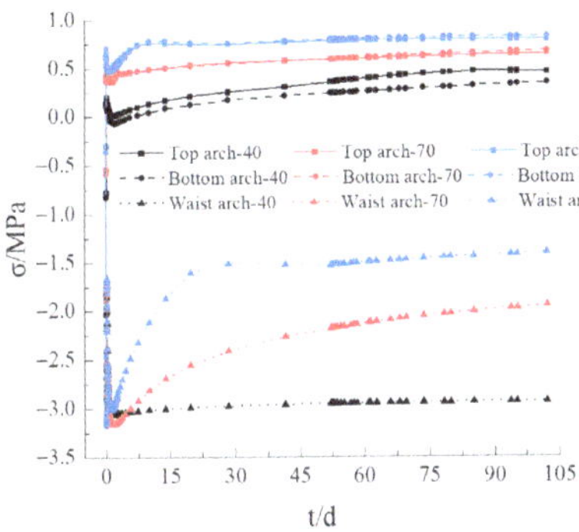

Figure 8. Curves of maximum principal stress variation at the vault and arch waist of the tunnel under different structure temperatures.

3.7. Settlement Analysis between Different High Structure Temperature Enclosures and Liners

The settlement curves for the top, bottom, and waist arches of the tunnel at 40 °C, 70 °C, and 100 °C were extracted to further analysis with the settlement pattern at the contact between the tunnel envelope and the lining at different high temperatures. As can be seen from Figure 9, the top, bottom, and waist arches followed the same trend at 40 °C, 70 °C, and 100 °C. The top arch and waist arch settlement, in general, showed a trend that first surged before slowing down, then decreasing, and finally growing slowly with time. The bottom arch rose on the same trend as the top and waist arches settled. During excavation, the settlement of the top and waist arches first increased sharply and then slowed down to a maximum value during excavation. In total, 7.71 mm and 1.38 mm were recorded for the top and waist arches, respectively, at 40 °C, and 5.35 mm for the waist arch bulge. The settlement of the top and waist arches at 70 °C was 7.80 mm and 1.42 mm, respectively, and the bulge value of the waist arch was 5.45 mm. The settlement of the top and waist arches at 100 °C was 7.89 mm and 1.48 mm, respectively, and the bulge value of the waist arch was 5.61 mm. At the completion of the tunnel excavation, the settlement of the top and waist arches at 40 °C was 7.64 mm and 1.33 mm, respectively, and the bulge value of the waist arch was 4.90 mm. The settlement of the top and waist arches at 70 °C was 7.74 mm and 1.37 mm, respectively, and the bulge value of the waist arch was 4.92 mm. The settlement of the top and waist arches at 100 °C was 7.90 mm and 1.43 mm, respectively, and the bulge value of the waist arch was 4.97 mm. After 100 days of operating, the settlement of the top and waist arches at 40 °C was 8.10 mm and 1.65 mm, respectively, and the bulge value of the waist arch was 4.72 mm. The settlement of the top and waist arches at 70 °C was 9.28 mm and 2.39 mm, respectively, with a bulge value of 4.34 mm for the waist arch. The settlement of the top and waist arches at 100 °C was 10.85 mm and 3.22 mm, respectively, with a bulge value of 4.14 mm for the waist arch. The top and waist arch settlement values gradually increased with time, and the bottom arch bulge value gradually decreased with time. The higher the temperature of the soil in the underwater shield tunnel, the greater the settlement fluctuations caused during excavation; the greater the variation in the settlement values of the top and waist arches, as well as the rise in the bottom arch over the same operating time, the more detrimental this was to the stability of the tunnel.

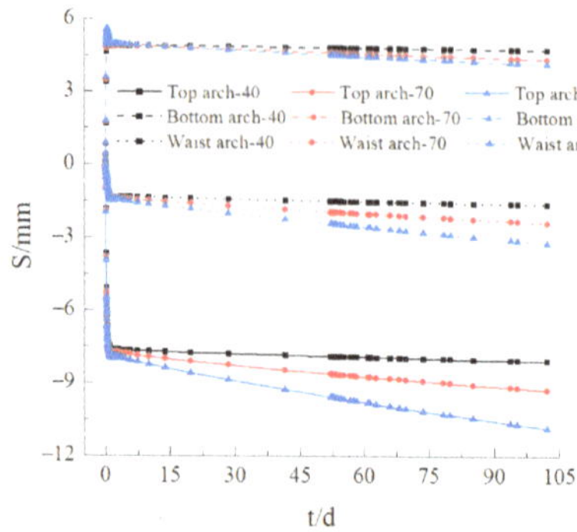

Figure 9. Variation in vault and waist settlement of the tunnel under different structure temperatures.

4. Conclusions

Through the numerical simulation of high structure temperatures in an underwater shield tunnel, the variation in the tunnel temperature circle was investigated. The variation patterns of the top arch, bottom arch, and waist arch temperatures, maximum principal stress, and settlement of the tunnel envelope were analyzed. The simulation results could be used as a reference for the design and construction of different structure temperature tunnels. The main conclusions of the summary of the laws of temperature, maximum principal stress, and settlement changes in underwater shield tunnels at different temperatures were as follows:

(1) The early excavation time of the underwater shield tunnel was short, and the temperature circle was small. The temperature circle expanded rapidly after 50 days of operation. The spread increased by 256.7%. The temperature change curves of the top, bottom, and waist arches decreased with time. The higher the temperature of the soil around the underwater shield tunnel, the greater the temperature drop.

(2) The process of the change in the maximum principal stress in the top, bottom, and waist arches could be divided into three phases: the period of sudden stress change, the period of stress fluctuation, and the period of stress stabilization. The higher the temperature in the soil, the more complex the temperature transfer between the soil and the lining was while generating greater temperature stresses and reducing the safety of the tunnel. When in high-temperature conditions, the temperature between the tunnel and the soil should be controlled to avoid creating additional temperature stresses that could affect the stability of the tunnel.

(3) Settlement changes could be divided into three phases: the abrupt settlement period and the settlement fluctuation period and settlement creep period. After the excavation, with a decrease in the temperature, the strength of the soil and lining increased. The settlement of the top arch, bottom arch, and waist arch increased slowly with time, and the growth rate decreased gradually. The higher the temperature of the tunnel structure, the greater the settlement and the more detrimental this was to the stability of the tunnel.

Author Contributions: L.Z.: Data curation, Resources; Q.W.: Project administration, Conceptualization; Y.J.: Methodology, Investigation; Z.L.: Writing—original draft, Validation; Y.W.: Supervision, Writing—review and editing. All authors have read and agreed to the published version of the manuscript.

Funding: This research received no external funding.

Data Availability Statement: Some or all of the data, models or codes that support the findings of this study are available from the corresponding author upon reasonable request.

Conflicts of Interest: The authors declare that they have no known competing financial interest or personal relationships that could have appeared to influence the work reported in this paper.

References

1. Lin, M.; Zhou, P.; Jiang, Y.; Zhou, F.; Lin, J.; Wang, Z. Numerical investigation on comprehensive control system of cooling and heat insulation for high geothermal tunnel: A case study on the highway tunnel with the highest temperature in China. *Int. J. Therm. Sci.* **2022**, *173*, 107385. [CrossRef]
2. Hu, Y.; Wang, M.; Wang, Q.; Liu, D.; Tong, J. Field test of thermal environment and thermal adaptation of workers in high geothermal tunnel. *Build. Environ.* **2019**, *160*, 106174. [CrossRef]
3. Zhang, G.; Cao, Z.; Wang, W.; Mei, X.; Zhao, X.; Shen, S.; Na, T. Field test and numerical investigation on thermal environment of tunnel with air layer structure. *Build. Environ.* **2021**, *203*, 108105. [CrossRef]
4. Chen, X.; Zhou, X.; Zhong, Z.; Liang, N.; Wang, Y.; Zhang, X. Study on temperature field and influencing factors of the high geothermal tunnel with extra-long one-end construction ventilation. *Int. J. Therm. Sci.* **2023**, *191*, 108322. [CrossRef]
5. Duan, L.; Zhang, Y.; Lai, J. Influence of Ground Temperature on Shotcrete-to-Rock Adhesion in Tunnels. *Adv. Mater. Sci. Eng.* **2019**, *2019*, 8709087. [CrossRef]
6. Luo, K.; Li, Y.; Zheng, C.; Gao, X.; Fan, J. Numerical simulation of temperature effect on particles behavior via electrostatic precipitators. *Appl. Therm. Eng.* **2015**, *88*, 127–139. [CrossRef]
7. Uejio, C.K.; Morano, L.H.; Jung, J.; Kintziger, K.; Jagger, M.; Chalmers, J.; Holmes, T. Occupational heat exposure among municipal workers. *Int. Arch. Occup. Environ. Health* **2018**, *91*, 705–715. [CrossRef] [PubMed]
8. Srinavin, K.; Mohamed, S. Thermal environment and construction workers' productivity: Some evidence from Thailand. *Build. Environ.* **2003**, *38*, 339–345. [CrossRef]
9. Zhao, Y.; Zhang, J.; Li, L.; Hu, T.; Zhao, Y. Analytical solution for the temperature field of an elliptical tunnel in cold regions. *Cold Reg. Sci. Technol.* **2023**, *208*, 103793. [CrossRef]
10. Zhou, P.; Feng, Y.; Zhou, F.; Wei, Z.; Gou, S.; Xu, H.; Wang, Z. Evaluation system of worker comfort for high geothermal tunnel during construction: A case study on the highway tunnel with the highest temperature in China. *Tunn. Undergr. Space Technol. Inc. Trenchless Technol. Res.* **2023**, *135*, 105028. [CrossRef]
11. John, H.; Kamelia, A. Fire-induced damage in tunnels: Thermo-mechanical modeling incorporating support system and geological conditions. *Tunn. Undergr. Space Technol. Inc. Trenchless Technol. Res.* **2023**, *135*, 105027.
12. Zhang, D.; Zhao, B.; Zhu, K. Mechanical characteristics analysis of horizontal lifting of subsea pipeline with different burial depths. *Front. Earth Sci.* **2022**, *10*, 1011291.
13. Wang, M.; Hu, Y.; Liu, D.; Jiang, C.; Wang, Q.; Wang, Y. A Study on the Heat Transfer of Surrounding Rock-Supporting Structures in High-Geothermal Tunnels. *Appl. Sci.* **2020**, *10*, 2307. [CrossRef]
14. Wei, Y.; Sun, K.; Zhong, X.; Jia, J.; Huang, Q.; Qin, J.; Xiong, Z. Study on effects of the train-induced airflow on the temperature field of high-speed railway tunnels in cold regions. *Therm. Sci. Eng. Prog.* **2023**, *41*, 101837. [CrossRef]
15. Kang, F.; Li, Y.; Tang, C. Numerical study on airflow temperature field in a high-temperature tunnel with insulation layer. *Appl. Therm. Eng.* **2020**, *179*, 115654. [CrossRef]
16. Zhou, X.; Zeng, Y.; Fan, L. Temperature field analysis of a cold-region railway tunnel considering mechanical and train-induced ventilation effects. *Appl. Therm. Eng.* **2016**, *100*, 114–124. [CrossRef]
17. Sun, B.; Xu, Z.D.; Zhou, H. A multiple back propagation neural network fusion algorithm for ceiling temperature prediction in tunnel fires. *Eng. Struct.* **2023**, *280*, 115601. [CrossRef]
18. Wan, H.; Xiao, Y.; Wei, S.; Zhang, Y.; Oka, Y. Experimental study on multiple fire hazards in a naturally ventilated tunnel: Assessment of the flame interaction and extension of two unequal fires. *Int. J. Therm. Sci.* **2023**, *187*, 108209. [CrossRef]
19. He, K.; Shi, L.; Zhang, S.; Cong, W.; Yang, H.; Cheng, X. Experimental study on temperature attenuation of smoke flow driven by dual fire sources in a tunnel. *Tunn. Undergr. Space Technol. Inc. Trenchless Technol. Res.* **2023**, *134*, 105004. [CrossRef]
20. Sun, S.; Yan, S.; Cao, X.; Zhang, W. Distribution Law of the Initial Temperature Field in a Railway Tunnel with High Rock Temperature: A Model Test and Numerical Analysis. *Appl. Sci.* **2023**, *13*, 1638. [CrossRef]
21. Wang, J.; Li, Z.; Li, G.; Xu, Y. Heat Hazard Control in High-Temperature Tunnels: Experimental Study of Coupled Cooling with Ventilation and Partial Insulation for Synergistic Geothermal Extraction. *Int. J. Environ. Res. Public Health* **2023**, *20*, 1941. [CrossRef] [PubMed]
22. Xu, D.; Zhang, B.; Zubin, A.I.; Bu, X.; Pan, H.; Chen, S. Spatial-temporal evolution principle of temperature field in a high-temperature geothermal highway tunnel. *Ain Shams Eng. J.* **2023**, *14*, 101965. [CrossRef]
23. Wang, T.; Zhang, C.; Xie, D. Study on creep deformation law of surrounding rock of slate tunnel considering temperature effect. *J. Hebei Univ. Sci. Technol.* **2022**, *43*, 90–98.

water

MDPI

Article

An Efficient Dynamic Coupling Calculation Method for Dam–Reservoir Systems Based on FEM-SBFEM

He Xu [1], Jianjun Xu [1,*], Dongming Yan [2], Kai Chen [3] and Degao Zou [3]

1 Power China Huadong Engineering Corporation Limited, Hangzhou 311122, China; bieshuxuhe@163.com
2 College of Civil Engineering and Architecture, Zhejiang University, Hangzhou 310058, China; dmyan@zju.edu.cn
3 School of Hydraulic Engineering, Dalian University of Technology, Dalian 116024, China; chenkai@dlut.edu.cn (K.C.); zoudegao@dlut.edu.cn (D.Z.)
* Correspondence: xu_jj@hdec.com; Tel.: +86-150-4243-1787

Abstract: In the dynamic analysis of dam–reservoir interactions, the computational efficiency of coupling system is relatively low. When numerical methods such as the scaled boundary finite element method (SBFEM) or the finite element method (FEM) are used to deal with hydrodynamic pressure, the additional mass matrix for the hydrodynamic pressure of incompressible reservoir water obtained is the full matrix. In this study, an efficient three dimensional (3D) dynamic fluid–solid coupling analysis method for dam–reservoir systems based on the FEM-SBFEM is proposed and applied to the dynamic calculation and analysis of an arch dam under seismic conditions, which adopts the SBFEM to solve the hydrodynamic pressure of the reservoir and employs the FEM to discretize the dam. In the proposed method, the hydrodynamic pressure additional mass matrix is simplified according to the physical meaning and distribution characteristics of the additional matrix with only a reduction coefficient α ($0 < \alpha \leq 1.0$), which is simple and easy to implement. The suggested value of the reduction coefficient α for the added mass matrix of the hydrodynamic pressure is selected to be 0.6 so as to ensure that the error of the maximum value of the dynamic response of the dam is limited within 5%, which is acceptable, and the elapsed time of calculation can be reduced to one twentieth of the accurate solution, which is a great jump in calculation efficiency. The proposed method provides a practical and effective process for the analysis of dam–reservoir dynamic interaction systems with a large computational scale and a fine grid scale.

Keywords: dam–reservoir interactions; computational efficiency; scaled boundary finite element method (SBFEM); hydrodynamic pressure; added mass matrix; simplification

Citation: Xu, H.; Xu, J.; Yan, D.; Chen, K.; Zou, D. An Efficient Dynamic Coupling Calculation Method for Dam–Reservoir Systems Based on FEM-SBFEM. *Water* **2023**, *15*, 3095. https://doi.org/10.3390/w15173095

Academic Editor: Ryan Bailey

Received: 28 April 2023
Revised: 14 August 2023
Accepted: 25 August 2023
Published: 29 August 2023

1. Introduction

The scaled boundary finite element method (SBFEM) [1] was recently developed as a semi-analytical numerical method to solve partial differential equations (PDEs). Since then, many researchers have made advancements in the application of the SBFEM or polyhedron SBFEM (PSBFEM) to model fluid–solid coupling problems [2–25]. Firstly, Deeks and Cheng [10] used SBFEM and potential flow theory to solve the problem of obstacle-induced fluid disturbance. They took the lead in extending the SBFEM to applications in computational fluid dynamics and opened up the possibility of the SBFEM-based study of dynamic structure–fluid interactions. Thereafter, Li et al. [11,12] used the SBFEM to analytically solve the two-dimensional Helmholtz equation and studied the dynamic coupling between waves of water and a floating structure. Furthermore, Teng et al. [13] and Cao et al. [14] used SBFEM to facilitate the solution of complex problems such as the dynamic coupling between ocean waves and coastal engineering structures as well as fluid sloshing in containers. Furthermore, Song and Tao [15–17] also researched ocean engineering problems, such as the solution of Poisson's equation and the interaction between waves and floating structures, and they improved the scaled boundary coordinate system to solve coupling problems involving circular geometry. Liu et al. [18] used the SBFEM to deal with short-crested wave interactions with a surface-piercing concentric cylindrical structure. Lin and Du [19] spearheaded research efforts in the computation of hydrodynamic pressure in front of dams based on the SBFEM. Until now, the SBFEM has made great progress and has been widely used in studying fluid–structure coupling analysis methods [2–9].

For the dynamic interactions between dams and reservoirs, the SBFEM possesses a unique advantage in that semi-infinite reservoir water can be simulated by discretizing only the interface between fluids and solids. Two main methods to model a semi-infinite reservoir water in front of the dam are as follows: (1) With the help of the SBFEM, the upstream surface of the dam body is directly discretized, and a prism-shaped, semi-infinite reservoir water is simulated in front of the dam [2,19–23]. The discretization of the fluid–solid interface not only reduces the dimension of the solution by one, but it also automatically meets the radiation condition at infinite distance in the fluid domain. (2) The finite element method (FEM) is used to discretize the near-field reservoir in front of the dam, and the SBFEM is used to simulate the prismatic semi-infinite reservoir to provide a truncated boundary condition for the reservoir tail, which can thusly exhibit characteristics of wave-free reflection [24,25]. Among the above two reservoir water simulation methods, the first method directly uses the SBFEM to calculate hydrodynamic pressure, while the second method uses the SBFEM to construct a truncated boundary at the end of the reservoir.

In the dynamic response analysis of dams under earthquake conditions, hydrodynamic pressure is one of the important factors that must be considered in order to reasonably evaluate the seismic safety of dams, so many researchers have done a lot of research work in this area. At present, the FEM [26–37], the boundary element method (BEM) [38–43], and the SBFEM [2,19–23] are the three commonly used methods to calculate the hydrodynamic pressure of reservoir water in front of a dam. In the BEM and SBFEM, the reservoir model is established based on the Eulerian approach [21,31], while both Eulerian and Lagrangian [31,33,35] approaches could be used to model the reservoir with the FEM. In the Eulerian approach, the hydrodynamic pressure of reservoir can be expressed as an additional mass matrix for incompressible reservoir water. For the Lagrangian approach, the Lagrangian fluid elements, such as acoustic elements in ABAQUS [44], are utilized to simulate the reservoir.

When the BEM is used to calculate the hydrodynamic pressure, the fundamental solution must be found first [19,21]. However, it is very difficult to obtain fundamental solutions for two-dimensional (2D) or three-dimensional (3D) reservoir water problems with complex geometric conditions that are often encountered in practical research work, which limits the wide application of the BEM in related fields. In contrast, the FEM and SBFEM are more widely applied and developed due to their good adaptability to complex geometric conditions for reservoir. However, for many dams that are built in valleys with complex 3D shapes, many degrees of freedom (DOFs) are required to model the complex shapes of reservoirs in order to accurately compute the hydrodynamic pressure when the FEM is used for elaborate reservoir modelling. By comparison, the SBFEM-based method [2,19–23] is very efficient and can precisely compute the hydrodynamic pressure induced by the vibration of the dam surface and the river valley with minimal DOFs for 3D reservoir water accurately simulated in 2D, which can significantly reduce the number of DOFs and improve computational efficiency.

In the analysis of dam–reservoir dynamic interactions, no matter what numerical analysis method is used for hydrodynamic pressure, there is an unavoidable problem, which is that the computational efficiency of the coupling system is relatively low. This is because when numerical methods such as the SBFEM, FEM, or BEM are used to deal with hydrodynamic pressure based on the Eulerian approach, the additional mass matrix for hydrodynamic pressure of incompressible reservoir water obtained is the full matrix [21]. When the number of free degrees of the coupling system reaches a certain scale, solving the coupled system equation will consume a lot of computing time, which may even reach an unacceptable degree. Especially for the numerical calculation of the coupling system with strong nonlinear (elastic–plastic) characteristics, such as the dynamic coupling system of a concrete-faced rockfill dam (CFRD) and reservoir water, the problem of time-consumption for calculation will become more prominent. The low efficiency of the coupling calculation affects the further development and application of dynamic interaction analysis between the dam and reservoir water, which is a key and difficult problem to be solved urgently.

In view of the key point faced by the further development and application of the dynamic interaction analysis method for dam–reservoir systems, as mentioned above, in this study, an efficient 3D dynamic fluid–solid coupling analysis method for dam–reservoir systems based on FEM-SBFEM is proposed and applied to the dynamic calculation and analysis of an arch dam under seismic conditions. By comparing the difference of the dynamic response of the arch dam under earthquake and calculation cost time, the calculation accuracy and efficiency of this efficient method are verified. The proposed efficient method is implemented in the program GEODYNA [45], which is also used for numerical coupling analysis.

Based on the previous research results, the coupling calculation method proposed in this paper adopts the SBFEM to solve the hydrodynamic pressure of reservoirs in front of dams, which makes it efficient to calculate and solve the additional mass matrix of hydrodynamic pressure, and it employs the FEM to discretize the dam. According to the physical meaning and distribution characteristics of the hydrodynamic pressure added mass matrix, the method realizes the efficient calculation and analysis of the dam–reservoir dynamic coupling systems by dealing with the added mass matrix. The proposed method provides a practical and effective way for the analysis of dam–reservoir dynamic interaction systems with a large computational scale and fine grid scale. In addition, this method is convenient for implementation and application on the software platform and has good prospects for promoting the development and application of a dam–reservoir dynamic coupling analysis method.

2. A Calculating Method for Hydrodynamic Pressure of Reservoir Based on SBFEM

The basic concept, equation, and derivation process of the solution for hydrodynamic pressure based on SBFEM are introduced below, which can also be found in the literature [2,19,21].

2.1. The Basic Equation and Boundary Conditions

In this study, the reservoir water is assumed to be an ideal fluid that is non-viscous, incompressible, and slightly disturbed. Under seismic loading, the dynamic water pressure in the reservoir will satisfy the Laplace equation:

$$\nabla^2 p = 0 \tag{1}$$

Neglecting the micro-amplitude gravity wave, the free surface S_0 boundary condition of the reservoir water is

$$p = 0 \tag{2}$$

Additionally, the boundary condition on the upstream face S_1 of the dam is as follows:

$$\partial p / \partial n = -\rho \ddot{u}_n \tag{3}$$

The boundary condition on the interface S_2 between the reservoir and bottom and bank slope is as follows:

$$\partial p / \partial n = -\rho \ddot{v}_n \tag{4}$$

In the above equations, ∇^2 is the Laplace operator, p is the reservoir hydrodynamic pressure, n is the normal direction of the interface between the solid and fluid, ρ is the water density, and $\ddot{u}_n$ and $\ddot{v}_n$ are normal acceleration values of the dam–reservoir interface and the river-valley interface, respectively.

When the SBFEM method is used to simulate the semi-infinite reservoir water in front of the dam, the analytical solution is obtained along the direction of the semi-infinite domain, so the non-reflection condition at the reservoir tail S_3 is automatically satisfied, which is illuminated below.

2.2. Solution for Hydrodynamic Pressure Based on SBFEM

Figure 1 shows a typical semi-infinite scale boundary finite element of the reservoir water in the front of the dam. According to the basic SBFEM theory, once the similarity center O is selected at infinity downstream of the dam, the prismatic fluid element of semi-infinite reservoir water is generated radially, with the similarity center acting as the origin by directly utilizing the 2D mesh on the dam upstream face. Furthermore, the solution can be analytically obtained along the radial direction without discretization. As shown in Figure 2, the 3D reservoir model, which is composed of a series of semi-infinite prismatic elements, can be derived from the 2D surface grid of the dam surface, which means there is no need to also divide the reservoir water grid, and the number of DOFs is limited.

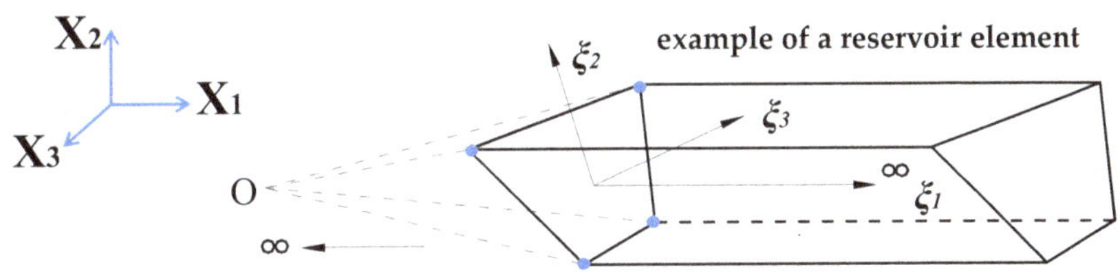

Figure 1. Typical scale boundary finite element of fluid.

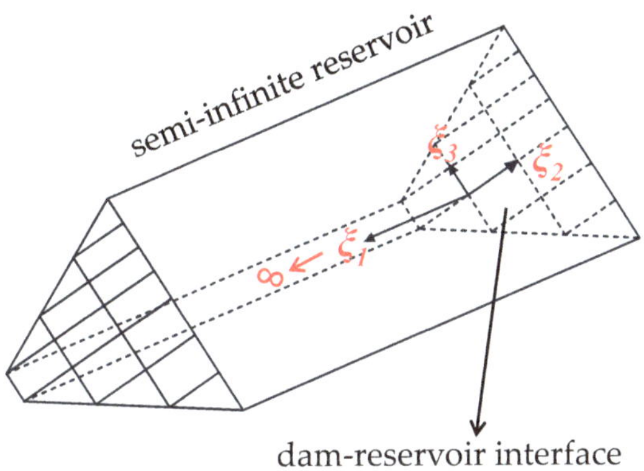

Figure 2. Reservoir model discretized by SBFEM.

The basic equation and boundary conditions (Equations (1)–(4)) of the hydrodynamic pressure of reservoir in front of the dam are solved using the weighted residual method. Through the weight function w, the following weak integral equation (Equation (5)) can be obtained:

$$\int_V \nabla w \nabla p dV + \rho \int_{S_1} w\ddot{u}_n dS + \rho \int_{S_2} w\ddot{v}_n dS = 0 \quad (5)$$

In order to derive the governing equations and boundary conditions of reservoir hydrodynamic pressure using SBFEM, it is necessary to transform the coordinates of the reservoir from the global Cartesian coordinate system to the local coordinate system of the scaled boundary. In the scaled boundary coordinate system, the radial local coordinate ξ_1 has the range [0, +∞]. As shown in Figure 2, $\xi_1 = 0$ is at the upstream face of the dam, and $\xi_1 = +\infty$ is at the infinity of the reservoir. The circumferential local coordinates ξ_2 and ξ_3 have the range [−1, 1]. The coordinates (X_1, X_2, X_3) of the global Cartesian coordinate system at any point in the reservoir area can be expressed in terms of the local coordinates

(ξ_1, ξ_2, ξ_3) of the scaled boundary. The radial local coordinate ξ_1 serves as a factor of proportionality, giving rise to the following set of equations:

$$\begin{aligned} X_1(\xi_1,\xi_2,\xi_3) &= x_1(\xi_2,\xi_3) + \xi_1 = [N(\xi_2,\xi_3)]\{x_1\} + \xi_1 \\ X_2(\xi_1,\xi_2,\xi_3) &= x_2(\xi_2,\xi_3) = [N(\xi_2,\xi_3)]\{x_2\} \\ X_3(\xi_1,\xi_2,\xi_3) &= x_3(\xi_2,\xi_3) = [N(\xi_2,\xi_3)]\{x_3\} \end{aligned} \tag{6}$$

where (x_1, x_2, x_3) are the nodal coordinates of the elements on the interface between the reservoir and dam. $[N(\xi_2,\xi_3)]$ represents the shape function of the elements, which is only related to the circumferential local coordinates but unrelated to the radial coordinate ξ_1.

According to the transformation formula, the differential operator ∇ can be expressed in terms of the scaled boundary coordinates through the Jacobian matrix $[J]$ as follows:

$$[J] = \begin{bmatrix} X_{1,\xi_1} & X_{2,\xi_1} & X_{3,\xi_1} \\ X_{1,\xi_2} & X_{2,\xi_2} & X_{3,\xi_2} \\ X_{1,\xi_3} & X_{2,\xi_3} & X_{3,\xi_3} \end{bmatrix} = \begin{bmatrix} 1 & 0 & 0 \\ [N(\xi_2,\xi_3)_{,\xi_2}]\{x_1\} & [N(\xi_2,\xi_3)_{,\xi_2}]\{x_2\} & [N(\xi_2,\xi_3)_{,\xi_2}]\{x_3\} \\ [N(\xi_2,\xi_3)_{,\xi_3}]\{x_1\} & [N(\xi_2,\xi_3)_{,\xi_3}]\{x_2\} & [N(\xi_2,\xi_3)_{,\xi_3}]\{x_3\} \end{bmatrix} \tag{7}$$

$$\left\{ \frac{\partial}{\partial X_1} \quad \frac{\partial}{\partial X_2} \quad \frac{\partial}{\partial X_3} \right\}^T = [J]^{-1}\left\{ \frac{\partial}{\partial \xi_1} \quad \frac{\partial}{\partial \xi_2} \quad \frac{\partial}{\partial \xi_3} \right\}^T = \left\{b^1\right\}\frac{\partial}{\partial \xi_1} + \left\{b^2\right\}\frac{\partial}{\partial \xi_2} + \left\{b^3\right\}\frac{\partial}{\partial \xi_3} \tag{8}$$

where $[J]^{-1} = [\{b^1\}\ \{b^2\}\ \{b^3\}]$.

By using the shape function $[N(\xi_2,\xi_3)]$ for coordinate transformation, the hydrodynamic pressure at any point in an element can be expressed as

$$p(\xi_1,\xi_2,\xi_3) = [N(\xi_2,\xi_3)]\{p(\xi_1)\} \tag{9}$$

where $\{p(\xi_1)\}$ is the hydrodynamic pressure at nodes of the fluid element.

Equations (6)–(9) are introduced into the SBFEM hydrodynamic pressure integral Equation (5), and the partial integration is carried out. Finally, the governing equation (Equation (10)) and boundary conditions (Equation (11)) of hydrodynamic pressure can be obtained in the frequency domain as follows:

$$\left[E^0\right]\{p(\xi_1)\}_{,\xi_1\xi_1} + \left(\left[E^1\right]^T - \left[E^1\right]\right)\{p(\xi_1)\}_{,\xi_1} - \left[E^2\right]\{p(\xi_1)\} - \rho\left[C^0\right]\{\ddot{v}_n\} = 0 \tag{10}$$

$$\left(\left[E^0\right]\{p(\xi_1)\}_{,\xi_1} + \left[E^1\right]^T p(\xi_1) + [M^1]\,\{\ddot{u}_n\}\right|_{\xi_1=0} = 0 \tag{11}$$

in which

$$\left[B^1\right] = \left\{b^1\right\}[N], \qquad \left[B^2\right] = \left\{b^2\right\}[N]_{,\xi_2} + \left\{b^3\right\}[N]_{,\xi_3} \tag{12}$$

$$\left[M^1\right] = \rho\int_{-1}^{1}\int_{-1}^{1}[N]^T[N]A\,d\xi_2 d\xi_3 \tag{13}$$

$$\left[E^0\right] = \int_{-1}^{1}\int_{-1}^{1}[B^1]^T[B^1]|J|d\xi_2 d\xi_3 \tag{14}$$

$$\left[E^1\right] = \int_{-1}^{1}\int_{-1}^{1}[B^2]^T[B^1]|J|d\xi_2 d\xi_3 \tag{15}$$

$$\left[E^2\right] = \int_{-1}^{1}\int_{-1}^{1}[B^2]^T[B^2]|J|d\xi_2 d\xi_3 \tag{16}$$

$$A = \sqrt{(x_{2,\xi_2}x_{3,\xi_3} - x_{3,\xi_2}x_{2,\xi_3})^2 + (x_{3,\xi_2}x_{1,\xi_3} - x_{1,\xi_2}x_{3,\xi_3})^2 + (x_{1,\xi_2}x_{2,\xi_3} - x_{2,\xi_2}x_{1,\xi_3})^2} \tag{17}$$

$$\left[C^0\right] = \int_\Gamma [N]^T[N]d\Gamma \tag{18}$$

$$d\Gamma = \sqrt{x_{2,\xi_2}{}^2 + x_{3,\xi_2}{}^2} d\xi_2 \Big|_{\xi_3=-1} \tag{19}$$

where $[N]$ denotes the interpolation shape function $[N(\xi_2, \xi_3)]$, and Γ in Equation (18) denotes the projection of the wet contour line of the reservoir at the interface between the dam and reservoir on the (X_2, X_3) plane.

It can be seen from Equations (12)–(19) that the coefficient matrices $[E^0]$, $[E^1]$, $[E^2]$, $[C^0]$, and $[M^1]$ are independent of the radial coordinate ξ_1 and can be straightforwardly obtained from geometry information of the grid on the dam upstream face. Then the coefficient matrices of elements can be integrated into the total SBFEM coefficient matrices, the process of which is similar to the FEM.

In order to analytically solve the governing equation (Equation (10)), it is necessary to introduce the nodal force matrix $\{q(\xi_1)\}$ of the hydrodynamic pressure.

$$\{q(\xi_1)\} = \left[E^0\right]\{p(\xi_1)\}_{,\xi_1} + [E^1]^T\{p(\xi_1)\} \tag{20}$$

Taking advantage of the new variables and expressions defined in Equation (21), the governing equation (Equation (10)) can be rewritten into a first-order ordinary differential equation (Equation (22)).

$$\{X(\xi_1)\} = \begin{Bmatrix} \{p(\xi_1)\} \\ \{q(\xi_1)\} \end{Bmatrix}, \quad \{F_0\} = \begin{Bmatrix} 0 \\ -\rho\left[C^0\right]\{\ddot{v}_n\} \end{Bmatrix} \tag{21}$$

$$\{X(\xi_1)\}_{,\xi_1} = [Z]\{X(\xi_1)\} + \{F_0\} \tag{22}$$

$$[Z] = \begin{bmatrix} -[E^0]^{-1}[E^1]^T & [E^0]^{-1} \\ [E^2] - [E^1]\,[E^0]^{-1}[E^1]^T & [E^1][E^0]^{-1} \end{bmatrix} \tag{23}$$

in which the coefficient matrix [Z] is the Hamilton matrix.

The eigenvalue problem, as shown in Equation (24), corresponding to the Hamilton matrix [Z] should be solved first.

$$[Z][\Phi] = [\Phi][\Lambda] \tag{24}$$

$$[\Lambda] = \begin{bmatrix} [\lambda_i] & 0 \\ 0 & [-\lambda_i] \end{bmatrix}, \ [\Phi] = \begin{bmatrix} [\Phi_{11}] & [\Phi_{12}] \\ [\Phi_{21}] & [\Phi_{22}] \end{bmatrix} \tag{25}$$

in which $[\Lambda]$ denotes the eigenvalue matrix, $[\Phi]$ denotes the eigenvector matrix, $[\lambda_i]$ is the diagonal matrix, and the real part of λ_i is positive.

The inverse matrix of the matrix $[\Phi]$, which is denoted with $[A]$, is solved and partitioned secondly.

$$[A] = [\Phi]^{-1}, \quad [A] = \begin{bmatrix} [A_{11}] & [A_{12}] \\ [A_{21}] & [A_{22}] \end{bmatrix} \tag{26}$$

In the end, by bringing in the boundary condition (Equation (11)) and carrying out a series of manipulations, the hydrodynamic pressure of the reservoir on the dam surface because of a seismic load can be expressed as

$$\{p(\xi_1 = 0)\} = -[\Phi_{12}]\left[\Phi_{22}\right]^{-1}[M^1]\{\ddot{u}_n\} - \left([\Phi_{12}]\left[\Phi_{22}\right]^{-1}[B_1] - [B_2]\right)\rho\left[C^0\right]\{\ddot{v}_n\} \tag{27}$$

in which

$$[B_1] = [\Phi_{21}]\left[\lambda_i^{-1}\right][A_{12}] + [\Phi_{22}]\left[-\lambda_i^{-1}\right][A_{22}] \tag{28}$$

$$[B_2] = [\Phi_{11}]\left[\lambda_i^{-1}\right][A_{12}] + [\Phi_{12}]\left[-\lambda_i^{-1}\right][A_{22}] \tag{29}$$

From Equation (27) above, the hydrodynamic pressure consists of two components: the hydrodynamic pressure caused by the vibration of the dam upstream face $\{\ddot{u}_n\}$, and that induced by the vibration of the river valley $\{\ddot{v}_n\}$ surrounding the reservoir area.

3. An Efficient Dynamic Coupling Calculation Method for Dam–Reservoir Systems

In Section 3.1, the conventional dynamic coupling method for dam and reservoir systems is described first, in which the hydrodynamic pressure is considered by means of the additional mass matrix [2,21–23]. In Section 3.2, the hydrodynamic pressure additional mass matrix, which is computed from Equation (27) as shown in Section 3.1, is further processed (in Section 3.2.2) according to the characteristics of the additional mass matrix as analyzed in Section 3.2.1. In Section 3.3, the efficient dynamic coupling calculation method based on the FEM-SBFEM is proposed with the hydrodynamic pressure additional mass matrix after treatment by way of Section 3.2.

3.1. *Conventional Dynamic Coupling Analysis Method for Dam–Reservoir Systems*

The FEM is employed for dam modeling, and the SBFEM is adopted for modeling the reservoir before the dam. The analysis equation for dynamic coupling of the dam and reservoir system is defined as follows:

$$[M_s]\{\ddot{u}_r(t)\} + [C_s]\{\dot{u}_r(t)\} + [K_s]\{u_r(t)\} = -[M_s]\{\ddot{u}_g(t)\} - (1/\rho)\left[L_1\right]^T[M^1]^T\{p(\xi_1 = 0)\} \tag{30}$$

in which $[M_s]$, $[C_s]$, and $[K_s]$ are mass, damping, and stiffness matrices of the dam, respectively. $\{\ddot{u}_r(t)\}$, $\{\dot{u}_r(t)\}$, and $\{u_r(t)\}$ are the relative acceleration, velocity, and displacement, respectively. $\{\ddot{u}_g(t)\}$ is the earthquake acceleration from the input system. $[L_1]$ is the conversion matrix for mapping global coordinates to the local coordinates of the dam upstream face.

After substituting Equation (27) into Equation (30), the dynamic coupling calculation equations are obtained as

$$([M_s] + [M_p])\{\ddot{u}_r(t)\} + [C_s]\{\dot{u}_r(t)\} + [K_s]\{u_r(t)\} = -([M_s] + [M_p])\{\ddot{u}_g(t)\} \tag{31}$$

$$[M_p] = (1/\rho)[L_1]^T([M^u][L_1] + [M^v][L_2]) \tag{32}$$

$$\left[M^u\right] = -[M^1]^T[\Phi_{12}][\Phi_{22}]^{-1}\left[M^1\right] \tag{33}$$

$$[M^v] = -[M^1]^T([\Phi_{12}]\left[\Phi_{22}\right]^{-1}[B_1] - [B_2])\rho\left[C^0\right] \tag{34}$$

$$[L_1]\{\ddot{u}_g(t) + \ddot{u}_r(t)\} = \{\ddot{u}_n\}, \qquad [L_2]\{\ddot{u}_g(t) + \ddot{u}_r(t)\} = \{\ddot{v}_n\} \tag{35}$$

where $[M^u]$ and $[M^v]$ are the dam upstream face and river valley components, respectively, of the hydrodynamic pressure additional mass matrix $[M_p]$. $[L_2]$ is the conversion matrix for mapping global coordinates to the local coordinates of the bank slope (river valley). As long as the additional mass matrix $[M_p]$ is added to the mass matrix of the finite element dynamic equation of the dam, the hydrodynamic pressure due to the ground motion input in different directions can be considered.

3.2. *Simplification of Hydrodynamic Pressure Additional Mass Matrix*

3.2.1. Physical Meaning and Distribution Characteristics of Matrix

The calculated additional mass matrix $[M^u]$ is a full matrix, and all elements are non-zero, that is, when calculating the hydrodynamic pressure caused by the excitation of the dam upstream face, the hydrodynamic pressure acting on a certain node is related to the acceleration excitations $\{\ddot{u}_n\}$ of all nodes on the dam upstream face below the water level. However, when calculating the hydrodynamic pressure caused by river valley excitation, the hydrodynamic pressure acting on a certain node on the dam upstream face is only related to the node acceleration excitations $\{\ddot{v}_n\}$ at the boundary between the dam surface and the river valley. Therefore, the additional mass matrix $[M^v]$ is a very sparse matrix containing a large number of zero elements. Obviously, due to the existence of the additional mass matrix $[M^u]$, the additional mass matrix $[M_p]$ is a full matrix, which greatly increases the time consumption of solving the equivalent stiffness matrix in dynamic analysis.

From the above analysis, it can be seen that the node degree of freedom correlation of the additional mass matrix $[M_p]$ can be reduced by the simplifying matrix $[M^u]$. When the additional mass matrix $[M^u]$ is of order n (i.e., there are n nodes below the water level line on the dam upstream face in total); the physical meaning of the element M^u_{ij} is the hydrodynamic pressure acting on the node i caused by the unit normal acceleration excitation of the node j on the water upstream face. The i-th row elements in matrix $[M^u]$, that is M^u_{i1} $M^u_{i2}\ldots$ $M^u_{ii}\ldots$ M^u_{in}, are selected, and then these row elements give the contribution of all nodes on the upstream face of the dam (below the water level) to the hydrodynamic pressure acting on node i.

The distribution characteristics of the hydrodynamic pressure added mass matrix $[M^u]$ are analyzed using the example of a vertical dam upstream face in a rectangular valley. The height of the dam is 200 m, and the width of the river valley is 400 m. The grid division of the reservoir is shown in Figure 3, and the water depth in front of the dam is 200 m (full reservoir).

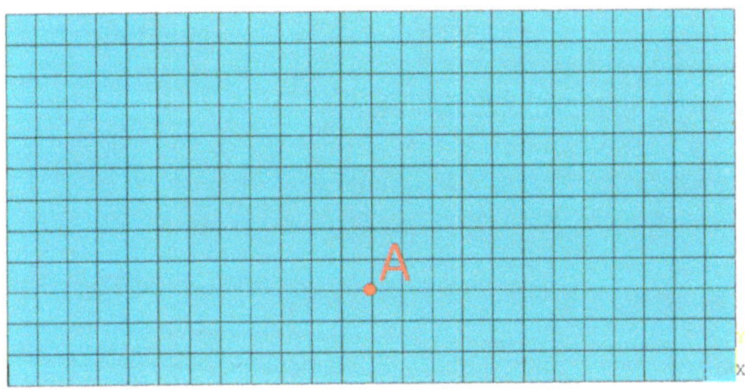

Figure 3. Mesh of dam upstream face (*A* represents a node of the mesh).

After solving the added mass matrix of hydrodynamic pressure based on the SBFEM, the row elements corresponding to node *A* (see Figure 3) in the added mass matrix $[M^u]$ are extracted, and the influence of all nodes at the dam upstream face on the hydrodynamic pressure of node *A* under unit normal acceleration excitation is shown in Figure 4 (normalized relative value). As shown in Figure 4, node *A* has the greatest impact on itself, and the closer the node is to node *A*, the greater the impact is on the hydrodynamic pressure of node *A*. Other nodes on the upstream face of the dam also conform to similar laws, so it is no longer described repetitively.

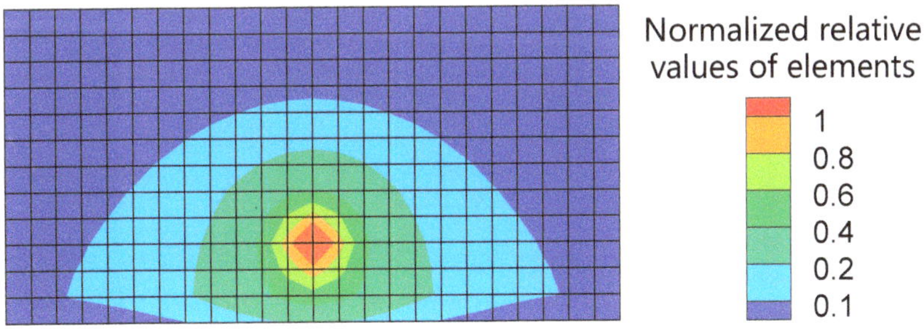

Figure 4. Effect of all nodes on node *A*.

3.2.2. Theoretical Analysis and Simplified Processing Method

For the additional mass matrix $[M^u]$, among all the elements in the i-th row, the element values corresponding to node i (M^u_{ii}) and its adjacent nodes are relatively large, and the acceleration values of these closer nodes in the dynamic coupling analysis are not much different. Therefore, when calculating the hydrodynamic pressure acting on node i, the added mass element of the i-th row can be directly superimposed on M^u_{ii}, and the corresponding elements at the original position of the row are directly taken as 0, which means that the hydrodynamic pressure acting on node i is only related to the acceleration of node i, and it has nothing to do with the acceleration of other nodes on the dam upstream face. At the same time, the instantaneous acceleration distribution of the dam upstream face is not consistent, which contains both positive and negative values. Therefore, the above treatment method for the additional mass matrix will cause the amplification of hydrodynamic pressure. Considering the amplification effect of the superposition of additional mass elements on the hydrodynamic pressure, the appropriate reduction treatment is needed in the process of element superposition.

According to the physical meaning of each element of the additional mass matrix $[M^u]$, a simple and easy simplification method is proposed by row processing, which is briefly described as follows:

(1) Extract the elements of the i-th row in the hydrodynamic pressure added mass matrix $[M^u]$. (2) Let the diagonal elements of the matrix $M^u_{ii} = \boldsymbol{\alpha} \sum_{j=1}^{n} M^u_{ij}$, where $\boldsymbol{\alpha}$ is the reduction coefficient ($0 < \boldsymbol{\alpha} \leq 1.0$); at the same time, set the value of other non-diagonal elements to zero. (3) According to this method, the additional mass matrix $[M^u]$ is processed from the first row until the last row.

After simplification, the additional mass matrix $[M^u]$ is transformed into a new diagonal matrix $[M^{u\alpha}]$. The additional mass matrix $[M^v]$ does not need to be simplified. Further, the overall additional mass matrix $[M_p]$ is updated to $[M_{p\alpha}]$ as follows:

$$[M_{p\alpha}] = (1/\rho)[L_1]^T([M^{u\alpha}][L_1] + [M^v][L_2]) \quad (36)$$

3.3. *Efficient Dynamic Coupling Calculation Method Based on FEM-SBFEM*

By replacing $[M_p]$ with $[M_{p\alpha}]$ in Equation (31), the efficient dynamic coupling calculation method for dam–reservoir systems based on the FEM-SBFEM is established:

$$([M_s] + [M_{p\alpha}])\{\ddot{u}_r(t)\} + [C_s]\{\dot{u}_r(t)\} + [K_s]\{u_r(t)\} = -([M_s] + [M_{p\alpha}])\{\ddot{u}_g(t)\} \quad (37)$$

The simplification method only needs to provide a reduction coefficient $\boldsymbol{\alpha}$ to realize the simplification of the additional mass matrix $[M_p]$ to a large extent, which is simple and easy to implement. The processed additional mass matrix $[M_{p\alpha}]$ contains many zero elements, so the computational efficiency of dynamic coupling systems is greatly improved. The practice shows that the value of reduction coefficient $\boldsymbol{\alpha}$ has an obvious influence on the calculation accuracy, which is presented in Section 4.

The efficient dynamic coupling calculation method based on the FEM-SBFEM is implemented on the strength of the Windows program GEODYNA [45], which was developed using object-oriented programming in Visual C++. Multicore parallel technology of the CPU is carried out in the GEODYNA software platform, by which the computational capacity of solving a large-scale nonlinear equation with millions of DOFs is enabled. The GEODYNA program has been widely used in the dynamic analysis of linear and nonlinear structures and fluid–solid coupling systems [2,21–23,46–51].

4. Dynamic Coupling Analysis of Arch Dam and Reservoir Systems

Based on the dynamic fluid–solid coupling analysis of a concrete arch dam and reservoir water under earthquake conditions, the sensitivity analysis for the value of reduction coefficient $\boldsymbol{\alpha}$ is carried out. Different reduction factors $\boldsymbol{\alpha}$ of additional mass matrix $[M_{p\alpha}]$ are selected to calculate and analyze the dynamic response of the arch dam. By comparing with the exact solution, which means the unsimplified full additional mass matrix $[M_p]$ is utilized in dynamic analysis, the recommended value of reduction factor α is discussed.

4.1. Calculation Model

4.1.1. Dam and Reservoir Model

The Morrow Point arch dam [19,20,36] was selected to establish the finite element model, as shown in Figure 5. The concrete arch dam is 141.73 m high and simulated by 3D eight-node or degenerate isoparametric elements. The dam FEM model contains 15,000 elements, of which 2500 elements are on the upstream face of the arch dam. Assuming that the arch dam is located on a rigid bedrock, the bottom of the arch dam and the boundary in contact with the river valley are both constrained in X, Y, and Z directions. The construction joints of the arch dam are not taken into account in the FEM model.

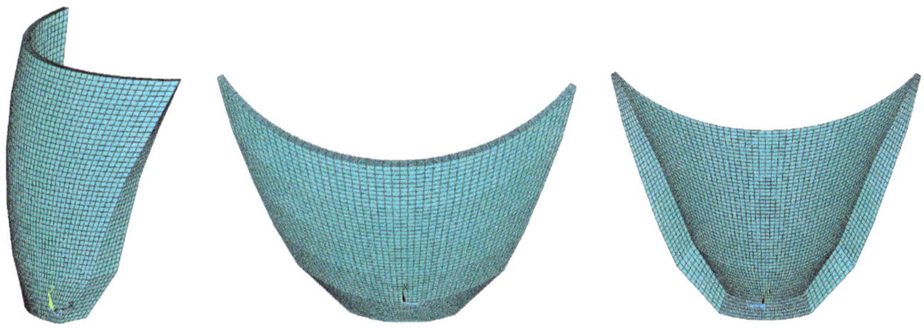

Figure 5. FEM model of Morrow Point arch dam.

The water depth of the reservoir at the upstream side of the dam is 133.23 m. When the SBFEM is used to simulate the reservoir water in front of the dam in the dynamic coupling analysis, the 2D finite element mesh of the dam upstream face is immediately used to generate the prismatic semi-infinite reservoir water, as indicated in Figure 2. The grid of the dam already contains the reservoir water grid information, so there is no need to divide the reservoir water grid separately, which improves the pre-processing efficiency. There are 2350 grids below the water level on the upstream face of dam, which means that there are 2350 scaled boundary finite elements for the reservoir water.

4.1.2. Material Parameters, Input Seismic Load, and Damping Methods

The constitutive model employed for the concrete arch dam is a linear elastic model (density ρ_d = 2.4 g/cm^3, elasticity modulus E = 25 GPa, Poisson's ratio ν = 0.167), which can fully show the advanced nature of the proposed efficient dynamic coupling calculation method. The density ρ_w of the reservoir water is 1.0 g/cm^3.

A set of earthquake waves, which were generated from the site spectrum of real engineering, were selected for dynamic analysis. As shown in Figure 6, the acceleration time history of three seismic waves was input from the bedrock with a consistent input method in up-downstream (X), vertical (Y), and dam axial (Z) directions. The peak ground acceleration (PGA) of horizontal bedrock motion, which is in the up-downstream and dam axial direction, was scaled to 1.5 m/s^2, and the PGA of vertical bedrock motion was 1.0 m/s^2.

The Rayleigh viscous damping method was adopted for the concrete arch dam in dynamic analysis [49]. A damping ratio of 5% was assumed for the concrete material.

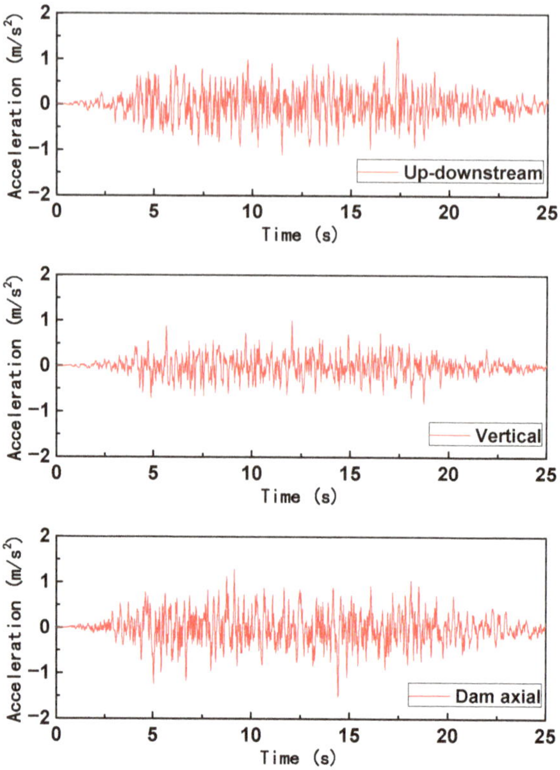

Figure 6. Input bedrock motion.

4.2. Effect of Additional Mass Matrix Simplification

In the finite element model of the arch dam, there are 2397 nodes below the water level line on the dam upstream face, that is, there are 2397 node degrees of freedom in the semi-infinite reservoir water model. As a consequence, the unsimplified additional mass matrix $[M^u]$ is a full matrix of order 2397, which contains 2397 × 2397 = 5,745,609 elements of the matrix and could provide an accurate dynamic coupling calculation result of dam and reservoir systems.

In the following numerical calculation and analysis, a total of 8 working conditions of hydrodynamic pressure additional mass matrix reduction coefficient α ($0 < \alpha \leq 1.0$) was selected for forming $[M^{u\alpha}]$, in which α = 1.0, 0.9, 0.8, 0.7, 0.6, 0.5, 0.4, and 0.3. α = 1.0 indicates that the additional mass matrix is simplified to a diagonal matrix without reduction, and $\alpha \approx 0$ indicates that the added mass matrix $[M^{u\alpha}]$ is close to the zero matrix, that is, the hydrodynamic pressure is almost not considered. When $0 < \alpha \leq 1.0$ in this numerical case, the simplified added mass matrix $[M^{u\alpha}]$ is a diagonal matrix, which contains only 2397 elements and can present an approximate analysis result of the dynamic interaction of the dam and reservoir. The simplification of the additional mass matrix can not only save a lot of memory occupied by the additional mass matrix of hydrodynamic pressure but also greatly reduce the computational time. However, only a reasonable value of the reduction coefficient α can ensure high calculation accuracy, as discussed below.

4.3. Results and Discussion

Under different reduction coefficient (α = 1.0, 0.9, 0.8, 0.7, 0.6, 0.5, 0.4, and 0.3) conditions, the effects of the hydrodynamic pressure added mass matrix $[M^{u\alpha}]$ ($[M_{p\alpha}]$) on the dynamic stress and acceleration extreme (maximum) value of the arch dam and their regularities of distribution and calculation time are studied. When calculating the error of dynamic stress and acceleration of the arch dam caused by the simplification of the additional mass matrix, the accurate calculation results of the unsimplified additional mass matrix $[M^u]$ ($[M_p]$) are taken as the benchmark. The results of the unsimplified additional mass matrix were verified to be of high accuracy [2,21]. By comparing the calculation accuracy and calculation time, the recommended value of the reduction coefficient α is given.

4.3.1. Acceleration of Arch Dam

Table 1 summarizes the maximum absolute values of the arch dam acceleration in up-downstream (a_x) and vertical (a_y) directions under different reduction coefficients α for simplification of additional mass matrix and unsimplified additional mass matrix conditions when the earthquake occurs. Table 2 shows the errors corresponding to the maximum dynamic acceleration caused by simplification of the added mass matrix of hydrodynamic pressure. It can be concluded from Tables 1 and 2 that the additional mass matrix reduction coefficient α has a big impact on the acceleration in the up-downstream direction (a_x) but has relatively little impact on the vertical acceleration (a_y). Compared with the accurate results in the condition of the unsimplified additional mass matrix, the maximum errors of acceleration, when the reduction coefficient α = 0.6 and 0.7, are 0.4% and 4.6%, respectively, which are acceptable from the perspective of calculation accuracy since the max errors are limited to less than 5.0%. When the reduction coefficient $\alpha < 0.6$ and $\alpha > 0.7$, the max errors of acceleration become bigger, which are from 9.0% to 22.7%.

Table 1. Max dynamic acceleration of arch dam.

Added Mass Matrix	Unsimplified (Accurate)	α = 1.0	α = 0.9	α = 0.8	α = 0.7	α = 0.6	α = 0.5	α = 0.4	α = 0.3
a_x (m/s^2)	6.583	5.518	5.596	5.989	6.280	6.554	7.209	7.926	8.075
a_y (m/s^2)	2.285	2.287	2.339	2.364	2.327	2.282	2.339	2.376	2.419

Table 2. Error of max dynamic acceleration of arch dam.

Added Mass Matrix	α = 1.0	α = 0.9	α = 0.8	α = 0.7	α = 0.6	α = 0.5	α = 0.4	α = 0.3
Error of a_x	16.2%	15.0%	9.0%	4.6%	0.4%	9.5%	20.4%	22.7%
Error of a_y	0.1%	2.4%	3.5%	1.8%	0.1%	2.3%	4.0%	5.9%

Figure 7 plots the distribution of the maximum absolute acceleration along the up-downstream direction (a_x) on the upstream face of the arch dam. As shown in Figure 7, the distribution laws of the maximum acceleration along the up-downstream direction (a_x) for each condition are similar, but the extent and area of the high acceleration response region are significantly different and gradually change with the decrease of the reduction coefficient α. In addition, the maximum acceleration distribution is most consistent with the accurate analysis solution (from unsimplified additional mass matrix) when the reduction coefficient α = 0.6, which also corresponds to the minimum error of acceleration in both vertical (a_y) and up-downstream (a_x) directions, as summarized in Table 2.

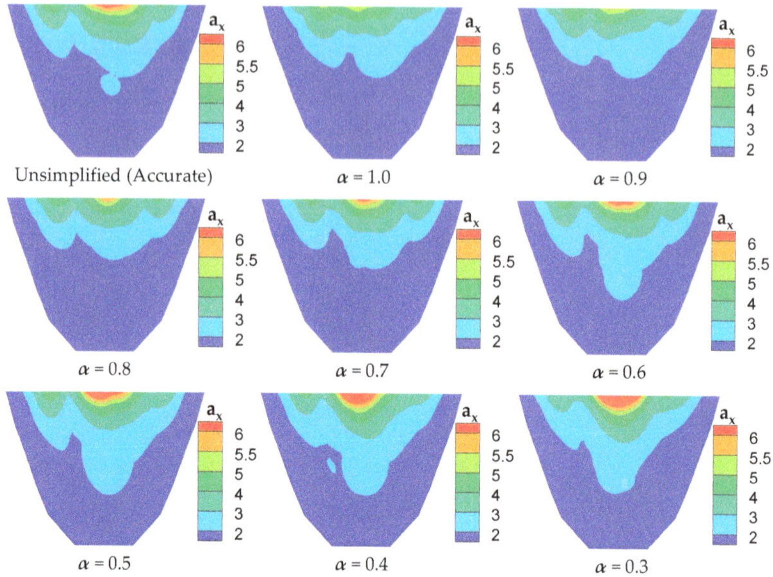

Figure 7. Distribution of maximum acceleration along the up-downstream direction (a_x) on upstream face of arch dam (m/s^2).

4.3.2. Stress of Arch Dam

Under different hydrodynamic pressure added mass matrix conditions, the maximum dynamic stresses of the arch dam concrete, including major principal stress (σ_1) and minor principal stress (σ_3), that occurred during the earthquake are collected in Table 3. The compressive stress of the dam concrete is set to be positive. Table 4 lists the errors corresponding to the maximum dynamic stress when the added mass matrix of hydrodynamic pressure is simplified with different reduction coefficients α. Tables 3 and 4 demonstrate that the reduction coefficient α has an obvious influence on both the major principal stress and minor principal stress of arch dam. As shown in Table 4, the maximum errors of dynamic stress are less than 5.0% when the reduction coefficient α = 0.4, 0.6, 0.7, and 0.8, which means the calculation accuracy of dynamic stress is effectively controlled. Furthermore, the maximum errors of dynamic stress are 2.2% and 1.3%, respectively, for reduction coefficient α = 0.6 and 0.8, the errors of which are relatively small in all reduction coefficients conditions. The maximum errors of dynamic stress vary from 5.0 to 11.2%, when the reduction coefficient α = 0.3, 0.5, 0.9, and 1.0, which are relatively big.

Table 3. Max dynamic stress of arch dam.

Added Mass Matrix	Unsimplified (Accurate)	α = 1.0	α = 0.9	α = 0.8	α = 0.7	α = 0.6	α = 0.5	α = 0.4	α = 0.3
σ_1 (MPa)	1.567	1.643	1.520	1.588	1.615	1.564	1.471	1.497	1.458
σ_3 (MPa)	−1.604	−1.783	−1.685	−1.617	−1.551	−1.568	−1.508	−1.555	−1.521

Table 4. Error of max dynamic stress of arch dam.

Added Mass Matrix	α = 1.0	α = 0.9	α = 0.8	α = 0.7	α = 0.6	α = 0.5	α = 0.4	α = 0.3
Error of σ_1	4.9%	3.0%	1.3%	3.1%	0.2%	6.1%	4.5%	7.0%
Error of σ_3	11.2%	5.0%	0.8%	3.3%	2.2%	6.0%	3.1%	5.2%

The distribution of the maximum dynamic minor principal stress (σ_3) on the upstream face of the arch dam for every condition of hydrodynamic pressure added mass matrix is depicted in Figure 8. Because seismic loads on the dam will eventually be transferred to the arch abutment, there are high maximum minor principal stress values on the contour of the dam, as shown in Figure 8. As seen in Figure 8, the dynamic minor principal stress (σ_3) distribution rules of each condition are basically consistent, whereas there are some differences in the location and range of the high stress zone of the arch dam. Although the calculation accuracy of the maximum dynamic stress is relatively high for the reduction coefficient α = 0.8, as mentioned above, the scope and area of the high stress zone are obviously different from the accurate solution for the unsimplified added mass matrix. For the condition of the reduction coefficient α = 0.4, the calculation error of maximum dynamic stress is restricted to less than 5%; however, the position of the high stress zone distinctly has a certain degree of deviation compared to the accurate result. When the reduction coefficient α = 0.6, the distribution law of the maximum dynamic stress of the arch dam, including the location of the high stress zone and the range of the high stress zone, is the most similar to the accurate solution, and the computational accuracy for maximum dynamic stress is also high in all the simplification conditions of the additional mass matrix reduction coefficient α, as seen in Table 4.

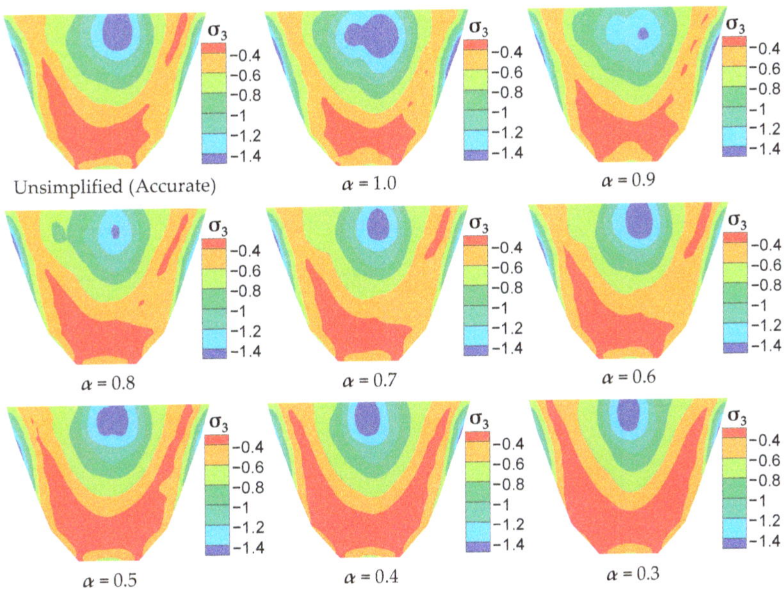

Figure 8. Distribution of maximum minor principal stress (σ_3) on upstream face of arch dam (MPa).

4.3.3. Computational Efficiency

The CPU model of the desktop computer used in the numerical calculation is an Intel (R) Xeon (R) Gold 6248R, whose main frequency is 3.00 Hz. Table 5 shows the consumption time and the corresponding time-consuming ratio of the dynamic fluid–solid coupling calculation for the dam–reservoir systems under different hydrodynamic pressure added mass matrix conditions. When calculating the time-consuming ratio of different working conditions, the accurate solution using the unsimplified additional mass matrix is used as the benchmark, of which the time-consuming ratio is 100%.

Table 5. Consumption time and the corresponding time-consuming ratio.

Added Mass Matrix	Unsimplified (Accurate)	$\alpha = 1.0$	$\alpha = 0.9$	$\alpha = 0.8$	$\alpha = 0.7$	$\alpha = 0.6$	$\alpha = 0.5$	$\alpha = 0.4$	$\alpha = 0.3$
consumption time (hours)	117.799	5.804	5.810	5.861	5.897	5.835	5.801	5.822	5.856
time-consuming ratio	100.0%	4.9%	4.9%	5.0%	4.9%	5.0%	4.9%	4.9%	5.0%

It can be seen from Table 5 that the consumption time of calculation is the most (117.799 h) when the unsimplified hydrodynamic pressure added mass matrix is used, which means there are accurate results of the dam dynamic response. When different reduction coefficients α are used to simplify the additional mass matrix, the consumption time of the dynamic coupling calculation is basically the same, which accounts for about 5% of that under the accurate solution condition. Table 5 shows that for the proposed efficient dynamic coupling calculation method of dam–reservoir systems, the elapsed time is sharply reduced, and the calculation efficiency has made a great jump. It can be seen that the additional mass matrix has a great influence on the computational efficiency of dynamic coupling analysis. Combining Table 2, Table 4, and Table 5, it can be seen that the dynamic coupling calculation efficiency of dam–reservoir systems has been greatly improved after the additional mass matrix is simplified, and the reasonable selection of reduction coefficient α can effectively ensure high calculation accuracy.

4.3.4. Suggested Value of Reduction Coefficient α

According to the analysis of the dam acceleration response results, when the reduction coefficient is 0.6, the calculation accuracy of the maximum acceleration of the dam is the highest (Table 2), and the distribution law of the maximum acceleration of the dam at this time is most similar to the accurate solution (Figure 7). In addition, by comprehensively analyzing the results of the maximum dynamic stress of the dam (Table 4) and its distribution law (Figure 8), it can be known that the reduction coefficient α of 0.6 cannot only ensure the accuracy of the distribution law of the maximum dynamic stress of the dam but also the high calculation accuracy of the maximum dynamic stress. At the same time, from the perspective of computational efficiency, different reduction coefficients almost do not affect the running time of the coupling calculation, and the calculation time can be reduced to about one-twentieth of the unsimplified condition of the additional mass matrix.

In summary, the suggested value of the reduction coefficient is 0.6, which can control the error of coupling calculation results of the dynamic response of the dam within 5% and reduce the elapsed calculation time by more than one order of magnitude. By simplifying the added mass matrix of hydrodynamic pressure, the proposed efficient analysis method can greatly reduce the calculation consumption time and improve the calculation efficiency under the premise of ensuring good accuracy. The efficient method proposed in this study can realize the simplified and efficient calculation and analysis of dynamic fluid–solid coupling between the dam and reservoir under earthquake conditions and is more suitable for engineering calculations with large degrees of freedom than traditional methods.

5. Conclusions

In this study, the hydrodynamic pressure of a reservoir with semi-infinite 3D shapes is directly solved using the SBFEM in 2D, which acting on the dam upstream face is expressed in the form of an additional mass matrix. Further, the dynamic coupling analysis method of dam–reservoir systems, in which the dam is simulated using the FEM, under earthquake conditions, is improved by processing the additional mass matrix of hydrodynamic pressure. Furthermore, the computational efficiency and accuracy of the improved analysis method are studied and verified with the seismic calculation of arch dam–reservoir systems. The following major conclusions are drawn:

1. An efficient 3D dynamic fluid–solid coupling calculation method for dam–reservoir systems based on the FEM-SBFEM is proposed by simplifying the hydrodynamic

pressure additional mass matrix according to the physical meaning and distribution characteristics of the additional matrix. The proposed method not only ensures the high accuracy of the numerical calculation results but also greatly reduces the consumption time of the dynamic coupling calculation.

2. The hydrodynamic pressure added mass matrix has a great influence on the computational efficiency of dynamic coupling analysis. The proposed method, which is simple and easy to implement, only needs to determine a reduction coefficient α ($0 < \alpha \leq 1.0$) to simplify the hydrodynamic pressure added mass matrix to a great extent and save a lot of memory occupied by the added mass matrix.
3. The suggested value of the reduction coefficient α for the added mass matrix of the hydrodynamic pressure is selected to be 0.6 so as to ensure that the distribution law of the dynamic response of the dam is consistent with the accurate solution, which means the unsimplified additional mass matrix condition. The error of the maximum value of the dynamic response of the dam is limited to within 5%, which is acceptable, and the elapsed time of calculation can be reduced to one twentieth of the accurate solution, which is a great jump in calculation efficiency.
4. The proposed method provides an accurate and efficient approach for dynamic fluid–solid coupling analysis and seismic safety evaluation of dam and reservoir systems and makes the application of dam–reservoir systems and a fluid–solid coupling analysis method in fine analysis with large-scale DOFs technically feasible.
5. The proposed dynamic coupling calculation method can also be further applied to the nonlinear numerical analysis of CFRD and the fine damage analysis of concrete dams under earthquake conditions. Furthermore, the additional mass matrix simplification method in the dynamic coupling analysis of dam and reservoir systems provided in this study is also applicable to the additional mass of hydrodynamic pressure calculated by other numerical methods (FEM, BEM, PSBFEM, etc.).

Author Contributions: Conceptualization, H.X. and D.Z.; methodology, H.X. and J.X.; software, H.X. and K.C.; validation, H.X. and D.Y.; resources, H.X., K.C., and D.Z.; writing—original draft preparation, H.X. and J.X.; writing—review and editing, D.Y. and K.C.; supervision, J.X., H.X, D.Y., K.C., and D.Z. All authors have read and agreed to the published version of the manuscript.

Funding: This research was supported by the Project funded by the China Postdoctoral Science Foundation (2021M690998), the National Natural Science Foundation of China (Grant Nos. 52192674, U2240211), and the Doctoral Research Start-up Fund Program Project (2021-BS-065).

Data Availability Statement: Data are not publicly available due to data privacy reasons.

Conflicts of Interest: The authors declare no conflict of interest.

References

1. Song, C.; Wolf, J.P. The scaled boundary finite-element method—Alias consistent infinitesimal finite-element cell method—For elastodynamics. *Comput. Methods Appl. Mech. Eng.* **1997**, *147*, 329–355. [CrossRef]
2. Xu, J.; Xu, H.; Yan, D.; Chen, K.; Zou, D. A Novel Calculation Method of Hydrodynamic Pressure Based on Polyhedron SBFEM and Its Application in Nonlinear Cross-Scale CFRD-Reservoir Systems. *Water* **2022**, *14*, 867. [CrossRef]
3. Zhang, G.; Zhao, M.; Zhang, J.; Du, X. Scaled Boundary Perfectly Matched Layer (SBPML): A novel 3D time-domain artificial boundary method for wave problem in general-shaped and heterogeneous infinite domain. *Comput. Methods Appl. Mech. Eng.* **2023**, *403*, 115738. [CrossRef]
4. Zhang, P.; Du, C.; Zhao, W.; Zhang, D. Hydraulic fracture simulation of concrete using the SBFEM-FVM model. *Struct. Eng. Mech.* **2021**, *80*, 553–562.
5. Li, J.; Shi, Z.; Liu, L.; Song, C. An efficient scaled boundary finite element method for transient vibro-acoustic analysis of plates and shells. *Comput. Struct.* **2020**, *231*, 106211. [CrossRef]
6. Fouladi, M.Q.; Badiei, P.; Vahdani, S. A study on full interaction of water waves with moored rectangular floating breakwater by applying 2DV scaled boundary finite element method. *Ocean Eng.* **2021**, *220*, 108450. [CrossRef]
7. Wang, P.; Lu, R.; Yan, Q.; Bao, X.; Zhang, J.; Liu, L. The internal substructure method for shock wave input in 2D fluid-structure interaction analysis with unbounded domain using doubly-asymptotic ABC. *Mech. Adv. Mater. Struct.* **2023**, *30*, 3111–3124. [CrossRef]

8. Pfeil, S.; Gravenkamp, H.; Duvigneau, F.; Woschke, E. Scaled boundary finite element method for hydrodynamic bearings in rotordynamic simulations. *Int. J. Mech. Sci.* **2021**, *199*, 106427. [CrossRef]
9. Zhao, M.; Wang, X.; Wang, P.; Du, X.; Liu, J. Seismic water-structure interaction analysis using a modified SBFEM and FEM coupling in a frequency domain. *Ocean Eng.* **2019**, *189*, 106374. [CrossRef]
10. Deeks, A.J.; Cheng, L. Potential flow around obstacles using the scaled boundary finite-element method. *Int. J. Numer. Methods Fluids* **2003**, *41*, 721–741. [CrossRef]
11. Li, B.; Cheng, L.; Deeks, A.J.; Teng, B. A modified scaled boundary finite-element method for problems with parallel side-faces. Part II. Application and evaluation. *Appl. Ocean Res.* **2005**, *27*, 224–234. [CrossRef]
12. Li, B.; Cheng, L.; Deeks, A.J.; Zhao, M. A semi-analytical solution method for two-dimensional Helmholtz equation. *Appl. Ocean Res.* **2006**, *28*, 193–207. [CrossRef]
13. Teng, B.; Zhao, M.; He, G.H. Scaled boundary finite element analysis of the water sloshing in 2D containers. *Int. J. Numer. Methods Fluids* **2006**, *52*, 659–678. [CrossRef]
14. Cao, F.S.; Teng, B. Scaled boundary finite element analysis of wave passing a submerged breakwater. *China Ocean Eng.* **2008**, *22*, 241–251.
15. Song, H.; Tao, L. Hydroelastic response of a circular plate in waves using scaled boundary FEM. In Proceedings of the ASME 2009 28th International Conference on Ocean, Offshore and Arctic Engineering, Honolulu, HI, USA, 31 May–5 June 2009.
16. Song, H.; Tao, L. Semi-analytical solution of Poisson's equation in bounded domain. *ANZIAM J.* **2009**, *51*, C169–C185. [CrossRef]
17. Song, H.; Tao, L. An efficient scaled boundary FEM model for wave interaction with a nonuniform porous cylinder. *Int. J. Numer. Methods Fluids* **2010**, *63*, 96–118. [CrossRef]
18. Liu, J.; Lin, G.; Li, J. Short-crested waves interaction with a concentric cylindrical structure with double-layered perforated walls. *Ocean Eng.* **2012**, *40*, 76–90. [CrossRef]
19. Lin, G.; Du, J.; Hu, Z. Dynamic dam-reservoir interaction analysis including effect of reservoir boundary absorption. *Sci. China Ser. E Technol. Sci.* **2007**, *50*, 1–10. [CrossRef]
20. Lin, G.; Wang, Y.; Hu, Z. An efficient approach for frequency-domain and time-domain hydrodynamic analysis of dam-reservoir systems. *Earthq. Eng. Struct. Dyn.* **2012**, *41*, 1725–1749. [CrossRef]
21. Xu, H.; Zou, D.; Kong, X.; Hu, Z. Study on the effects of hydrodynamic pressure on the dynamic stresses in slabs of high CFRD based on the scaled boundary finite-element method. *Soil Dyn. Earthq. Eng.* **2016**, *88*, 223–236. [CrossRef]
22. Xu, H.; Zou, D.; Kong, X.; Su, X. Error study of Westergaard's approximation in seismic analysis of high concrete-faced rockfill dams based on SBFEM. *Soil Dyn. Earthq. Eng.* **2017**, *94*, 88–91. [CrossRef]
23. Xu, H.; Zou, D.; Kong, X.; Hu, Z.; Su, X. A nonlinear analysis of dynamic interactions of CFRD-compressible reservoir system based on FEM-SBFEM. *Soil Dyn. Earthq. Eng.* **2018**, *112*, 24–34. [CrossRef]
24. Wang, X.; Jin, F.; Prempramote, S.; Song, C. Time-domain analysis of gravity dam–reservoir interaction using high-order doubly asymptotic open boundary. *Comput. Struct.* **2011**, *89*, 668–680. [CrossRef]
25. Gao, Y.; Jin, F.; Wang, X.; Wang, J. Finite Element Analysis of Dam–Reservoir Interaction Using High-Order Doubly Asymptotic Open Boundary. In *Seismic Safety Evaluation of Concrete Dams*; Butterworth-Heinemann: Oxford, UK, 2013; pp. 173–198.
26. Das, S.K.; Mandal, K.K.; Niyogi, A.G. Finite element-based direct coupling approach for dynamic analysis of dam–reservoir system. *Innov. Infrastruct. Solut.* **2023**, *8*, 44. [CrossRef]
27. Gorai, S.; Maity, D. Seismic Performance Evaluation of Concrete Gravity Dams in Finite-Element Framework. *Pract. Period. Struct. Des. Constr.* **2022**, *27*, 04021072. [CrossRef]
28. Wang, C.; Zhang, H.; Zhang, Y.; Guo, L.; Wang, Y.; Thira Htun, T.T. Influences on the seismic response of a gravity dam with different foundation and reservoir modeling assumptions. *Water* **2021**, *13*, 3072. [CrossRef]
29. Pelecanos, L.; Kontoe, S.; Zdravković, L. The effects of dam–reservoir interaction on the nonlinear seismic response of earth dams. *J. Earthq. Eng.* **2020**, *24*, 1034–1056. [CrossRef]
30. Sharma, V.; Fujisawa, K.; Murakami, A. Space–time FEM with block-iterative algorithm for nonlinear dynamic fracture analysis of concrete gravity dam. *Soil Dyn. Earthq. Eng.* **2020**, *131*, 105995. [CrossRef]
31. Karabulut, M.; Kartal, M.E. Seismic analysis of Roller Compacted Concrete (RCC) dams considering effect of viscous boundary conditions. *Comput. Concr. Int. J.* **2019**, *27*, 255–266.
32. Wang, J.T.; Lv, D.D.; Jin, F.; Zhang, C.H. Earthquake damage analysis of arch dams considering dam–water–foundation interaction. *Soil Dyn. Earthq. Eng.* **2013**, *49*, 64–74. [CrossRef]
33. Bayraktar, A.; Kartal, M.E. Linear and nonlinear response of concrete slab on CFR dam during earthquake. *Soil Dyn. Earthq. Eng.* **2010**, *30*, 990–1003. [CrossRef]
34. Wang, J.T.; Chopra, A.K. Linear analysis of concrete arch dams including dam-water-foundation rock interaction considering spatially varying ground motions. *Earthq. Eng. Struct. Dyn.* **2010**, *39*, 731–750. [CrossRef]
35. Bayraktar, A.; Dumanoğlu, A.A.; Calayir, Y. Asynchronous dynamic analysis of dam-reservoir-foundation systems by the Lagrangian approach. *Comput. Struct.* **1996**, *58*, 925–935. [CrossRef]
36. Fok, K.L.; Chopra, A.K. Earthquake analysis of arch dams including dam–water interaction, reservoir boundary absorption and foundation flexibility. *Earthq. Eng. Struct. Dyn.* **1986**, *14*, 155–184. [CrossRef]
37. Fenves, G.; Chopra, A.K. Earthquake analysis of concrete gravity dams including reservoir bottom absorption and dam-water-foundation rock interaction. *Earthq. Eng. Struct. Dyn.* **1984**, *12*, 663–680. [CrossRef]

38. Bouaanani, N.; Miquel, B. An error estimator for transmitting boundary conditions in fluid-structure interaction problems. *WIT Trans. Built Environ.* **2011**, *115*, 169–178.
39. Küçükarslan, S.; Coşkun, S.B. Transient dynamic analysis of dam-reservoir interaction by coupling DRBEM and FEM. *Eng. Comput.* **2004**, *21*, 692–707. [CrossRef]
40. Fahjan, Y.M.; Börekçi, O.S.; Erdik, M. Earthquake-induced hydrodynamic pressures on a 3D rigid dam–reservoir system using DRBEM and a radiation matrix. *Int. J. Numer. Methods Eng.* **2003**, *56*, 1511–1532. [CrossRef]
41. Tsai, C.S.; Lee, G.C.; Yeh, C.S. Time-domain analyses of three-dimensional dam-reservoir interactions by BEM and semi-analytical method. *Eng. Anal. Bound. Elem.* **1992**, *10*, 107–118. [CrossRef]
42. Tsai, C.S. Analyses of three-dimensional dam—Reservoir interactions based on bem with particular integrals and semi-analytical solution. *Comput. Struct.* **1992**, *43*, 863–872. [CrossRef]
43. Tsai, C.S.; Lee, G.C. Arch dam–fluid interactions: By FEM–BEM and substructure concept. *Int. J. Numer. Methods Eng.* **1987**, *24*, 2367–2388. [CrossRef]
44. Degao, Z.; Xianjing, K.; Bin, X. *User Manual for Geotechnical Dynamic Nonlinear Analysis*; Institute of Earthquake Engineering, Dalian University of Technology: Dalian, China, 2005.
45. Dassault Systemes Simulia Corp. *ABAQUS Analysis User's Guide*. 2016. Available online: http://130.149.89.49:2080/v2016/books/usb/default.htm?startat=pt04ch18s01aus106.html (accessed on 24 August 2023).
46. Chen, K.; Zou, D.; Tang, H.; Liu, J.; Zhuo, Y. Scaled boundary polygon formula for Cosserat continuum and its verification. *Eng. Anal. Bound. Elem.* **2021**, *126*, 136–150. [CrossRef]
47. Qu, Y.; Zou, D.; Kong, X.; Yu, X.; Chen, K. Seismic cracking evolution for anti-seepage face slabs in concrete faced rockfill dams based on cohesive zone model in explicit SBFEM-FEM frame. *Soil Dyn. Earthq. Eng.* **2020**, *133*, 106106. [CrossRef]
48. Pang, R.; Xu, B.; Zhou, Y.; Song, L. Seismic time-history response and system reliability analysis of slopes considering uncertainty of multi-parameters and earthquake excitations. *Comput. Geotech.* **2021**, *136*, 104245. [CrossRef]
49. Zou, D.; Xu, B.; Kong, X.; Liu, H.; Zhou, Y. Numerical simulation of the seismic response of the Zipingpu concrete face rockfill dam during the Wenchuan earthquake based on a generalized plasticity model. *Comput. Geotech.* **2013**, *49*, 111–122. [CrossRef]
50. Chen, K.; Zou, D.; Liu, J.; Zhuo, Y. A high-precision formula for mixed-order polygon elements based on SBFEM. *Comput. Geotech.* **2023**, *155*, 105209. [CrossRef]
51. Nie, X.; Chen, K.; Zou, D.; Kong, X.; Liu, J.; Qu, Y. Slope stability analysis based on SBFEM and multistage polytree-based refinement algorithms. *Comput. Geotech.* **2022**, *149*, 104861. [CrossRef]

MDPI AG
Grosspeteranlage 5
4052 Basel
Switzerland
Tel.: +41 61 683 77 34

Water Editorial Office
E-mail: water@mdpi.com
www.mdpi.com/journal/water

www.ingramcontent.com/pod-product-compliance
Lightning Source LLC
Chambersburg PA
CBHW040747250926
51850CB00007BB/1429